COLLEGE ALGEBRA
AND TRIGONOMETRY

SECOND EDITION

STEVEN J. BRYANT CALIFORNIA STATE UNIVERSITY, SAN DIEGO

JACK KARUSH LAWRENCE RADIATION LABORATORY

LEON NOWER CALIFORNIA STATE UNIVERSITY, SAN DIEGO

DANIEL SALTZ CALIFORNIA STATE UNIVERSITY, SAN DIEGO

GOODYEAR PUBLISHING COMPANY, INC.
SANTA MONICA, CALIFORNIA

Library of Congress Cataloging in Publication Data

College algebra and trigonometry, 2nd Ed.

1. Algebra. 2. Trigonometry. I. Bryant, Steven Jerome, 1923–

QA154.2.C645 1974   512′.13   73-81071

ISBN 0-87620-171-0

© 1974 by Goodyear Publishing Company, Inc.
Santa Monica, California

Current printing (last digit): 10 9 8 7 6 5 4 3

ISBN 0-87620-171-0

Library of Congress Catalog Card Number 73-81071

Y-1710-6

Printed in the United States of America

Design: Einar Vinje

Editorial and Production: Project Publishing and Design

Contents

1

NUMBERS AND FUNCTIONS

2
POLYNOMIAL AND RATIONAL FUNCTIONS

3
TRANSCENDENTAL FUNCTIONS

4
LINEAR SYSTEMS AND AN INTRODUCTION TO ANALYTIC GEOMETRY

5
ADDITIONAL TOPICS

Preface

Preface to the First Edition

Mathematicians nearly always associate mathematical ideas with pictures; beginners seldom do. This is unfortunate, because it is almost impossible to study many important mathematical functions without also studying their graphs.

Since the main purpose of this book is to prepare the reader for the study of calculus, the analysis of *functions and their graphs* is emphasized throughout. The reader will find that to come up with the correct picture he will not only exercise his skills in arithmetic and algebra, but will also acquire and strengthen his intuition of *continuity*—so indispensable for a further study of mathematics. To help the growth of this intuition, many problems in this book simply ask for the graph of a given function, and the answer (a picture) is given in the answer section.

The basic review (Chapter 1) provides the less well prepared reader with an opportunity to strengthen his techniques in arithmetic and algebra. However, it should be remembered that this material is *preliminary* to the subject matter of the book.

The early occurrence of the chapter on sequences makes it possible to discuss the completeness of the real number system at the right place: *before* the topics of continuity and irrational exponents. Furthermore, the brief exposition of *limits* in this chapter is good preparation for the discussion of asymptotic behavior in rational functions.

The core of the book is contained in Parts Two and Three. Part Two deals in detail with *algebraic* (polynomial and rational) functions, Part Three with *transcendental* (exponential, logarithmic, and trigonometric) functions. Part Four is a self-contained treatment of *linear systems, matrices, and determinants;* however, Chapter 11 (Analytic Geometry) provides the geometric background for the study of linear systems as well as an introduction to conics and three-dimensional coordinate geometry. Lastly, in Part Five the reader will find an introduction to several important areas of modern mathematics: *sets, combinatorics,* and *probability.*

January, 1971

Preface to the Second Edition

A new feature of this revision is the inclusion of sets of miscellaneous problems for Chapters 2 through 12. These problems range from routine to difficult and can be used for tests or review assignments. With this in mind, we have not included answers to these miscellaneous problems in the text.

The chapter on Trigonometry has been expanded and reorganized.

Also a section on Linear Programming has been added to Chapter 12.

There is no change in basic philosophy in this revised edition. But substantial changes have been made in response to those who have used the previous edition.

We want to express our gratitude to the three teams of graduate students* who, working independently, did all the problems in the book and eliminated errors that have appeared in the previous edition.

*Barbara Leake, UCLA
Thomas Stewart, UCLA
Chris Francis, UCLA
Dale Larson, UCLA
Le Anne Cooney, SDSU
Allan Witcher, SDSU

We also extend thanks for helpful comments and suggestions from those who reviewed the manuscript.

Prof. Gerald Bradley, Claremont Men's College
Prof. Grant A. Fraser, University of Santa Clara
Prof. Bill New, Cerritos College
Prof. Tom Reimer, Lane Community College

STEVEN BRYANT
JACK KARUSH
LEON NOWER
DANIEL SALTZ

January, 1974

1

NUMBERS
AND
FUNCTIONS

The four chapters of Part I develop the basic concepts and tools for the analysis of functions in Parts II and III.

Chapter 1 is a review of the properties of real numbers. Chapter 2 introduces the coordinate plane and graphs, thus providing the geometric background for the study of functions.

Chapter 3 introduces functions and their graphs, as well as some aspects of the algebra of functions including sums, products, shifts, and stretches.

Chapter 4 deals with sequences and progressions, and then introduces limits, convergence, and the completeness of the real number system. It includes a short section on induction.

1 Basic Review

1-1 INTRODUCTION

In addition to providing a review of the properties of real numbers, this chapter is designed to help the reader increase the skills he needs in studying algebra and to indicate without undue emphasis how the subject of algebra is developed from just a few fundamental properties (axioms).

By the end of the chapter, the reader will have acquired competence in solving problems that involve order, powers, roots, fractions, factoring, and related equations.

1-2 BASIC PROPERTIES

We shall assume that the reader is familiar with the interpretation of real numbers as points on a line (see Figure 1-1) and with the usual computations of arithmetic. In this section, we shall show how these computations, as well as certain much-used theorems, are derived from basic properties (or axioms) of the real number system.

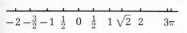

Figure 1-1

While the results of this section are merely a review, the proofs may be new to the reader; and although the remainder of the book does not depend on these proofs, some students do benefit from this type of activity (following a proof step by step), since it provides a secure basis for deciding whether or not a given computation is valid.

In the following list of basic properties, and in the remainder of this section, a, b, c, and d are *any* real numbers, unless otherwise specified. Furthermore, a statement such as "$a = b$" means "a *is* b" or, equivalently, that "a" and "b" are merely different symbols for the *same* number (just as H_2O and water are merely different names—i.e., symbols—for the same thing).

Name of property	Addition	Multiplication
1. **Commutative:**	$a + b = b + a$	$ab = ba$
2. **Associative:**	$(a + b) + c = a + (b + c)$	$(ab)c = a(bc)$
3. **Distributive:**	$a(b + c) = ab + ac$	

and, in view of the commutative property,

$$(b + c)a = ba + ca$$

4. **The property of identities:***	$a + 0 = 0 + a = a$	$a \cdot 1 = 1 \cdot a = a$

(i.e., 0 and 1 are, respectively, *additive* and *multiplicative identities*)

5. **The property of inverses:†** $a + (-a) = (-a) + a = 0$

$$a \cdot \frac{1}{a} = \frac{1}{a} \cdot a = 1$$

(if $a \neq 0$)

(i.e., a has $-a$ as *additive inverse* and, if $a \neq 0$, $1/a$ as *multiplicative inverse*)

Recall at this point that, *by definition*,

$$a - c = a + (-c)$$

and, for $c \neq 0$,

$$\frac{a}{c} = a \cdot \frac{1}{c}$$

That is, *to subtract c means to add $-c$, and to divide by c means to multiply by $1/c$.* Note also that, since $a = b$ means

---

*It turns out that 0 and 1 are *the only* (i.e., *unique*) identities, respectively, for addition and multiplication: for if $0'$ is an *additive identity*, then $0 + 0' = 0$; and we also have $0 + 0' = 0'$ because of the property of 0. Hence, $0' = 0$. An analogous argument shows that 1 is the only *multiplicative identity.*
†The *uniqueness* of inverses will be shown shortly.

that a and b are the same number, it follows that for *any* number c, if

$$a = b$$

then

$$a + c = b + c$$

and

$$ac = bc$$

That is, addition and multiplication (and hence also subtraction and division) "preserve equality." * Conversely,

if $a + c = b + c$,   then $a = b$

and, for $c \neq 0$,

if $ac = bc$,   then $a = b$

To see this, suppose that $a + c = b + c$; then, since addition preserves equality, $(a + c) + (-c) = (b + c) + (-c)$,   and hence $a + [c + (-c)] = b + [c + (-c)]$. Since $c + (-c) = 0$, it follows that $a = b$. A similar argument shows that if $ac = bc$ (with $c \neq 0$) then $a = b$.

Summarizing the foregoing observations, we see that

$a = b$ if and only if $a + c = b + c$

and for $c \neq 0$,

$a = b$ if and only if $ac = bc$

Remark When "*if and only if*" connects two statements, it means that these two statements are *logically equivalent*; that is, if either one of them is true, then so is the other.

We shall now prove several important theorems, using the properties discussed above.

---

*It is now possible to show why *inverses are unique*. Suppose x is an additive inverse of a: $a + x = 0$; then, since addition preserves equality,

$$(-a) + (a + x) = (-a) + 0$$
$$(-a + a) + x = (-a) + 0 \qquad \textit{[by associativity]}$$
$$0 + x = (-a) + 0 \qquad \textit{[since } -a + a = 0]$$

and finally

$$x = -a \qquad \textit{[since } 0 + x = x \textit{ and}$$
$$(-a) + 0 = -a]$$

The uniqueness of multiplicative inverses is shown analogously.

Theorem 1

For any real number a, $-(-a) = a$; and if $a \neq 0$, then $\dfrac{1}{1/a} = a$.

$-(-a) = a$

$-(a) + a = 0$

Proof   Note that (by definition) $-(-a)$ is the additive inverse of $-a$; and that a is the additive inverse of $-a$ [since $(-a) + a = 0$]. Since $-a$ has *only one* additive inverse, it follows that $-(-a) = a$. An analogous proof shows that if $a \neq 0$, then $\dfrac{1}{1/a} = a$, and is left as an exercise.

Theorem 2

For any real numbers a and b, $-(a + b) = -a - b$; and if $a \neq 0$ and $b \neq 0$, then $\dfrac{1}{ab} = \dfrac{1}{a} \cdot \dfrac{1}{b}$.

Proof   Note that

$$(-a - b) + (a + b) = (-a) + (-b) + a + b$$
$$= (-a) + a + (-b) + b$$
$$= 0 + 0 = 0$$

while by definition we also have $-(a + b) + (a + b) = 0$. Since $a + b$ has only one additive inverse, it follows that $-(a + b) = -a - b$. The proof for multiplicative inverses is similar, and is left as an exercise.

Theorem 3

For any real number a, $a \cdot 0 = 0$.

Proof   We have

$$a \cdot 0 = a(0 + 0) \qquad [\text{since } 0 + 0 = 0]$$
$$= a \cdot 0 + a \cdot 0 \qquad [\text{by the distributive property}]$$

Hence,

$$a \cdot 0 - a \cdot 0 = a \cdot 0 + a \cdot 0 - a \cdot 0 \qquad [\text{since subtraction preserves equality}]$$

and therefore

$$0 = a \cdot 0$$

Theorem 4

If a and b are real numbers and $ab = 0$, then $a = 0$ or $b = 0$.

Proof If $a = 0$, the conclusion holds; if $a \neq 0$ and $ab = 0$, then

$$\frac{1}{a}(ab) = \frac{1}{a} \cdot 0 \qquad \text{[since multiplication preserves equality]}$$

$$\left(\frac{1}{a}a\right)b = 0 \qquad \text{[associative property and Theorem 3]}$$

$$1 \cdot b = 0 \qquad \left[\text{since } \frac{1}{a} \cdot a = 1\right]$$

and finally

$$b = 0 \qquad \text{[since } 1 \cdot b = b]$$

Example 1

If x is a real number for which $(x - 2)(x + 1) = 0$, then either $x - 2 = 0$ or $x + 1 = 0$; that is, either $x = 2$ or $x = -1$.
□

Theorems 3 and 4 yield the following important result. For any real numbers a and b,

$$ab = 0 \quad \text{if and only if} \quad a = 0 \text{ or } b = 0$$

Theorem 5

If a and b are real numbers, then $(-a)b = -ab = a(-b)$.

Proof On the one hand, $-ab$ is the additive inverse of ab by definition; on the other hand, we also have $(-a)b + ab = (-a + a)b = 0 \cdot b = 0$. Since ab has but one additive inverse, it follows that $(-a)b = -ab$. A similar argument shows that $a(-b) = -ab$.

Theorem 6

If a and b are real numbers, then $(-a)(-b) = ab$.

Proof

$$(-a)(-b) = -[a(-b)] \qquad \text{[Theorem 5]}$$

$$= -(-ab) \qquad \text{[Theorem 5]}$$

$$= ab \qquad \text{[Theorem 1]}$$

Theorem 7

If a, b, and c are real numbers, then $a(b - c) = ab - ac$.

Proof

$$a(b - c) = a[b + (-c)] = ab + a(-c)$$
$$= ab + (-ac) = ab - ac$$

The following theorem provides a basis for computations that involve fractions.

Theorem 8

If a, b, c, and d are real numbers with $b \neq 0$ and $d \neq 0$, then

(1) $\dfrac{a}{b} \cdot \dfrac{c}{d} = \dfrac{ac}{bd}$

(2) $\dfrac{a}{b} + \dfrac{c}{b} = \dfrac{a + c}{b}$

(3) $\dfrac{a}{b} + \dfrac{c}{d} = \dfrac{ad + bc}{bd}$

Proof

1. $\dfrac{a}{b} \cdot \dfrac{c}{d} = \left(a \cdot \dfrac{1}{b}\right) \cdot \left(c \cdot \dfrac{1}{d}\right) = (ac) \cdot \left(\dfrac{1}{b} \cdot \dfrac{1}{d}\right) = ac \cdot \dfrac{1}{bd} = \dfrac{ac}{bd}$

2. $\dfrac{a}{b} + \dfrac{c}{b} = a \cdot \dfrac{1}{b} + c \cdot \dfrac{1}{b} = (a + c) \cdot \dfrac{1}{b} = \dfrac{a + c}{b}$

3. $\dfrac{a}{b} + \dfrac{c}{d} = \dfrac{a}{b} \cdot \dfrac{d}{d} + \dfrac{c}{d} \cdot \dfrac{b}{b} = \dfrac{ad}{bd} + \dfrac{bc}{bd} = \dfrac{ad + bc}{bd}$

The reader is asked to justify the intermediate steps and to verify the following additional relationships:

4. $\dfrac{a}{b} - \dfrac{c}{d} = \dfrac{ad - bc}{bd}$

5. If also $c \neq 0$, then $\dfrac{a/b}{c/d} = \dfrac{ad}{bc}$

6. $-\dfrac{a}{b} = \dfrac{-a}{b} = \dfrac{a}{-b}$

An important consequence of (1) is the *cancellation property* of division: if a, b, and c are real numbers with $b \neq 0$ and $c \neq 0$, then

$$\frac{ac}{bc} = \frac{a}{b}$$

This is so because

$$\frac{ac}{bc} = \frac{a}{b} \cdot \frac{c}{c} = \frac{a}{b} \cdot 1 = \frac{a}{b}$$

Example 2

(a) $\dfrac{15}{21} = \dfrac{5 \cdot 3}{7 \cdot 3} = \dfrac{5}{7}$

(b) If $x \neq 0$, then $\dfrac{x^2}{x} = \dfrac{x \cdot x}{1 \cdot x} = x$ and $\dfrac{x}{x^2} = \dfrac{1 \cdot x}{x \cdot x} = \dfrac{1}{x}$

(c) If $x \neq 0$ and $x \neq 1$, then

$$\frac{x^2(x-1)}{x(x-1)^2} = \frac{x^2}{x} \cdot \frac{x-1}{(x-1)^2}$$

$$= x \cdot \frac{1}{x-1}$$

$$= \frac{x}{x-1}$$

(d) If a and b are numbers, with $a \neq b$ and $a \neq 2b$, then

$$\frac{(a-2b)}{(a-b)} \cdot \frac{(a-b)^2}{(a-2b)^2} = \frac{(a-b)^2(a-2b)}{(a-b)(a-2b)^2}$$

$$= \frac{(a-b)^2}{a-b} \cdot \frac{(a-2b)}{(a-2b)^2}$$

$$= (a-b) \cdot \frac{1}{(a-2b)}$$

$$= \frac{a-b}{a-2b}$$

□

The following examples use the foregoing properties and theorems as well as some of their consequences.

Example 3

(a) $-(-5) = 5$

(b) $4 - (-3) = 4 + (-(-3)) = 4 + 3 = 7$

(c) $\dfrac{3}{7} + \dfrac{4}{11} = \dfrac{3 \cdot 11 + 4 \cdot 7}{7 \cdot 11} = \dfrac{61}{77}$

(d) $\dfrac{5}{7} \cdot \dfrac{2}{3} = \dfrac{5 \cdot 2}{7 \cdot 3} = \dfrac{10}{21}$

(e) $\dfrac{-3}{5} = -\dfrac{3}{5} = \dfrac{3}{-5}$

(f) $\dfrac{3}{14} - \dfrac{4}{9} = \dfrac{3 \cdot 9 - 4 \cdot 14}{14 \cdot 9} = \dfrac{-29}{126} = -\dfrac{29}{126}$

(g) $\dfrac{24}{76} = \dfrac{4 \cdot 6}{4 \cdot 19} = \dfrac{4}{4} \cdot \dfrac{6}{19} = 1 \cdot \dfrac{6}{19} = \dfrac{6}{19}$

Example 4

Let x and y be real numbers; then

(a) $3(x - 2) = 3x - 3 \cdot 2 = 3x - 6$

(b) $3(5x) = (3 \cdot 5)x = 15x$

(c) $5x + 7x = (5 + 7)x = 12x$

(d) $x^2 - xy = x(x - y)$

(e) $-[3y + 5(y - 2)] = -[3y + 5y - 10]$
$$= -[8y - 10] = -8y + 10$$

(f) $(x + y)^2 = (x + y)(x + y) = (x + y)x + (x + y)y$
$$= x^2 + yx + xy + y^2 = x^2 + 2xy + y^2$$

Example 5

Find all real numbers x for which $5x + 7 = 0$.

Solution

x is such a number if and only if

$(5x + 7) + (-7) = 0 + (-7)$ [*since addition preserves equality*]

$5x + [7 + (-7)] = -7$ [*associative law for addition*]

$5x + 0 = -7$ [*additive inverse*]

$5x = -7$ [*additive identity*]

$\dfrac{1}{5}(5x) = \dfrac{1}{5}(-7)$ [*since multiplication preserves equality*]

$\left(\dfrac{1}{5} \cdot 5\right)x = \dfrac{-7}{5}$ $\left[\begin{array}{l}\textit{associative law and the} \\ \textit{definition that } \dfrac{1}{a}b = \dfrac{b}{a}\end{array}\right]$

$1 \cdot x = -\dfrac{7}{5}$ $\left[\textit{since } \dfrac{-a}{b} = -\dfrac{a}{b}\right]$

or, finally

$$x = -\dfrac{7}{5}$$

Since 0 does not have a multiplicative inverse, $\frac{1}{0}$ does not denote a number. And, in general, if a is *any* number, $\frac{a}{0}$ does not denote a number. For example, $\frac{1}{x-2}$ does not denote a number if x is 2. In each of the following examples the fact that $\frac{a}{0}$ does not denote a number restricts the number of possibilities for solutions.

Example 6

Find all numbers x for which

$$\frac{2}{3x-1} + \frac{4}{5x-3} = \frac{7}{1-3x}$$

Solution

First note that the only numbers x under consideration are those for which $3x - 1 \neq 0$ and $5x - 3 \neq 0$. Therefore,

$$\frac{2}{3x-1} + \frac{4}{5x-3} = \frac{7}{1-3x} = \frac{-7}{3x-1}$$

if and only if

$$\left(\frac{2}{3x-1} + \frac{4}{5x-3}\right)(3x-1)(5x-3)$$

$$= \frac{-7}{3x-1}(3x-1)(5x-3)$$

(we are multiplying by a nonzero number); that is,

$$10x - 6 + 12x - 4 = -35x + 21$$

$$57x = 31$$

or, finally

$$x = \tfrac{31}{57}$$

Exercises 1-2

In Exercises 1–8, compute

1. $-[-(-27)]$ -27
2. $5 + [-(-7)]$
3. $(-5)(-2)(-1)$
4. $7 \cdot 0 \cdot 4$
5. $\frac{2}{11} - \frac{1}{5}$
6. $\frac{1}{4} + \frac{3}{8} - \frac{7}{3}$
7. $\frac{5}{4} \cdot \frac{2}{15}$
8. $\left(-\frac{4}{49}\right)\left(\frac{5}{28}\right)\left(\frac{-14}{125}\right)$

In Exercises 9–13, compute as follows:

$$\frac{1 + \frac{2}{3}}{3 - \frac{2}{5}} = \frac{1 + \frac{2}{3}}{3 - \frac{2}{5}} \cdot \frac{3 \cdot 5}{3 \cdot 5} = \frac{15 + \frac{2}{3} \cdot 3 \cdot 5}{3 \cdot 15 - \frac{2}{5} \cdot 3 \cdot 5} = \frac{15 + 10}{45 - 6} = \frac{25}{39}$$

9. $\dfrac{\frac{5}{8} - 7}{\frac{9}{16} + \frac{2}{7}}$

10. $\dfrac{15 - \frac{3}{11}}{\frac{7}{13} + 5}$

11. $\dfrac{5\frac{3}{7} - 2\frac{2}{3}}{3\frac{7}{9} + 5\frac{3}{7}}$

12. $\dfrac{4}{4 + \dfrac{3}{3 + \frac{2}{3}}}$

13. $\dfrac{5}{\dfrac{3}{8} - \dfrac{1}{\dfrac{2}{3} + \dfrac{6}{1 - \frac{3}{7}}}}$

Let x and y be real numbers. Compute as in Exercise 14.

14. $5(y - 3) = 5y - 5 \cdot 3 = 5y - 15$

15. $x(3y - 4)$

16. $-xy(x + y)$

17. $(x + 1)^2$

18. $(x - 1)^2$

19. $5x\left(x + \dfrac{1}{x}\right), (x \neq 0)$

20. $xy\left(\dfrac{1}{x} + \dfrac{1}{y}\right), (x \neq 0, y \neq 0)$

21. $(x + y)\left(\dfrac{x - y}{x + y} + 1\right), (x + y \neq 0)$

22. $xy\left(\dfrac{x - y}{3} - \dfrac{x + y}{2}\right)$

In Exercises 23–35, find all real numbers x for which:

23. $3x + 4 = 0$

24. $\frac{2}{3}x + \frac{1}{4} = 0$

25. $6x + 4 - [-(3x - 7)] = 15$

26. $\dfrac{x}{3} + \dfrac{2}{7} - \left(\dfrac{4x}{7} - \dfrac{5}{11}\right) = 0$

27. $(x - 1)(x - \pi)(x + 3) = 0$

28. $(2x - 3)(7 - 4x) = 0$

29. $\left(3x - \dfrac{1}{5}\right)\left(\dfrac{x}{5} - 3\right) = 0$

30. $2x + 7 = -4x + 2$

31. $-\left(\frac{2}{3}x + \frac{1}{5}\right) = \frac{1}{3}x - \frac{1}{10}$

32. $\frac{5}{3}(6 - x) = \frac{1}{5}x + \frac{8}{3}(3 - x)$

33. $\frac{1}{2}(x - 1) = \frac{1}{3}x - \frac{2}{3}(1 + x)$

34. $\dfrac{4}{x + 1} = \dfrac{3}{x - 2} + \dfrac{5}{x + 1}$

35. $\dfrac{7}{3x - 4} + \dfrac{1}{x - 1} - \dfrac{3}{4 - 3x} = 0$

In Exercises 36–43, simplify. (All letters denote real numbers for which all denominators are not zero.)

36. $\dfrac{a}{a + b} \cdot \dfrac{5}{3b} \cdot \dfrac{7b(a + b)}{2a} = \dfrac{35}{6} \cdot \dfrac{a}{a} \cdot \dfrac{b}{b} \cdot \dfrac{a + b}{a + b} = \dfrac{35}{6}$

37. $\dfrac{5(a + b)}{ab} \cdot \dfrac{2b}{5} \cdot \dfrac{a}{2a + 2b}$

38. $\dfrac{x + 3a}{a - 2b} \cdot \dfrac{2b - a}{bx + 3ab}$

39. $\dfrac{\dfrac{a + b}{a - b}}{1 + \dfrac{a + b}{a - b}} = \dfrac{\left(\dfrac{a + b}{a - b}\right)(a - b)}{\left(1 + \dfrac{a + b}{a - b}\right)(a - b)}$

$$= \dfrac{a + b}{a - b + a + b} = \dfrac{a + b}{2a}$$

40. $\dfrac{\dfrac{2x + y}{x + y} + 1}{\dfrac{2x + y}{x + y} - 1}$

41. $\dfrac{\dfrac{a + 2b}{a - b} + 3}{\dfrac{1}{a - b} + 7}$

◊ **42.** $\dfrac{a}{a - \dfrac{b}{b - \dfrac{c}{1 - \dfrac{d}{d - x}}}}$

◊ **43.** $\dfrac{x}{x + \dfrac{y}{y + \dfrac{z}{1 - \dfrac{1}{1 - a}}}}$

In Exercises 44–49, indicate whether true or false.

44. "Subtraction" is associative; that is, $a - (b - c) = (a - b) - c$ for any real numbers a, b, and c.

45. If a, b, and c are real numbers and $abc = 0$, then $ab = 0$ or $a(-c) = 0$.

46. If $a(b + c) = 0$ and $a \neq 0$, then $b = -c$.

47. There is a real number x for which $\dfrac{2x}{x + 1} = 2$.

48. Every real number has a multiplicative inverse.

49. If a and b are real numbers and if $4(a + b) = 1$, then $a + b - x = -(x - \frac{1}{4})$ for any real number x.

50. Show that every nonzero number has precisely one multiplicative inverse.

51. Show that if a is any nonzero number, then $\dfrac{1}{1/a} = a$.

52. If a and b are nonzero numbers, show that $1/ab = (1/a) \cdot (1/b)$.

1-3 ORDER

Since real numbers are regarded as points on a line, there is a natural way of ordering them: If a is a real number to the left of the real number b, then a is **less than** b, written

$$a < b$$

or equivalently, b is **greater than** a, written

$$b > a$$

Also, $a \leq b$ means a *is less than or equal to b,* and $a \geq b$ means a *is greater than or equal to b.*

By definition, if a is a real number, then

a is **positive** if and only if $a > 0$

and

a is **negative** if and only if $a < 0$

We see that *every real number is either positive* (to the right of zero) *or negative* (to the left of zero) *or zero.* This is a fundamental property of order, known as **trichotomy.** Another fundamental property, known as **transitivity,** is also self-evident from the geometric interpretation: *if a, b, and c are real numbers with $a > b$ and $b > c$, then $a > c$.*

The following three properties of order are also fundamental. Let a, b, and c be any real numbers.

1. $a < b$ if and only if $a + c < b + c$

2. For $c > 0$, $a < b$ if and only if $ac < bc$

3. For $c < 0$, $a < b$ if and only if $ac > bc$

(1) states that "addition preserves inequality"; (2) states that "multiplication by positive numbers preserves inequality"; and (3) states that "multiplication by negative numbers *reverses* inequality."

Example 1

(a) $5 > 3$ and $(-7) \cdot 5 < (-7) \cdot 3$; (i.e., $-35 < -21$)

(b) $-3 < -2$, and $(-8)(-3) > (-8)(-2)$; (i.e., $24 > 16$)

(c) If x and y are numbers and $x < y$, then $(-1)x > (-1)y$; that is, $-x > -y$.

(d) If x and y are numbers and $x > -2y$, then
   (i) $(-4)x < (-4)(-2y)$; that is, $-4x < 8y$,
   (ii) $7x > 7(-2y)$; that is, $7x > -14y$.

The foregoing definitions and fundamental properties will now be used to prove a theorem that establishes additional important properties of order.

Theorem 1

(a) Let a and b be real numbers. Then $b > a$ **if and only if** $b - a > 0$.
(b) The sum of positive numbers is positive.
(c) The product of positive numbers is positive, and the product of a positive and a negative number is negative.
(d) The product of two negative numbers is positive.

Proof

(a) By property (1), $b > a$ if and only if $b + (-a) > a + (-a)$; that is, if and only if $b - a > 0$.
(b) If $a > 0$ then, by property (1), $a + b > 0 + b = b$ for any real number b. Hence, if also $b > 0$, then $a + b > 0$ by transitivity.
(c) Let $a > 0$. If $b > 0$, then by property (2), $ab > 0 \cdot b = 0$; and if $b < 0$, then by property (3), $ab < 0 \cdot b = 0$.
(d) If $a < 0$ and $b < 0$, then, since multiplying by the negative number b reverses the inequality $a < 0$, we have $ab > 0 \cdot b = 0$.

Example 2

(a) $-5 > -8$, and $(-5) - (-8) = (-5) + 8 = 3 > 0$
(b) Let x and y be numbers. Then
    (i) $(-3x)(2y) = (-3)(2)xy = -6xy$, and
    (ii) $(-5x)(-8y) = (-5)(-8)xy = 40xy$

□

The following two corollaries are frequently used:

Corollary 1

If a is a nonzero real number, then $a^2 > 0$. Also, $a^4 = (a^2)^2 > 0$, $a^6 > 0$, $a^8 > 0$, and so on.

Proof This follows directly from parts (c) and (d) of Theorem 1 (and trichotomy).

Corollary 2

If a is a nonzero real number, then a and $1/a$ are both positive or both negative.

Proof Since $a \cdot (1/a) = 1$ is not negative, it cannot be the product of two numbers one of which is positive and the other negative [by part (c) of Theorem 1].

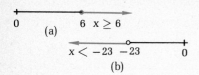

Figure 1-2

Example 3

(a) Let x be a real number. Then, $x - 6 \geq 0$ if and only if $(x - 6) + 6 \geq 0 + 6$; that is, $x \geq 6$. See Figure 1-2(a).

(b) Let x be a real number. Then, $x + 8 < -15$ if and only if $(x + 8) - 8 < -15 - 8$; that is, $x < -23$. See Figure 1-2(b).

Example 4

Find all real numbers x for which $3x - 5 < -7$.

Solution

x is such a number if and only if $3x - 5 + 5 < -7 + 5$, $3x < -2$, and finally, $x < -\frac{2}{3}$ [since $\frac{1}{3} > 0$ and $\frac{1}{3}(-2) = -\frac{2}{3}$]. Thus the required real numbers are those less than $-\frac{2}{3}$.

Example 5

Find all real numbers x for which $(4x - 7)^4 = 0$.

Solution

Since $(4x - 7)^4 \geq 0$ for all numbers x, we see that $(4x - 7)^4 = 0$ if and only if $4x - 7 = 0$, or, equivalently, $x = \frac{7}{4}$.

Example 6

Find all real numbers x and y for which $(x - 1)^2 + (3y - 2)^4 = 0$.

Solution

Since $(x - 1)^2 \geq 0$ and $(3y - 2)^4 \geq 0$ for all numbers x and y, the sum $(x - 1)^2 + (3y - 2)^4 = 0$ if and only if $x - 1 = 0$ and $3y - 2 = 0$; that is, if and only if $x = 1$ and $y = \frac{2}{3}$.

Example 7

Find all real numbers x for which $(x - 1)(3x + 2) > 0$.

Solution

Note that $(x - 1)(3x + 2) = 0$ for $x = -\frac{2}{3}$ and $x = 1$ (and also that $x - 1 < 0$ if and only if $x < 1$, while $3x + 2 < 0$ if and only if $x < -\frac{2}{3}$). Refer now to Figure 1-3. We shall consider what happens to the left of $-\frac{2}{3}$, between $-\frac{2}{3}$ and 1, and to the right of 1:

1. When $x < -\frac{2}{3}$, then also $x < 1$, and hence $x - 1 < 0$ and $3x + 2 < 0$; therefore, $(3x + 2)(x - 1) > 0$.
2. When $-\frac{2}{3} < x < 1$, then $x - 1 < 0$, but $3x + 2 > 0$, and hence $(3x + 2)(x - 1) < 0$.

$x - 1 < 0$	$x - 1 < 0$	$x - 1 > 0$
$3x + 2 < 0$	$3x + 2 > 0$	$3x + 2 > 0$

$-\frac{2}{3}$ 0   1

Figure 1-3

3. When $x > 1$, then both factors are positive, and hence $(3x + 2)(x - 1) > 0$.

We see therefore, that $(3x + 2)(x - 1) > 0$ if and only if $x < -\frac{2}{3}$ or $x > 1$.

Example 8

Find all real numbers x for which $(x + 1)(2 - 3x) > 0$.

Solution

First note that

$$x + 1 < 0 \text{ for } x < -1$$
$$x + 1 > 0 \text{ for } x > -1$$
$$2 - 3x < 0 \text{ for } x > \tfrac{2}{3}$$
$$2 - 3x > 0 \text{ for } x < \tfrac{2}{3}$$

Also note that $-1 < \frac{2}{3}$. Now consider the following cases (see Figure 1-4). For $x < -1$, $x + 1 < 0$ and $2 - 3x > 0$; hence $(x + 1)(2 - 3x) < 0$. For $-1 < x < \frac{2}{3}$, $x + 1 > 0$ and $2 - 3x > 0$; hence $(x + 1)(2 - 3x) > 0$. For $\frac{2}{3} < x$, $x + 1 > 0$ and $2 - 3x < 0$; hence $(x + 1)(2 - 3x) < 0$. Thus the set of all numbers x for which $(x + 1)(2 - 3x) > 0$ is the set of all numbers between -1 and $\frac{2}{3}$.

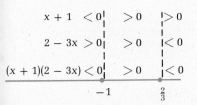

Figure 1-4

Example 9

Find all real numbers x for which $\dfrac{3x + 5}{2x + 1} \geq 3$.

Solution

x is such a number if and only if

$$\frac{3x + 5}{2x + 1} - 3 \geq 0$$

$$\frac{-3x + 2}{2x + 1} \geq 0$$

or, equivalently,

$$\frac{-3x + 2}{2x + 1} = 0 \quad \text{or} \quad \frac{-3x + 2}{2x + 1} > 0$$

Now

$$\frac{-3x + 2}{2x + 1} = 0 \quad \text{if and only if} \quad x = \frac{2}{3}$$

It remains to find all x for which $\dfrac{-3x + 2}{2x + 1} > 0$. First note that

$$-3x + 2 > 0 \quad \text{for } x < \tfrac{2}{3}$$
$$-3x + 2 < 0 \quad \text{for } x > \tfrac{2}{3}$$
$$2x + 1 > 0 \quad \text{for } x > -\tfrac{1}{2}$$

Hence (see Figure 1-5),

$$\dfrac{-3x + 2}{2x + 1} > 0 \quad \text{for } \quad -\tfrac{1}{2} < x < \tfrac{2}{3}$$

Combining all our observations, we see that

$$\dfrac{3x + 5}{2x + 1} \geq 3 \quad \text{if and only if} \quad -\tfrac{1}{2} < x \leq \tfrac{2}{3}$$

Figure 1-5

Exercises 1-3

In Exercises 1–11, answer true or false (Exercises 7–11 are stated for all real numbers a, b, and c, with a negative, b positive, and c ≠ 0).

1. $-4 < -5$ **2.** $[-(-2)] > -6$

3. $(-3)[2 - (-5)] > 4[3 - (-4)](-1)$

4. $2(4 - 3) \leq \dfrac{-4}{-2}$ **5.** $\dfrac{-4}{2} \leq \dfrac{4}{-2}$

6. $\dfrac{-3}{2} > \dfrac{3}{-2}(4 - 2)$

7. $\dfrac{a}{-b} \geq \dfrac{-a}{b}$ and $\dfrac{-a}{b} \geq \dfrac{a}{-b}$

8. $6ab(-a)(-b) > 0$ **9.** $(ab)(b - a) < 0$

10. $12c > 2(b - 1)c$ **11.** $abc > a(b - 1)c$

In Exercises 12–25, find all real numbers x for which

12. $x + 7 < 15$ **13.** $2x + 9 \geq 15$

14. $3x + 4 < 2x + 9$ **15.** $-5(x + 2) \geq 3x + 4$

16. $\dfrac{3}{x - 1} \geq 0$ **17.** $\dfrac{7}{6x - 9} > 0$

18. $1 < 3x + 2 < 4$ **19.** $4 < 3x + 9 < 5$

20. $\dfrac{5}{3} + 2 < \dfrac{3x + 6}{3} < \dfrac{7}{3} + (-9)$ **21.** $6 - x < x < 10 - x$

22. $3 - 2x < 4 - x < 2 - 2x$ **23.** $2 + 5x < 10x < 4 + 5x$

24. $3x + [2 + (-3)] \leq 6x - 3 < 3x$

25. $-1 + \tfrac{2}{3}x \leq x \leq 1 + \tfrac{2}{3}x$

In Exercises 26–33, find all real numbers x and y for which

26. $(3x - 9)^2 + (y - 1)^2 = 0$ **27.** $(2x - 4)^4 + (y - 3)^2 = 0$

28. $(x - 4)(x - 3) \geq 0$

29. $(2x - 8)[-(4x + 12)] < 0$

30. $\dfrac{3x + 4}{2x + 9} \leq 1$ **31.** $\dfrac{-(1 + 3x)}{(x - 2)} \geq 0$

32. $(x - 1)(x - 2)(x - 3) > 0$

33. $(x - 2)(x - 3)(2x + 6) \leq 0$

In Exercises 34–40, answer true or false (a, b, c, and d are real numbers).

34. $(-a)b$ is negative.

35. If c is positive and $a > b$, then $(a - b)(-c) < 0$.

36. If $a > 0$, $b > 0$, $c > 0$, and $d < 0$, then $(a - c)b > d(-b)$.

37. If $a < 1$, then $a^2 < a$.   **38.** If $a > 1$, then $a^2 > a$.

39. If $0 < a < 1$, then $a^2 < a$.

40. If $0 < a < b$ and $c \geq 0$, then $\dfrac{a}{b} \leq \dfrac{a + c}{b + c}$.

1-4 INTEGER EXPONENTS

Recall that if x is a real number and n is a natural number (that is, a positive integer) then x^n (read "x to the nth power" or "the nth power of x") is the product $\underbrace{x \cdot x \cdots x}_{n \text{ times}}$.

In other words, $x^1 = x$ and $x^n = x^{n-1} \cdot x$ for $n > 1$. The reader will have no difficulty in verifying that, more generally, if m is also a natural number, then

$$x^n \cdot x^m = x^{n+m}, \quad (x^n)^m = x^{n \cdot m}$$

and

$$\frac{x^n}{x^m} = \begin{cases} x^{n-m} & \text{if } n > m \\ \dfrac{1}{x^{m-n}} & \text{if } n < m \end{cases} \quad x \neq 0$$

Furthermore, if y is also a real number, then

$$(xy)^n = x^n \cdot y^n$$

To make computations with integer exponents consistent with the foregoing facts, we now define

$$x^0 = 1 \quad \text{and} \quad x^{-n} = \frac{1}{x^n}$$

for any nonzero real number x and natural number n. It can be shown that with these definitions, the "rules of exponents" still hold. They are summarized below for reference:

If x and y are nonzero real numbers, and m and n are any integers, then

(a) $x^m \cdot x^n = x^{m+n}$ (b) $(x^m)^n = x^{m \cdot n}$

(c) $(xy)^n = x^n \cdot y^n$ (d) $\dfrac{x^n}{x^m} = x^{n-m}$

The proofs are left as an exercise.

Example 1

If x, y, and z are real numbers, then

(a) $5^9 \cdot 5^7 = 5^{9+7} = 5^{16}$
(b) $2^5 \cdot 4^3 = 2^5 \cdot (2^2)^3 = 2^5 \cdot 2^6 = 2^{11}$
(c) $x^3 x^8 = x^{3+8} = x^{11}$
(d) $(xy)^2(x^3y) = x^2y^2x^3y = x^{2+3}y^{2+1} = x^5y^3$
(e) $(x^3y^2)^8 = (x^3)^8(y^2)^8 = x^{24}y^{16}$
(f) $(x^2y^3z^4)^5 = x^{10}y^{15}z^{20}$

Example 2

(a) $\dfrac{5^9}{5^7} = 5^{9-7} = 5^2 = 25$

(b) $\dfrac{5^7}{5^9} = \dfrac{1}{5^{9-7}} = \dfrac{1}{25}$

(c) $\dfrac{5^4}{25^3} = \dfrac{5^4}{(5^2)^3} = \dfrac{5^4}{5^6} = \dfrac{1}{5^2} = \dfrac{1}{25}$

(d) If $x \neq 0$, $y \neq 0$, then $\dfrac{(x^2y)^5}{(xy)^7} = \dfrac{x^{10}y^5}{x^7y^7} = \dfrac{x^3}{y^2}$

Example 3

(a) $10^0 = 1$

$10^1 = 10$ and $10^{-1} = \dfrac{1}{10} = .1$

$10^2 = 100$ and $10^{-2} = \dfrac{1}{10^2} = .01$

$10^3 = 1000$ and $10^{-3} = \dfrac{1}{10^3} = .001$

(b) $2^0 = 1$

$2^1 = 2$ and $2^{-1} = \frac{1}{2} = .5$

$2^2 = 4$ and $2^{-2} = \frac{1}{4} = .25$

$2^3 = 8$ and $2^{-3} = \frac{1}{8} = .125$

Example 4

If x is a nonzero real number, then

(a) $\dfrac{x^5}{x^7} = x^5 \cdot \dfrac{1}{x^7} = x^5 \cdot x^{-7} = x^{5-7} = x^{-2} = \dfrac{1}{x^2}$

(b) $(x^{-3})^2 = x^{-6} = \dfrac{1}{x^6}$

(c) $(x^{-2})^{-3} = x^6$, or $(x^{-2})^{-3} = \left(\dfrac{1}{x^2}\right)^{-3} = \dfrac{1}{x^{-6}} = x^6$,

or $(x^{-2})^{-3} = \dfrac{1}{(x^{-2})^3} = \dfrac{1}{x^{-6}} = x^6$

(d) If y is also a nonzero real number, then

$$(x^{-3}y^2)^{-4} = x^{12}y^{-8} = \dfrac{x^{12}}{y^8}$$

(e) $(x + y)^{-1} = \dfrac{1}{x + y}$ for $x + y \neq 0$

Exercises 1-4

In Exercises 1–13, compute (express answers in exponential form as in Exercise 1).

1. $3^4 \cdot 9 = 3^4 \cdot 3^2 = 3^{4+2} = 3^6$

2. $2^2 \cdot 2^4$
3. $(2^3)^2$

4. $2^3 \cdot 4^2$
5. $36^2 \cdot 6^{-4}$

6. $2^{-2}(2^2 + 2)$
7. $3^3(6 + 3 \cdot 8^0)$

8. $\dfrac{2^7}{2^6}$
9. $\dfrac{6^4}{6^2}$

10. $\left(\dfrac{6^2}{36^3}\right)^{-3}$
11. $\left(\dfrac{3^3}{9}\right)^{-1}$

12. $(3^{-1} + 2^{-1})^{-2}$
13. $\dfrac{4^{-1} + 3^{-2}}{2}$

In Exercises 14–34, compute (x, y, and z are nonzero real numbers).

14. $(x^4y^3)^2x^{-1}$
15. $(x^{-2}y^4)^3(xy)^3$

16. $\dfrac{x^4}{x^5 y^{-1}}$ **17.** $(x^2 y^2)^{-3}$

18. $\dfrac{24 x^4 y^3}{8 y^3 x}$ **19.** $\dfrac{27 x^4 y^3}{18 x^5 y^2}$

20. $\dfrac{4 x^4 y^6}{6 z^5} \cdot \dfrac{9 x^3 z^6}{12 y^7}$ **21.** $\dfrac{7 x^2 y^2}{10 y z^4} \cdot \dfrac{5 z^2}{14 x^2 z^2}$

(In Exercises 22–29, n is a natural number.)

22. $\left(x^{n+1} \right) \dfrac{1}{x^{1-n}}$ **23.** $(x^2 y^{2n})^3 (x^{-5} y^{-7n})$

24. $(x^{2+n})(x^{1-2n})$ **25.** $(x^{2+4n})(x^{-3n})$

26. $\dfrac{x^{n+1} y^{n+1}}{x^{2-n} y^{2n-1}}$ **27.** $\dfrac{x^{2n-1} y^{n+2}}{x^{n+4} y^{2n-1}}$

28. $(x^{2n+6} y^{n-4})^4 (x^{3n-1} y^{2n-3})^{-1}$ **29.** $\left(\dfrac{12 x^{2n-2}}{6 x^{n-2}} \right)^4 \left(\dfrac{1}{2 x^{2n}} \right)^2$

[Specify "necessary" restrictions for x and y in each step of your calculations in Exercises 30–34. For example, in Exercise 30, since $(x + y^{-1})^{-1} = \dfrac{1}{x + (1/y)}$, we must have $y \neq 0$ and $x + (1/y) \neq 0$.]

30. $(x + y^{-1})^{-1}$ **31.** $xy(x^{-1} + y^{-1})$

32. $\dfrac{x^{-2} - y^{-2}}{x^{-2} y^{-2}}$ **33.** $\dfrac{(x^{-1} + y^{-1})(x + y)^{-1}}{(xy)^{-1}}$

34. $\dfrac{x^{-2} - y^{-2}}{(x - y)^2}$

In Exercises 35–39, answer true or false (a and b are real numbers).

35. If $0 < a < 1$, then $a^n < a$ for every natural number n.

36. If $a < b$, then $a^2 < b^2$.   **37.** If $a < b$, then $a^3 < b^3$.

38. If $0 < a < b$, then $a^n < b^n$ for every natural number n.

39. If $a < 0 < b$, then $1/a^n > 1/b^n$ for every natural number n.

1-5 ROOTS AND ABSOLUTE VALUE

Let n be a natural number and x a real number. Then an **nth root of *x*** is any real number whose nth power is x. In other words, the real number y is an nth root of x if and only if $y^n = x$.

Case I Since an even power of a real number cannot be negative, only nonnegative real numbers have even roots. If

n is even and $x > 0$, we denote by $\sqrt[n]{x}$ the **positive nth root of x**; thus,

$$y = \sqrt[n]{x} \quad \textbf{if and only if} \quad y > 0 \text{ and } y^n = x$$

Furthermore, if n is even then $(-y)^n = y^n$, and so we see that every positive real number x has two nth roots: one positive ($\sqrt[n]{x}$) and one negative ($-\sqrt[n]{x}$).

Case II If n is odd, then any real number x has precisely one nth root, which is positive if x is positive and negative if x is negative; in either case, it is denoted by $\sqrt[n]{x}$.

Also, $\sqrt[n]{x} = 0$ if and only if $x = 0$, for any natural number n.

It is general usage to replace the symbol $\sqrt[2]{}$ with $\sqrt{}$. It follows from Case I that if x is a real number, then $\sqrt{x^2}$ is nonnegative, regardless of whether x is positive, negative, or zero; more precisely,

$$\sqrt{x^2} = \begin{cases} x & \text{if } x \geq 0 \\ -x & \text{if } x < 0 \end{cases}$$

This number $\sqrt{x^2}$ is called the **absolute value of x** and is usually denoted by $|x|$. Thus,

$$|x| = \begin{cases} x & \textbf{if } x \geq 0 \\ -x & \textbf{if } x < 0 \end{cases}$$

Example 1

(a) $|-3| = |3| = 3$
(b) If x is a real number, then $|-x| = |x| \geq 0$
□

It is useful to observe that $|x|$ is the *distance* between x and 0. Thus, for example, the distance between -5 and 0 is $|-5| = 5$. With this in mind, the following two properties of absolute value are easily verified (see Figure 1-6):

If b is a nonnegative real number, then:

1. The numbers x for which $|x| = b$ are b and $-b$; that is, $|x| = b$ if and only if $x = b$ or $x = -b$;
2. x is a real number for which $|x| < b$ if and only if $-b < x < b$; and x is a real number for which $|x| > b$ if and only if $x > b$ or $x < -b$.

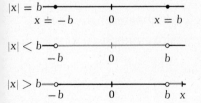

$|x| = b$
$x = -b \quad\quad 0 \quad\quad x = b$

$|x| < b$
$-b \quad\quad 0 \quad\quad b$

$|x| > b$
$-b \quad\quad 0 \quad\quad b \quad x$

Figure 1-6

Example 2

Find all real numbers x for which (a) $|2x - 3| = 5$
(b) $|2x - 3| < 5$ (c) $|2x - 3| \leq 5$ (d) $|2x - 3| > 5$

Solution

(a) x is a real number for which $|2x - 3| = 5$ if and only if $2x - 3 = 5$ or $2x - 3 = -5$; that is, $x = 4$ or $x = -1$.

(b) x is a real number for which $|2x - 3| < 5$ if and only if

$$-5 < 2x - 3 < 5$$
$$-2 < 2x < 8$$

or, finally,

$$-1 < x < 4$$

(c) combining (a) and (b), we see that the solution to (c) is the set of all real numbers x for which $-1 \leq x \leq 4$.

(d) x is a real number for which $|2x - 3| > 5$ if and only if

$$2x - 3 > 5, \text{ or, } -5 > 2x - 3$$

that is,

$$2x > 8 \quad \text{or} \quad -2 > 2x$$

or, finally,

$$x > 4 \quad \text{or} \quad x < -1$$

Example 3

(a) $\sqrt[3]{-8} = -2$ since $(-2)^3 = -8$

(b) $\sqrt[3]{27} = 3$ since $3^3 = 27$

(c) $\sqrt[4]{16} = 2$ since $2^4 = 16$

(d) $\sqrt[n]{0} = 0$ since $0^n = 0$ for any positive integer n

(e) $\sqrt[n]{1} = 1$ since $1^n = 1$ for any positive integer n

(f) $\sqrt[5]{\dfrac{243}{3125}} = \dfrac{3}{5}$ since $\left(\dfrac{3}{5}\right)^5 = \dfrac{3^5}{5^5} = \dfrac{243}{3125}$

(g) $\sqrt[3]{-\dfrac{27}{8}} = -\dfrac{3}{2}$ since $\left(-\dfrac{3}{2}\right)^3 = -\dfrac{27}{8}$

Example 4

$$|5| = 5, \quad |-5| = 5, \quad |0| = 0, \quad \sqrt{(-3)^2} = |-3| = 3$$

Two important properties or roots are

$$\sqrt[n]{xy} = \sqrt[n]{x}\,\sqrt[n]{y}$$

and

$$\sqrt[n]{\frac{x}{y}} = \frac{\sqrt[n]{x}}{\sqrt[n]{y}}$$

whenever all the roots are defined. Proof of these properties is left as an exercise.

Example 5

$$\sqrt[3]{(-8)\cdot 27} = \sqrt[3]{-8}\,\sqrt[3]{27} = -2\cdot 3 = -6$$

$$\sqrt[4]{\frac{16}{81}} = \frac{\sqrt[4]{16}}{\sqrt[4]{81}} = \frac{2}{3}$$

Example 6

Let x and y be positive real numbers, then

$$\sqrt{25x^8y^4} = \sqrt{25}\,\sqrt{x^8y^4} = \sqrt{25}\,\sqrt{x^8}\,\sqrt{y^4}$$
$$= 5x^4y^2 \text{ (since } 5^2 = 25, \quad (x^4)^2 = x^8 \text{ and } (y^2)^2 = y^4)$$

Example 7

Let x and y be positive real numbers, then

(a) $\sqrt{\dfrac{x^6}{4y^2}} = \dfrac{x^3}{2y}$

(b) $\sqrt[3]{\dfrac{x^{12}y^7}{27}} = \sqrt[3]{\dfrac{(x^4)^3(y^2)^3y}{27}} = \dfrac{x^4y^2\,\sqrt[3]{y}}{3}$

(c) $\sqrt[5]{x^{11}y^{17}} = \sqrt[5]{(x^2)^5x\cdot(y^3)^5y^2} = x^2y^3\,\sqrt[5]{xy^2}$

Example 8

$$\sqrt{8} + \sqrt{32} - \sqrt{18} = \sqrt{4\cdot 2} + \sqrt{16\cdot 2} - \sqrt{9\cdot 2}$$
$$= 2\sqrt{2} + 4\sqrt{2} - 3\sqrt{2}$$
$$= 3\sqrt{2}$$

Exercises 1-5

In Exercises 1–47, simplify.

1. $\sqrt{144}$ 2. $\sqrt{169}$

3. $\sqrt[4]{81}$ 4. $\sqrt[6]{64}$

5. $\sqrt[7]{-128}$ 6. $\sqrt[3]{-216}$

7. $\sqrt[3]{\dfrac{-64}{125}}$ 8. $\sqrt[4]{\dfrac{16}{625}}$

9. $\sqrt{24}$ 10. $\sqrt{54}$

11. $\sqrt{75(16)^0}$ 12. $\sqrt{36(25)}$

13. $\sqrt[3]{-108(64)}$ 14. $\sqrt[3]{-\frac{16}{81}(-216)}$

15. $\sqrt[n+1]{1}$ 16. $\sqrt[14]{2^2 + 2(2)(-2) + (-2)^2}$

17. $\sqrt{3 \cdot 18^6}$ 18. $\sqrt{27^4 \cdot 2^2}$

19. $\sqrt{(-3)^6 \cdot 5}$ 20. $\sqrt[4]{16^7 \cdot 18^5}$

21. $\sqrt[5]{6^8 \cdot 9^{10} \cdot 4^7}$ 22. $\sqrt[3]{(-3)^{15}(-4)^{21}}$

23. $\sqrt{|-4|^2}$ 24. $\sqrt[3]{8 \cdot |-12|^3 \cdot 6^{-6}}$

25. $3\sqrt{32} - 2\sqrt{50}$

26. $2\sqrt{218} - 4\sqrt{96}$

27. $\sqrt{96} + \sqrt{150}$

28. $3\sqrt{6} + 4\sqrt{54} - \sqrt{600}$

29. $\sqrt{8} + \sqrt{16} - 2\sqrt{50}$

30. $-\sqrt[3]{16} + \sqrt[3]{375} - \sqrt[3]{54}$

31. $5\sqrt{48} + \sqrt{162} - 3\sqrt{375}$

32. $3 \cdot \sqrt{\frac{32}{9}} - 4 \cdot \sqrt{\frac{18}{25}} + 2 \cdot \sqrt{50}$

33. $2 \cdot \sqrt{\frac{3}{4}} - 3 \cdot \sqrt{\frac{4}{9}} + 2 \cdot \sqrt{\frac{27}{16}}$

In Exercises 34–47, x, y, and z are positive real numbers. Simplify.

34. $\sqrt{x^4}$

35. $\sqrt{x^5 y^3}$

36. $\sqrt{x^{-4} y^5}$

37. $\sqrt[7]{x^9(-y)^{11} z^{-14}}$

38. $\sqrt[5]{\dfrac{x^6}{y^7 z^8}}$

39. $\sqrt[16]{\dfrac{x^{32} z^{-17}}{y^{20}}}$

40. $\sqrt{(x - y)^2}$

41. $\sqrt{(x - y)^4}$

42. $\sqrt[3]{(x - y)^3(x - z)^4(y - z)^7}$

43. $\sqrt[7]{(-3x)^9(yz)^{-8}}$

44. $\sqrt[19]{(x^3y^4)^7(-x)^2y^{-28}}$

45. $\sqrt{x^3} - 2 \cdot \sqrt{x^5}$

46. $3 \cdot \sqrt{\dfrac{x^3}{y^2}} - 4 \cdot \sqrt{\dfrac{x^5}{y^2}}$

47. $3 \cdot \sqrt[3]{x^5y} - 4 \cdot \sqrt[3]{x^7yx} + 5 \cdot \sqrt[3]{x^9yx^{-1}}$

In Exercises 48–53, find all real numbers x for which

48. $|x| = 5$ **49.** $|x - 2| = 7$

50. $|x - 3| < 1$ **51.** $|x + 4| \geq 2$

52. $|3x - 1| = 3$ **53.** $|2x + 1| \leq 7$

In Exercises 54–60, answer true or false and justify your answer (a and b are real numbers).

54. $|a \cdot b| = |a| \cdot |b|$

55. If $a < 0$ and $b < 0$, then $|a + b| = |a| + |b|$

56. If $a < 0$, then $\sqrt[3]{a^3} = -a$

57. $|a| < 4$ if and only if $-4 < a < 4$

58. $|a| \leq |b|$ if and only if $-|b| \leq a \leq |b|$

59. $|a - b| \leq |a| - |b|$

60. $||a| - |b|| \leq |a + b| \leq |a| + |b|$

In Exercises 61 and 62, find all real numbers x for which

61. $|x| + |x - 2| = 5$ **62.** $|3x - 2| + |5x + 7| = 1$

1-6 RATIONAL EXPONENTS

To extend the notion of exponents to rational numbers so that the laws of exponents still hold, we first observe that for any positive real number x and natural number n we must have $(x^{1/n})^n = x^{n/n} = x^1 = x$. In other words, $x^{1/n}$ must be the number whose nth power is x; thus, we define

$$x^{1/n} = \sqrt[n]{x}$$

The extension that immediately suggests itself, if the rules are to hold, is

$$x^{m/n} = (\sqrt[n]{x})^m$$

for any positive real number x, natural number n, and integer m; for then we have

$$x^{m/n} = (x^{1/n})^m = (\sqrt[n]{x})^m$$

It might appear that there is an ambiguity in this definition. For example, should $7^{.4}$ mean $(\sqrt[10]{7})^4$ or, since $.4 = \frac{4}{10} = \frac{2}{5}$, should it mean $(\sqrt[5]{7})^2$? In fact, there is no ambiguity since $(\sqrt[5]{7})^2 = (\sqrt[10]{7})^4$, and, in general, $(\sqrt[nk]{x})^{mk} = (\sqrt[n]{x})^m$ for any positive integer k.*

Example 1

$$8^{2/3} = (\sqrt[3]{8})^2 = 2^2 = 4, \qquad (\tfrac{9}{4})^{5/2} = (\sqrt{\tfrac{9}{4}})^5 = (\tfrac{3}{2})^5 = \tfrac{243}{32}$$

$$10^{.2} = 10^{2/10} = 10^{1/5} = \sqrt[5]{10}$$

Because of the way rational exponents have been defined, the laws of exponents still hold:

1. $x^{-r} = \dfrac{1}{x^r}$

2. $x^{r+s} = x^r \cdot x^s$

3. $(x^r)^s = x^{rs}$

4. $(xy)^r = x^r \cdot y^r$

for any positive real numbers x and y and rational numbers r and s. (In case x or y is negative, these rules apply whenever the corresponding conditions regarding roots are satisfied. See Section 1–5.) We shall prove (2) and (3) and leave the proofs of (1) and (4) as an exercise.

Let $r = m/n$ and $s = p/q$, where n and q are natural numbers and m and p any integers; also, let $y = \sqrt[nq]{x} = x^{1/nq}$, where x is a positive real number, so that $x = y^{nq}$. Then

$$x^r = (x^{1/n})^m = (y^q)^m = y^{qm}, \qquad x^s = (x^{1/q})^p = (y^n)^p = y^{np}$$

and so

$$x^r \cdot x^s = y^{qm} \cdot y^{np} = y^{qm+np} = (x^{1/nq})^{qm+np} = x^{qm+np/nq} = x^{r+s}$$

Also,

$$(x^r)^s = (y^{mq})^s = [(y^m)^q]^{p/q} = y^{mp} = x^{mp/nq} = x^{rs}$$

Example 2

(a) $(64)^{1/2}(64)^{2/3} = 8 \cdot 4^2 = 8 \cdot 16 = 128$ or

$(64)^{1/2} \cdot (64)^{2/3} = (64)^{1/2+2/3} = (64)^{7/6} = [(64)^{1/6}]^7 = 2^7 = 128$

(b) $[(32)^{1.8}]^{1/3} = (32)^{1/3(1.8)} = (32)^{.6} = (32)^{3/5} = [(32)^{1/5}]^3 = 2^3 = 8$

---

*Let $y = \sqrt[nk]{x}$, so that $x = y^{n \cdot k}$; then $\sqrt[n]{x} = \sqrt[n]{(y^k)^n} = y^k$ and $(\sqrt[n]{x})^m = (y^k)^m = y^{k \cdot m} = (\sqrt[nk]{x})^{mk}$. Thus, $(\sqrt[4]{3})^8 = 3^{8/4} = 3^2 = 9$.

Example 3

(a) $(27)^{2/3} \cdot 4^{-5/2} = (\sqrt[3]{27})^2 \cdot \dfrac{1}{(\sqrt[2]{4})^5} = \dfrac{3^2}{2^5} = \dfrac{9}{32}$

(b) $\dfrac{\sqrt[3]{4}}{\sqrt{2}} = 4^{1/3} \cdot 2^{-1/2} = (2^2)^{1/3} \cdot 2^{-1/2} = 2^{2/3 - 1/2} = 2^{1/6} = \sqrt[6]{2}$

Example 4

If x, y, and z are positive real numbers, then

(a) $x^{2/5} \cdot x^{3/5} \cdot y^{1/2} \cdot y^{-1/3} \cdot z^{-1/2} \cdot z^{1/4} = x^{2/5 + 3/5} y^{1/2 - 1/3} z^{-1/2 + 1/4}$

$$= xy^{1/6} z^{-1/4} = \dfrac{x \cdot \sqrt[6]{y}}{\sqrt[4]{z}}$$

(b) $\left(\dfrac{32x^{15}y^{10}}{z^{20}}\right)^{1/5} = \dfrac{32^{1/5}x^{15/5}y^{10/5}}{z^{20/5}} = \dfrac{2x^3y^2}{z^4}$

(c) $(x^{5/6}y^{10}z^{-1/3})^{-3/5} = x^{-1/2}y^{-6}z^{1/5} = \dfrac{\sqrt[5]{z}}{y^6 \cdot \sqrt{x}}$

Exercises 1-6

In Exercises 1–8, compute.

1. $5^{1/2} \cdot 5^{3/2}$

2. $4^{5/2} \cdot 4^{-2}$

3. $32^{.2}$

4. $1^{-2.4}$

5. $(-64)^{5/3}$

6. $[(1024)^{-(.3)}]^{1/3}$

7. $\left(\dfrac{64^{-2/5}}{2^{-2}}\right)^{1/2}$

8. $\dfrac{5^{1/3} \cdot 4^{1/2}}{5^{4/3} \cdot 4^{1/4}}$

In Exercises 9–20, simplify (x, y, and z are positive real numbers).

9. $\left(\dfrac{64x^{36}(yz)^{24}}{3^{-4}}\right)^{1/6}$

10. $(27x^{3/5}y^{-3})^{1/3}$

11. $(16x^{-6}y^{-2/3})^{-1/2}$

12. $\dfrac{32^{3/5}x^{3/7}y^{-2/3}}{4x^{-3/7}y}$

13. $\dfrac{16^{3/4}x^{-2/3}y}{8^{2/3}x^{-1}y^{-1}}$

14. $(9x^4y^6)^{1/2}(27x^3y^9)^{-1/3}$

15. $(4x^{-2}y^4)^{1/2}(8x^3y^{-6})^{1/3}$

16. $\left(\dfrac{32x^3z^{5/3}}{x^0 \cdot 3^{5/6}}\right)^{-1/5}$

17. $\left(\dfrac{x^{-6}y^{6/5}}{64^{-1}x^{-6/7}}\right)^{-1/6}$

In Exercises 18–20, m, n, and k, are natural numbers.

18. $\left(\dfrac{x^{1/m-2}}{x^{1/m+2}}\right)^{(m^2-4)/m}$

19. $\left(\dfrac{x^{m+3n}}{x^{2n}}\right)^{m/(m+n)}$

20. $\left(\dfrac{x^{k-3}}{x^{-4}}\right)^{1/(k^2-1^2)}$

In Exercises 21–24, answer true or false (m and n are natural numbers; a and b are real numbers).

21. $\sqrt[m]{\sqrt[n]{a}} = \sqrt[mn]{a}$

22. $(a^{2n})^{1/2n} = |a|$

23. $\sqrt[n]{a+b} = \sqrt[n]{a} + \sqrt[n]{b}$

24. $(1+a)^n = 1 + a^n$

1-7 FACTORING AND SPECIAL PRODUCTS

Many computations employ repeated application of the distributive property. For example, if a and b are real numbers, then

$$(a+b)^2 = (a+b)(a+b) = (a+b)a + (a+b)b$$
$$= aa + ba + ab + bb$$
$$= a^2 + 2ab + b^2$$

That is,

$$(a+b)^2 = a^2 + 2ab + b^2$$

Similarly, if a, b, c, d, and e are real numbers, then

$$(a+b)(c+d+e) = (a+b)c + (a+b)d + (a+b)e$$
$$= ac + bc + ad + bd + ae + be$$

Example 1

Let x be a real number.

(a) $(2x+1)(5x-3) = (2x+1)(5x) + (2x+1)(-3)$
$$= 10x^2 - x - 3$$

(b) $(x-3)(x^3-2x+5) = (x-3)x^3 + (x-3)(-2x) + (x-3)5$
$$= x^4 - 3x^3 - 2x^2 + 11x - 15$$

□

Thus, in particular, for any real number x,

$$x^2 + 2x + 1 = (x+1)^2$$

This fact has the interesting application shown in the following example.

Example 2

Find all real numbers x for which $x^2 + 2x + 1 = 0$.

Solution

For any real number x, $x^2 + 2x + 1 = (x + 1)^2$. Hence, $x^2 + 2x + 1 = 0$ if and only if $(x + 1)^2 = 0$, and from this we see that $x + 1 = 0$. (Why?) Hence, $x = -1$.

□

The following "special products" occur frequently enough to be worth listing. Each can be verified using the basic properties of real numbers. In each case, x and y are real numbers.

1. $x^2 - y^2 = (x + y)(x - y)$
   Proof: $(x + y) \cdot (x - y) = (x + y) \cdot x - (x + y) \cdot y$
   $$= x^2 + yx - xy - y^2$$
   $$= x^2 - y^2$$

2. $x^2 + (a + b)x + ab = (x + a)(x + b)$ for any real numbers a and b

3. $x^3 - y^3 = (x - y)(x^2 + xy + y^2)$

4. $x^3 + y^3 = (x + y)(x^2 - xy + y^2)$

5. $x^n - y^n = (x - y)(x^{n-1} + x^{n-2}y + x^{n-3}y^2 + \cdots + y^{n-1})$
   for any positive integer n

6. $x^n + y^n = (x + y)(x^{n-1} - x^{n-2}y + x^{n-3}y^2 - \cdots + y^{n-1})$
   for any *odd* positive integer n

7. $(x + y)^2 = x^2 + 2xy + y^2$

8. $(x + y)^3 = x^3 + 3x^2y + 3xy^2 + y^3$

9. $(x + y)^4 = x^4 + 4x^3y + 6x^2y^2 + 4xy^3 + y^4$

Example 3

For every real number x

(a) $x^2 - 4 = (x + 2)(x - 2)$

(b) $x^2 - 5 = (x + \sqrt{5})(x - \sqrt{5})$

(c) $x^2 - 2x - 15 = (x - 5)(x + 3)$

(d) $x^3 - 1 = (x - 1)(x^2 + x + 1)$

Example 4

Let x and y be real numbers.

(a) $x^6 - y^4 = (x^3)^2 - (y^2)^2 = (x^3 + y^2)(x^3 - y^2)$

(b) $x^5y^{12} - 9x^3y^8 = x^3y^8(x^2y^4 - 9)$
   $$= x^3y^8[(xy^2)^2 - 3^2]$$
   $$= x^3y^8(xy^2 + 3)(xy^2 - 3)$$

(c) $x^2 - xy + 3x - 3y = x(x - y) + 3(x - y)$
   $$= (x + 3)(x - y)$$

Example 5

Show that $a^2 + \dfrac{1}{a^2} \geq 2$ for any nonzero real number a.

Solution

Let a be any nonzero real number; then,

$$a^2 + \frac{1}{a^2} \geq 2$$

if and only if

$$a^2 \left(a^2 + \frac{1}{a^2} \right) \geq 2a^2$$

if and only if

$$a^4 - 2a^2 + 1 \geq 0$$

if and only if

$$(a^2 - 1)^2 \geq 0$$

Since $(a^2 - 1)^2$ is indeed greater than or equal to zero for any real number a and the argument was *if and only if* at each stage, it follows that

$$a^2 + \frac{1}{a^2} \geq 2 \quad \text{for any nonzero real number}$$

Exercises 1-7

1. Apply the basic properties of real numbers to verify the special products (2) − (4) and (7) − (9).

In Exercises 2–11, x and y are real numbers; apply the distributive property of real numbers to compute.

2. $(x + 4)(x - 3)$
3. $(x^3 - y^3)(x^3 + y^3)$
4. $(3x + \sqrt{3})(x - \sqrt{3})$
5. $(x - \sqrt{x^2})^2$
6. $(x - 4)(x^2 + 4x + 16)$
7. $(x + 3)(x^2 - 3x + 9)$
8. $(x - 2)(x^3 + 2x^2 + 4x + 8)$
9. $(x + 3)(x^3 - 3x^2 + 9x - 27)$
10. $(2x + 4)^3$
11. $(3x - 9)^3$

In Exercises 12-24, let x and y be real numbers; factor as in Exercise 12.

12. $x^2 + 5x - 14 = (x - 2)(x + 7)$

13. $4x^2 - 14x + 12$ **14.** $2x^2 + \frac{13}{2}x + \frac{15}{4}$

15. $6x^2 - 5x - 6$ **16.** $6x^2 + 11x - 2$

17. $x^2 - 7$ **18.** $13 - 4x^2$

19. $2x^3 - 4x^2 - 3x + 6$ **20.** $49y^2 - 3x^2$

21. $x^4y^2 - x^2y^4$ **22.** $x^2 - 2xy^2 + y^4$

23. $x^3y - x^2 + 3xy^3 - 3y^2$ **24.** $x^4 - y^2$

In Exercises 25-27, answer true or false; justify your answer.

◊**25.** If $x > 0$ and $y > 0$, then $\dfrac{x}{y} + \dfrac{y}{x} \geq 2$

◊**26.** If $x \neq y$, then $x^2 + 2xy \geq 2x^2 + y^2$

◊**27.** For any real numbers a and b, $a^2 + b^2 \geq 2ab$

1-8 FRACTIONS

We now apply the results of the preceding sections to more sophisticated problems involving fractions.

Example 1

If $x \neq 0$ and $y \neq 0$, then

$$\frac{3}{2x^2} - \frac{7}{5xy^2} = \frac{3}{2x^2} \cdot \frac{5y^2}{5y^2} - \frac{7}{5xy^2} \cdot \frac{2x}{2x}$$

$$= \frac{15y^2}{10x^2y^2} - \frac{14x}{10x^2y^2} = \frac{15y^2 - 14x}{10x^2y^2}$$

Example 2

If $x \neq 3$ or 5, then

$$\frac{6}{x^2 + 2x - 15} + \frac{x}{x - 3} = \frac{6}{(x + 5)(x - 3)} + \frac{x}{x - 3}$$

$$= \frac{6}{(x + 5)(x - 3)} + \frac{x}{x - 3} \cdot \frac{x + 5}{x + 5}$$

$$= \frac{6 + x^2 + 5x}{(x - 3)(x + 5)}$$

$$= \frac{(x + 3)(x + 2)}{(x - 3)(x + 5)}$$

Example 3

If $x \neq 0$, 1, or -1, then

$$\frac{1 + \dfrac{1}{x}}{x - \dfrac{1}{x}} = \frac{\dfrac{x}{x} + \dfrac{1}{x}}{\dfrac{x^2}{x} - \dfrac{1}{x}} = \frac{\dfrac{x+1}{x}}{\dfrac{x^2-1}{x}}$$

$$= \frac{x+1}{x} \cdot \frac{x}{x^2-1} = \frac{(x+1)}{(x+1)(x-1)}$$

$$= \frac{1}{x-1}$$

Example 4

If a and b are real numbers for which $a + 2b \neq 0$, $b^2 - a^2 \neq 0$, and $a^2 + ab \neq 0$, then

$$\frac{\dfrac{a^2 + 2ab + b^2}{a + 2b}}{\dfrac{b^2 - a^2}{a^2 + 2ab}} = \frac{a^2 + 2ab + b^2}{a + 2b} \cdot \frac{a^2 + 2ab}{b^2 - a^2}$$

$$= \frac{(a+b)^2}{a+2b} \cdot \frac{a(a+2b)}{(b-a)(b+a)}$$

$$= \frac{a(a+b)}{b-a}$$

Exercises 1–8

In Exercises 1–6, simplify (x and y are nonzero real numbers) as in Exercise 1.

1. $\dfrac{3}{x^2y} - \dfrac{4}{y^2} = \dfrac{3}{x^2y} \cdot \dfrac{y}{y} - \dfrac{4}{y^2} \cdot \dfrac{x^2}{x^2} = \dfrac{3y}{x^2y^2} - \dfrac{4x^2}{y^2x^2} = \dfrac{3y - 4x^2}{x^2y^2}$

2. $\dfrac{2}{x^3y} - \dfrac{7}{3xy}$

3. $\dfrac{4}{5x^5y} - \dfrac{6}{3x^3y^2}$

4. $\dfrac{x^3y^2}{xy^2} + \dfrac{xy^4}{xy^3}$

5. $\dfrac{1}{xy} - \dfrac{y}{x^2} + \dfrac{y^2}{x^2} + \dfrac{1}{y^2}$

6. $\dfrac{x^2}{y^6} - \dfrac{1}{x^2y^4} + \dfrac{x^4}{y^{10}}$

In Exercises 7–19, simplify; assume that x, y, z, a, and b are real numbers with all denominators not zero.

7. $\dfrac{6}{1 - 3x} - \dfrac{1}{2x - 1} + \dfrac{3}{x} + \dfrac{x}{5x - 6x^2 - 1}$

8. $\dfrac{1}{x + 2} - \dfrac{2x + 9}{6 + x - x^2} - \dfrac{2x}{x^2 - 2x - 3}$

9. $\dfrac{3}{x^2 - 1} + \dfrac{4x}{2x^2 + x - 3} - \dfrac{3x + 2}{2x^2 + 5x + 3}$

10. $\dfrac{2 - x}{x^2 - 1} + \dfrac{8}{1 - x^2} - \dfrac{5 + x}{1 - x^2}$

11. $\dfrac{x + 3}{4x^2 + 4x + 1} - \dfrac{2x + 1}{4x^2 - 4x + 1} - \dfrac{16x^4 + 3}{16x^4 - 8x^2 + 1}$

12. $\dfrac{\dfrac{3x - 1}{9x^2 - 1}}{4x + 5}$

13. $\dfrac{\dfrac{x^2 + 14x - 15}{x^2 + 4x - 5}}{\dfrac{x^2 + 12x - 45}{x^2 + 6x - 27}}$

14. $\dfrac{\dfrac{x^2 - 11x + 18}{x^2 - 9x + 8}}{\dfrac{x^2 - 7x + 10}{x^2 - 6x - 16}}$

15. $\dfrac{1 + \dfrac{2}{x - 1}}{\dfrac{x^2 + x}{x^2 + x - 2}}$

16. $\dfrac{(x + 1)^{-1} - (x - 1)^{-1}}{(x - 1)^{-1} + (x + 1)^{-1}}$

17. $\dfrac{(x - 2)^{-1} - (x - 3)^{-1}}{1 + (x^2 - 5x + 6)^{-1}}$

18. $\dfrac{\dfrac{a + 1}{a - 1} - \dfrac{a - 1}{a + 1}}{\dfrac{a + 1}{a - 1} + \dfrac{a - 1}{a + 1}}$

19. $\dfrac{\dfrac{1 - a^2}{1 + b} \cdot \left(\dfrac{a}{1 + a} - 1\right)}{1 - \left(\dfrac{1}{1 - b} - \dfrac{a^2 + b^2 - a + b}{1 - b^2}\right)}$

20. $1 - \dfrac{\dfrac{x^2}{y^2} - \dfrac{27y}{x}}{\dfrac{x}{y} + 3 + \dfrac{9y}{x}}$

1-9 SOLVING EQUATIONS

So far, we have solved many simple equations. We shall now solve more complicated equations, using the results of preceding sections.

Example 1

Find all real numbers x for which $x^2 - x - 6 = 0$; i.e., solve the equation $x^2 - x - 6 = 0$.

Solution

Since, for any real number x, $x^2 - x - 6 = (x - 3)(x + 2)$, we see that $x^2 - x - 6 = 0$ if and only if $x - 3 = 0$ or $x + 2 = 0$. Hence, the solutions are 3 and -2. (Here we once more use the fact that if a and b are real numbers, then $ab = 0$ if and only if $a = 0$ or $b = 0$.)

Example 2

Let a be a positive real number. Find all real numbers x for which $x^2 = a$.

Solution

Let x be a real number. Note that $x^2 = a$ if and only if $x^2 - a = 0$. Now, for every real number x,

$$x^2 - a = (x + \sqrt{a})(x - \sqrt{a})$$

and so x is a real number for which $x^2 = a$ if and only if x is a real number for which $x + \sqrt{a} = 0$ or $x - \sqrt{a} = 0$. Thus, *the real numbers x for which $x^2 = a$ are $\sqrt{a}$ and $-\sqrt{a}$.*

Example 3

Find all real numbers x for which $(x^2 - 2)^{-1} = 1$.

Solution

Let x be a real number for which $x^2 - 2 \neq 0$. Then

$$(x^2 - 2)^{-1} = \frac{1}{x^2 - 2} = 1$$

if and only if (multiplying by $x^2 - 2$)

$$1 = x^2 - 2$$
$$3 = x^2$$

Therefore, the solutions are $\sqrt{3}$ and $-\sqrt{3}$.

□

Let a and b be real numbers, with $b \neq 0$. Recall that $a/b = 0$ if and only if $a = 0$. The next two examples illustrate an application of this fact.

Example 4

The real numbers x for which

$$\frac{(x - 3)(x - 5)(x + 2)}{(x + 7)(x + 3)(x + 2)} = 0$$

are those numbers for which $(x - 3)(x - 5)(x + 2) = 0$ and $(x + 7)(x + 3)(x + 2) \neq 0$. Thus, the only solutions are 3 and 5. (Why is -2 not a solution?)

Example 5

Find all real numbers x for which

$$\frac{x + 2}{x - 1} + 2 \cdot \frac{x - 7}{x^2 + 2x - 3} = 0$$

Solution

First, we find the least common denominator as follows. Notice that for all real numbers x, $x^2 + 2x - 3 = (x - 1)(x + 3)$. Thus, if $x \neq 1$, $x \neq -3$, we see that

$$\begin{aligned}
\frac{x + 2}{x - 1} + \frac{2(x - 7)}{(x - 1)(x + 3)} &= \frac{x + 2}{x - 1} \cdot \frac{x + 3}{x + 3} + \frac{2x - 14}{(x - 1)(x + 3)} \\
&= \frac{x^2 + 5x + 6}{(x - 1)(x + 3)} + \frac{2x - 14}{(x - 1)(x + 3)} \\
&= \frac{x^2 + 7x - 8}{(x - 1)(x + 3)} \\
&= \frac{(x - 1)(x + 8)}{(x - 1)(x + 3)}
\end{aligned}$$

Thus, the solutions are those real numbers x for which $(x - 1)(x + 8) = 0$ but $(x - 1)(x + 3) \neq 0$. The only solution is, therefore, -8.

Exercises 1-9

Find all real numbers x for which

1. $2x^2 - 15x = 27$
2. $6x^2 - 5x - 6 = 0$
3. $x^2 + 7x + 12 = 0$
4. $2x^2 + 5x + 2 = 0$
5. $x^2 - 2x = 0$
6. $3x^2 - 4x = 0$
7. $x^4 = x^3$
8. $3x^3 + 2x^2 = 0$
9. $9x^2 + 12x + 4 = 0$
10. $4x^2 - 36x = -81$
11. $3x^3 - 3x^2 - 18x = 0$
12. $x^2(x - 1) - x(x - 1) = 12(x - 1)$
13. $x^2(x^2 - 4) - 9(x^2 - 4) = 0$
14. $x^3 - 100x^2 - x + 100 = 0$
15. $x^5 - 16x - 3x^4 + 48 = 0$
16. $\dfrac{2}{4x - 1} + \dfrac{7x}{4x - 1} = 0$
17. $\dfrac{2x^2 - 1}{x - 3} = x + 3 + \dfrac{17}{x - 3}$
18. $\dfrac{x}{x + 3} + \dfrac{5x^2}{x^2 - 9} = 0$

19. $\dfrac{2}{x + 1} - \dfrac{3}{x - 1} = \dfrac{2x + 5}{x^2 - 1}$

20. $\dfrac{5}{x - 3} + \dfrac{2}{x + 1} = \dfrac{-7x + 3}{x^2 - 2x - 3}$

21. $\dfrac{x + 2}{2x + 5} + \dfrac{x - 3}{2x^2 - x - 15} = 0$

22. $\dfrac{x + 2}{12 - 3x} = \dfrac{x - 3}{3x^2 - 21x + 36}$

23. $(x^2 - 5)^{-1} = 2$

24. $(2x^2 - 3)^{-1} = \frac{1}{2}$

25. $\dfrac{(x - 4)(x + 6)(x - 3)}{(2x + 6)(x + 5)(3x + 2)} = 0$

26. $\dfrac{(x^2 - 14x - 15)(3x - 1)}{(x^2 + 6x + 9)(x + 1)} = 0$

◊27. $|x^2 - 4| \leq 3$

◊28. $|3x^2 + 7x + 3| > 1$

1-10 COMPLETING THE SQUARE.
THE QUADRATIC FORMULA

Our previous success in solving equations such as $x^2 - x - 6 = 0$ rested on our ability to factor by inspection (with some trial-and-error) to see that for every real number x, $x^2 - x - 6 = (x - 3)(x + 2)$. We next investigate a more systematic approach to factoring. This procedure, called **completing the square,** depends on the following observation about numbers.

Let b and x be real numbers. We can always obtain from the number $x^2 + bx$ a "perfect square" by adding $(b/2)^2$ to it. Specifically, for any real numbers b and x,

$$x^2 + bx + \left(\frac{b}{2}\right)^2 = \left(x + \frac{b}{2}\right)^2$$

This is so because

$$\left(x + \frac{b}{2}\right)^2 = x^2 + 2\left(\frac{b}{2}\right)x + \left(\frac{b}{2}\right)^2 = x^2 + bx + \left(\frac{b}{2}\right)^2$$

Example 1

Find all real numbers x for which $x^2 - x - 6 = 0$.

Solution

This problem has already been solved by factoring. We shall now solve this problem again to illustrate the procedure of completing the square. Note that for every real number x

$$x^2 - x - 6 = x^2 - 1 \cdot x + (-\tfrac{1}{2})^2 - (-\tfrac{1}{2})^2 - 6$$
$$= (x - \tfrac{1}{2})^2 - \tfrac{1}{4} - 6$$
$$= (x - \tfrac{1}{2})^2 - \tfrac{25}{4}$$
$$= [(x - \tfrac{1}{2}) + \tfrac{5}{2}][(x - \tfrac{1}{2}) - \tfrac{5}{2}] \quad [\textit{difference of squares}]$$
$$= (x + 2)(x - 3)$$

Thus, -2 and 3 are the real numbers x for which $x^2 - x - 6 = 0$.

Another way to approach this problem is to note that for every real number x,

$$x^2 - x - 6 = (x - \tfrac{1}{2})^2 - \tfrac{25}{4}$$

and thus $x^2 - x - 6 = 0$ if and only if

$$(x - \tfrac{1}{2})^2 - \tfrac{25}{4} = 0, \qquad (x - \tfrac{1}{2})^2 = \tfrac{25}{4}$$

and so

$$x - \tfrac{1}{2} = \sqrt{\tfrac{25}{4}} = \tfrac{5}{2} \quad \text{or} \quad x - \tfrac{1}{2} = -\tfrac{5}{2}$$

that is,

$$x = 3 \quad \text{or} \quad x = -2$$

Example 2

Find all real numbers x for which $2x^2 - 6x - 5 = 0$.

Solution

For every real number x,

$$2x^2 - 6x - 5 = 2(x^2 - 3x - \tfrac{5}{2})$$
$$= 2[x^2 - 3x + (\tfrac{3}{2})^2 - (\tfrac{3}{2})^2 - \tfrac{5}{2}]$$
$$= 2[(x - \tfrac{3}{2})^2 - \tfrac{9}{4} - \tfrac{5}{2}]$$
$$= 2[(x - \tfrac{3}{2})^2 - \tfrac{19}{4}]$$

Thus, x is a real number for which $2x^2 - 6x - 5 = 0$ if and only if x is a real number for which

$$2[(x - \tfrac{3}{2})^2 - \tfrac{19}{4}] = 0$$
$$(x - \tfrac{3}{2})^2 = \tfrac{19}{4}$$
$$x - \frac{3}{2} = \sqrt{\frac{19}{4}} \quad \text{or} \quad x - \frac{3}{2} = -\frac{\sqrt{19}}{2}$$

and, finally,

$$x = \frac{3}{2} + \frac{\sqrt{19}}{2} = \frac{3 + \sqrt{19}}{2}$$

$$x = \frac{3}{2} - \frac{\sqrt{19}}{2} = \frac{3 - \sqrt{19}}{2}$$

□

In general, let a, b, and c be real numbers, with $a \neq 0$. Then

$$ax^2 + bx + c = a\left(x^2 + \frac{b}{a}x + \frac{c}{a}\right)$$

$$= a\left[x^2 + \frac{b}{a}x + \left(\frac{b}{2a}\right)^2 - \left(\frac{b}{2a}\right)^2 + \frac{c}{a}\right]$$

$$= a\left[\left(x + \frac{b}{2a}\right)^2 + \frac{-b^2 + 4ac}{4a^2}\right]$$

$$= a\left[\left(x + \frac{b}{2a}\right)^2 - \frac{b^2 - 4ac}{4a^2}\right]$$

Thus, if x is a real number and $b^2 - 4ac \geq 0$,* then

$$ax^2 + bx + c = 0 \quad \text{if and only if}$$
$$x + \frac{b}{2a} = \frac{\sqrt{b^2 - 4ac}}{\sqrt{4a^2}} \quad \text{or} \quad x + \frac{b}{2a} = -\frac{\sqrt{b^2 - 4ac}}{\sqrt{4a^2}}$$

or, finally,

$$x = \frac{-b + \sqrt{b^2 - 4ac}}{2a} \quad \text{or} \quad x = \frac{-b - \sqrt{b^2 - 4ac}}{2a}$$

This is called the **quadratic formula,** and is often summarized by writing

$$x = \frac{-b \pm \sqrt{b^2 - 4ac}}{2a}$$

---

*The case where $b^2 - 4ac < 0$ is discussed in the section on complex numbers (Section 2-4).

Example 3

Find all real numbers x for which $3x^2 - 2x - 4 = 0$.

Solution

With $a = 3$, $b = -2$, and $c = -4$, we have

$$\frac{-(-2) \pm \sqrt{(-2)^2 - 4(3)(-4)}}{6} = \frac{2 \pm \sqrt{52}}{6} = \frac{2 \pm 2\sqrt{13}}{6}$$

$$= \frac{1 \pm \sqrt{13}}{3}$$

Hence, the solutions are

$$\frac{1 + \sqrt{13}}{3} \quad \text{and} \quad \frac{1 - \sqrt{13}}{3}$$

Example 4

Find all real numbers x for which

$$x^2 - (4 + 2\sqrt{3})x + 7 + 4\sqrt{3} = 0$$

Solution

With $a = 1$, $b = -(4 + 2\sqrt{3})$, and $c = 7 + 4\sqrt{3}$, we have

$$\frac{(4 + 2\sqrt{3}) \pm \sqrt{(4 + 2\sqrt{3})^2 - 4(7 + 4\sqrt{3})}}{2}$$

$$= \frac{4 + 2\sqrt{3} \pm \sqrt{(16 + 16\sqrt{3} + 12) - (28 + 16\sqrt{3})}}{2}$$

$$= \frac{4 + 2\sqrt{3} \pm \sqrt{0}}{2}$$

$$= 2 + \sqrt{3}$$

Thus, there is precisely one solution, and it is the number $2 + \sqrt{3}$.

Example 5

Find all real numbers x for which $2x^2 + 3x + 2 = 0$.

Solution

With $a = 2$, $b = 3$, and $c = 2$, we see that

$$b^2 - 4ac = 9 - 4 \cdot 2 \cdot 2 = 9 - 16 < 0$$

Hence, there are no solutions to the problem.

Example 6

Find all real numbers x for which

$$3x^4 - 5x^2 + 1 = 0$$

Solution

Let x be such a number and let $t = x^2$; then $x^4 = t^2$, and hence $3t^2 - 5t + 1 = 0$. Now using the quadratic formula

$$t = \frac{5 \pm \sqrt{25 - 12}}{6} = \frac{5 \pm \sqrt{13}}{6}$$

then

$$x^2 = \frac{5 \pm \sqrt{13}}{6}$$

Hence there are only four solutions

$$\sqrt{\frac{5 + \sqrt{13}}{6}} \quad \text{and} \quad -\sqrt{\frac{5 + \sqrt{13}}{6}}$$

Exercises 1-10

By completing the square in Exercises 1–13, find all real numbers x for which

1. $x^2 - 4x - 2 = 0$
2. $x^2 - 4x + 2 = 0$
3. $2x^2 - 2x + 1 = 0$
4. $-4x^2 + 3x + 2 = 0$
5. $4x^2 - 6x - 7 = 0$
6. $6x + 4 = 9x^2$
7. $3x^2 = 15 - 4x$
8. $9x^2 = 6x - \frac{1}{4}$
9. $x^2 + 5x + \frac{25}{4} = 0$
10. $2x^2 + 3x + 9 = 0$
11. $x^2 - \frac{1}{3}x + \frac{1}{36} = 0$
12. $4x^2 - 4x + 1 = 0$
13. $3x^2 + 6x + 9 = 0$

In Exercises 14–23, use the quadratic formula to find all real numbers x for which

14. $x^2 + x - 7 = 0$
15. $2x^2 + x - 11 = 0$
16. $x^2 - \sqrt{2} \cdot x - 1 = 0$
17. $x^2 + x + 1 = 0$
18. $\sqrt{3}x^2 + \sqrt{2}x - 2 = 0$
19. $3x^2 - 2\sqrt{6}x + 2 = 0$
20. $x^2 + 4x + 5 = 0$
21. $x^2 + 10x + 25 = 0$
22. $6x^4 - 17x^2 - 45 = 0$
23. $15x^4 - 26x^2 + 8 = 0$

24. Suppose a, b, c, x_1, and x_2 are real numbers for which $x_1 \neq x_2$ or

$$ax_1^2 + bx_1 + c = 0 \quad \text{and} \quad ax_2^2 + bx_2 + c = 0$$

Compute $x_1 + x_2$ and $x_1 \cdot x_2$.

25. The sum of two real numbers is 14 and their product is 45. Find real numbers a, b, and c so that if x_1 and x_2 are the above-mentioned numbers, then

$$ax_1^2 + bx_1 + c = 0 \quad \text{and} \quad ax_2^2 + bx_2 + c = 0$$

2 The Coordinate Plane

2-1 INTRODUCTION

Just as real numbers have been identified with points on a line, so *pairs* of real numbers may be identified with points on a plane. In this chapter, such identification will provide us with a tool for applying arithmetic and algebra to plane geometry and, conversely, for applying geometric intuition to the study of algebra. It is natural, therefore, to include complex numbers in this chapter, since after an algebraic development they are identified with points in a plane. Other topics covered in this chapter include (1) the construction of a plane rectangular coordinate system, (2) graphs, and (3) relations.

2-2 RECTANGULAR COORDINATES AND DISTANCE

A basic connection between algebra and geometry is made through the construction of a *Cartesian* (or *rectangular*) *coordinate system*: two perpendicular *number lines,* one called the **horizontal** axis (sometimes called the x-axis) and the other the **vertical** axis (sometimes called the y-axis) are constructed so that their origins coincide. Each point in the plane determined by these axes is then assigned a pair of real numbers, its **coordinates,** as indicated in Figure 2-1. Thus, for example, if a and b are real numbers, the (ordered) pair (a, b) is assigned to the point whose perpendicular projections on the horizontal and

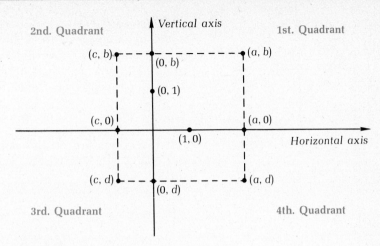

Figure 2-1

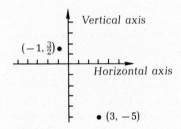

Figure 2-2

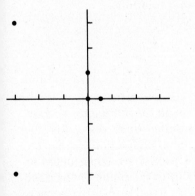

Figure 2-3

vertical axes are a and b, respectively. For convenience, we shall refer to the point corresponding to (a, b) as "the point (a, b)." Here a is called the **first coordinate** and b the **second coordinate** of (a, b). A plane in which a coordinate system has been constructed is called a **coordinate plane.** In Figure 2-1, notice that the coordinate axes divide the plane into 4 **quadrants.**

Example 1

$(3, -5)$ is the point 3 units to the right of the vertical axis and 5 units below the horizontal axis. $(-1, \frac{3}{2})$ is 1 unit to the left of the vertical axis and $\frac{3}{2}$ units above the horizontal axis. See Figure 2-2.

Example 2

The points $(\frac{1}{2}, 0)$, $(0, 0)$, $(0, 1)$, $(-3, -3)$ and $(-3, 3)$ are **plotted** in Figure 2-3.

□

We are now in a position to calculate the distance between any two points in a coordinate plane:*

If (x_1, y_1) and (x_2, y_2) are any two points and if d is the distance between them, then

$$d = \sqrt{(x_1 - x_2)^2 + (y_1 - y_2)^2}$$

To see this, observe first that if the points are not on the

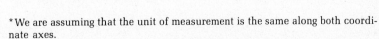

*We are assuming that the unit of measurement is the same along both coordinate axes.

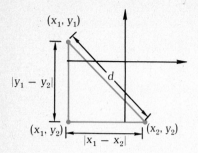

Figure 2-4

same horizontal or vertical line, then the segment joining them is the hypotenuse of the right triangle with vertices (x_1, y_1), (x_1, y_2), and (x_2, y_2). See Figure 2-4. The lengths of the horizontal and vertical legs of this triangle are $|x_1 - x_2|$ and $|y_1 - y_2|$, respectively. Hence, by the theorem of Pythagoras,

$$d = \sqrt{|x_1 - x_2|^2 + |y_1 - y_2|^2} = \sqrt{(x_1 - x_2)^2 + (y_1 - y_2)^2}$$

If the points are on the same horizontal line, then this is still true; for in this case $y_1 = y_2$, and so

$$d = \sqrt{(x_1 - x_2)^2 + (y_1 - y_2)^2} = \sqrt{(x_1 - x_2)^2 + 0}$$
$$= \sqrt{(x_1 - x_2)^2} = |x_1 - x_2|$$

A similar argument holds for vertical distance.

Example 3

Compute the distance between $(1, \sqrt{2})$ and $(-5, -3)$.

Solution

The distance is

$$\sqrt{[1 - (-5)]^2 + [\sqrt{2} - (-3)]^2} = \sqrt{6^2 + (\sqrt{2} + 3)^2}$$
$$= \sqrt{47 + 6\sqrt{2}} \approx \sqrt{55.484}$$

Example 4

If a point (x, y) is equidistant from $(0, 1)$ and $(1, 0)$, then we have

$$x^2 + (y - 1)^2 = (x - 1)^2 + y^2$$

or, simplifying, $y = x$. In other words, all points (x, y) equidistant from $(0, 1)$ and $(1, 0)$ satisfy $y = x$. See Figure 2-5.

Example 5

Show that the triangle whose vertices are $(2, 1)$, $(-1, 4)$, and $(-2, -3)$ is a right triangle.

Solution

Recall that a triangle is a right triangle if and only if the square of the length of one of its sides is the sum of the squares of the lengths of the other two sides. To see if this is the case here, we proceed as follows.

1. Let d_1 be the distance between $(2, 1)$ and $(-1, 4)$; then $d_1^2 = 3^2 + 3^2 = 18$.

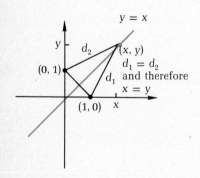

Figure 2-5

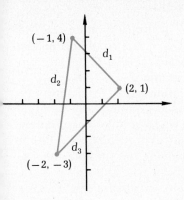

Figure 2-6

2. Let d_2 be the distance between $(-1, 4)$ and $(-2, -3)$; then
$d_2^2 = 1^2 + 7^2 = 50.$
3. Let d_3 be the distance between $(2, 1)$ and $(-2, -3)$; then
$d_3^2 = 4^2 + 4^2 = 32.$

Since $d_2^2 = d_1^2 + d_3^2$, it follows that the triangle *is* a right triangle. See Figure 2-6.

Exercises 2-2

In Exercises 1–7, find the distance between the points.

1. $(0, 5)$ and $(1, 4)$ **2.** $(6, \sqrt{3})$ and $(2, 2\sqrt{3})$

3. $(3 + \sqrt{3}, 6 + \sqrt{2})$ and $(3 - \sqrt{3}, 6 - \sqrt{2})$

4. $(\sqrt{2}, \sqrt{3})$ and $(1, -1)$ **5.** (a, b) and (b, a)

6. $(a, b + 1)$ and $(a + 1, b)$

7. $(2a - 1, (a - 1)^2)$ and $(a^2, 0)$

8. Find all real numbers x for which the distance between $(3, x)$ and $(1, -1)$ is 5.

9. Find all real numbers x for which the distance between $(x, -1)$ and $(2, x)$ is 3.

In Exercises 10 and 11, show that the triangle with the given vertices is a right triangle.

10. $(0, 6)$, $(9, -6)$, and $(-3, 0)$ **11.** $(2, 1)$, $(-1, 4)$,   and $(-2, -3)$

12. Show that the triangle whose vertices are $(2, 1)$ $(-1, 2)$, and $(2, 6)$ is an isosceles triangle.

13. Find a point (x, y) for which the triangle with vertices (x, y), $(1, 2)$, and $(3, 6)$ is an equilateral triangle.

14. If x and y are real numbers for which $x^2 + y^2 = 20$, show that the distance from (x, y) to $(1, 2)$ is half the distance from (x, y) to $(4, 8)$.

15. A point (x, y) is equidistant from $(0, 3)$ and $(3, 1)$. Find an equation relating the first coordinate of the point to the second.

◊**16.** Show that the midpoint of the segment joining two points (x_1, y_1) and (x_2, y_2) is $\left(\dfrac{x_1 + x_2}{2}, \dfrac{y_1 + y_2}{2}\right)$.

17. Given a triangle ABC, with vertices $A(1,2)$, $B(1,6)$, and $C(5,4)$; if the midpoint of $\overline{AB}$ is D, prove that $\overline{BD}$ is perpendicular to $\overline{DC}$.

◊**18.** Let $P_1 = (x_1, y_1)$ and $P_2 = (x_2, y_2)$, and let P be the point between P_1 and P_2 for which $P_1P = r\overline{PP_2}$. Show that $P = \left(\dfrac{x_1 + rx_2}{1 + r}, \dfrac{y_1 + ry_2}{1 + r}\right)$.

2-3 GRAPHS OF RELATIONS

An equation or inequality relating arbitrary numbers, x and y, is an example of a **relation;** that is, it is a condition on the ordered pair (x, y). Such an algebraic relation (or condition) determines a geometric object, its **graph:** that is, the set of all points, P = (x, y) for which the relation holds. Frequently we go the other way: a graph is described geometrically in terms of a condition on the point P and we obtain a corresponding algebraic relation between x and y by expressing the condition in terms of x and y, typically employing the formula for distance in the plane.

In general, *any* set of ordered pairs is called a relation. For example, the set of all ordered pairs (x, y) where x is a parent and y is a child of x is a relation. In what follows, however, the relations considered are subsets of the plane.

Example 1

A point (x, y) is on the circle with center (−5, 7) and radius 2 if and only if the distance between (x, y) and (−5, 7) is 2; that is, if and only if

$$[x - (-5)]^2 + (y - 7)^2 = 2^2$$

In other words,

$$(x + 5)^2 + (y - 7)^2 = 4$$

is an equation of the circle with center (−5, 7) and radius 2. See Figure 2-7.

□

The previous example may be generalized as follows: Let a, b, and r be numbers, with r > 0. A point (x, y) is on the circle with center (a, b) and radius r if and only if the distance between (x, y) and (a, b) is r; or, equivalently, if and only if $(x - a)^2 + (y - b)^2 = r^2$. That is,

$$(x - a)^2 + (y - b)^2 = r^2$$

is an equation of the circle with **center** *(a, b) and* **radius** *r.*

Example 2

Find an equation of the circle with center at $(\frac{1}{2}, -\frac{2}{3})$ and radius $\sqrt{7}$.

Solution

$$(x - \tfrac{1}{2})^2 + (y + \tfrac{2}{3})^2 = 7$$

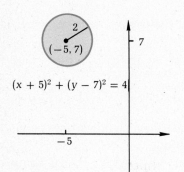

$(x + 5)^2 + (y - 7)^2 = 4$

Figure 2-7

Example 3

Describe the curve whose equation is $x^2 - 2x + y^2 + 4y = -3$.

Solution

If x and y are numbers, then

$$x^2 - 2x + y^2 + 4y = -3$$

if and only if

$$(x^2 - 2x + 1) + (y^2 + 4y + 4) = -3 + 1 + 4$$

or, equivalently,

$$(x - 1)^2 + (y + 2)^2 = 2$$

Hence, the curve is the circle with center at $(1, -2)$ and radius $\sqrt{2}$.

Example 4

Sketch the set of all points (x, y) for which $(x + 3)^2 + (y - 1)^2 \leq 2$.

Solution

This is the set of all points whose distance from $(-3, 1)$ is less than or equal to $\sqrt{2}$. Equivalently, it is the set of all points *interior to* or *on* the circle with radius $\sqrt{2}$ and center $(-3, 1)$. See Figure 2-8.

Example 5

Sketch the set of all points (x, y) for which $|x| \leq 1$ and $|y| \leq 1$.

Solution

A point (x, y) is in this set if and only if $-1 \leq x \leq 1$ and $-1 \leq y \leq 1$. See Figure 2-9.

Example 6

Sketch the graph of the equation $(y - x) \cdot (x - 5) = 0$.

Solution

A point (x, y) is on this graph if and only if $y = x$ or $x = 5$. The set of all points (x, y) for which $y = x$ is the line L_1 and the set of all points (x, y) for which $x = 5$ is the vertical line L_2, sketched in Figure 2-10.

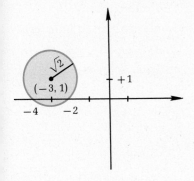

Figure 2-8

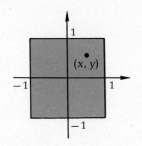

Figure 2-9

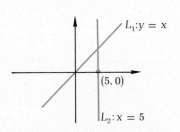

Figure 2-10

Example 7

Find an equation for the set of all points (x, y) that are equidistant from the points $(-1, 2)$ and $(3, -2)$. (The reader may recall from geometry that this set is simply the perpendicular bisector of the segment joining the two given points.)

Solution

(x, y) is such a point if and only if

$$(x + 1)^2 + (y - 2)^2 = (x - 3)^2 + (y + 2)^2$$

(since the distances are equal if and only if their squares are equal) or, equivalently,

$$x^2 + 2x + 1 + y^2 - 4y + 4 = x^2 - 6x + 9 + y^2 + 4y + 4$$

Simplifying, we obtain the equation

$$x - y = 1$$

Example 8

Sketch the set of all points (x, y) for which $x^2 + y^2 + 2x - 6y = -10$.

Solution

Completing the square, we see that a point (x, y) is on the graph if and only if

$$x^2 + 2x + 1 + y^2 - 6y + 9 = -10 + 1 + 9$$

or, equivalently,

$$(x + 1)^2 + (y - 3)^2 = 0$$

For any real numbers a and b, $a^2 + b^2 = 0$ if and only if $a = 0$ and $b = 0$. Hence, we have $x + 1 = 0$ and $y - 3 = 0$, and so the only point on the graph is $(-1, 3)$.

Example 9

Sketch the graph of the equation $y^2 = x^2$.

Solution

A point (x, y) is on this graph if and only if

$$y^2 - x^2 = 0$$
$$(y + x)(y - x) = 0$$

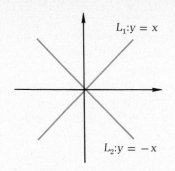

Figure 2-11

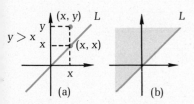

Figure 2-12

or, equivalently,

$$y = x \quad \text{or} \quad y = -x$$

The set of all points for which $y = x$ is the line L_1, and the set of all points for which $y = -x$ is the line L_2, as shown in Figure 2-11.

Example 10

Sketch the set of points (x, y) for which $y \geq x$.

Solution

Let L be the line whose equation is $y = x$. Notice in Figure 2-12(a) that the points (x, y) for which $y \geq x$ are precisely the points on or above the line L. A sketch of the required set is given in Figure 2-12(b).

Example 11

Find an equation for the set of all points (x, y) that are equidistant from the point $(0, \frac{1}{4})$ and the horizontal line L whose equation is $y = -\frac{1}{4}$ (see Figure 2-13).

Solution

Let (x, y) be any point in this set. The distance d_1 from (x, y) to $(0, \frac{1}{4})$ is $\sqrt{x^2 + (y - \frac{1}{4})^2}$, and the distance d_2 from (x, y) to the line L is $|y + \frac{1}{4}|$ (see Figure 2-13). Since $d_1 = d_2$, we have

$$\sqrt{x^2 + (y - \tfrac{1}{4})^2} = y + \tfrac{1}{4}$$

Squaring and simplifying, we obtain the equation

$$y = x^2$$

This curve is called a *parabola** and is sketched in Figure 2-13.

*A more complete discussion of parabolas appears in Sections 10-2 and 10-3.

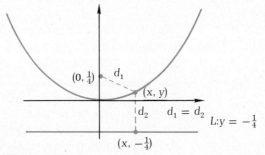

Figure 2-13

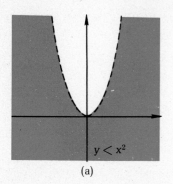

$y < x^2$

(a)

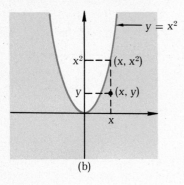

$y = x^2$

$x^2 \cdots (x, x^2)$

$y \cdots (x, y)$

x

(b)

Figure 2-14

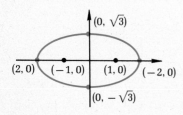

$(0, \sqrt{3})$

$(2, 0)$ $(-1, 0)$ $(1, 0)$ $(-2, 0)$

$(0, -\sqrt{3})$

Figure 2-15

Example 12

Sketch the set of points (x, y) for which $y < x^2$.

Solution

The sketch of the graph is given in Figure 2-14(a). To see this is the graph, observe [as in Figure 2-14(b)] that (x, y) is a point for which $y < x^2$ if and only if (x, y) is below the parabola whose equation is $y = x^2$.

Example 13

Sketch the set of all points the sum of whose distances from $(1, 0)$ and $(-1, 0)$ is 4.

Solution

A point (x, y) is in this set if and only if

$$\sqrt{(x - 1)^2 + y^2} + \sqrt{[x - (-1)]^2 + y^2} = 4$$

or, equivalently,

$$\sqrt{(x - 1)^2 + y^2} = 4 - \sqrt{(x + 1)^2 + y^2}$$

Squaring and simplifying, we have

$$x^2 - 2x + 1 + y^2 = 16 - 8\sqrt{(x + 1)^2 + y^2}$$
$$+ x^2 + 2x + 1 + y^2$$

and

$$x + 4 = 2\sqrt{(x + 1)^2 + y^2}$$

Squaring and simplifying again, we obtain

$$3x^2 + 4y^2 = 12$$

or

$$\frac{x^2}{4} + \frac{y^2}{3} = 1$$

The graph of this equation is called an *ellipse* and is given in Figure 2-15.

□

Examples 11 and 13 indicate a method for obtaining equations for curves that are defined in terms of distance. In particular, curves known as the *conic sections* (circles, ellipses, parabolas, and hyperbolas) can be defined in this way. See Chapter 10 for a more complete discussion of these curves.

Exercises 2-3

In Exercises 1-14, sketch the set of all points (x, y) for which

1. $x = 2$
2. $y = 3$
3. $y = x + 1$
4. $y = 3x$
5. $y = 3x + 1$
6. $y = -x$
7. $y = -x + 1$
8. $y = -3x + 1$
9. $y = x^2$
10. $y = x^3$
11. $y = -x^2$
12. $y = -x^3$
13. $x + y = 1$
14. $x - y = 1$

In Exercises 15–18, sketch the set of all points (x, y) for which

15. $(x - 1)^2 + (y + 3)^2 = 4$
16. $(x + \frac{1}{2})^2 + (y + \frac{1}{2})^2 = \frac{1}{4}$
17. $(x + 2)^2 + (y - 3)^2 \leq 2$
18. $(x + 2)^2 + (y - 3)^2 \geq 2$

In Exercises 19–26, describe the curve whose equation is

19. $x^2 + 6x + y^2 - 4y = 5$
20. $x^2 - 8x + y^2 - 6y = 0$
21. $x^2 - 3x + y^2 - 2y = -\frac{13}{4}$
22. $x^2 - 5x + y^2 - 10y = 0$
23. $4x^2 - 4x + 4y^2 + 2y = 1$
24. $5x^2 - 3x + 5y^2 + y = -\frac{1}{100}$
25. $5x^2 - 3x + 5y^2 + y = \frac{37}{100}$
26. $5x^2 - 3x + 5y^2 + y = -\frac{38}{100}$

In Exercises 27–30, find an equation for the circle with

27. center $(0, 1)$ and radius 5
28. center $(-2, 3)$ and radius $\sqrt{2}$
29. center $(1, 0)$ and radius 4
30. center $(-\frac{1}{2}, \sqrt{2})$ and radius $\sqrt{3}$

In Exercises 31–33, sketch the set of points (x, y) for which

31. $y \geq x^2$
32. $y \leq x$
33. $y \geq x^2$ and $y \leq x$

Before doing the next four Exercises, refer to Section 1-5 for the definition of absolute value. In Exercises 34–39, sketch the set of all points (x, y) for which

34. $y \leq |x|$
35. $y \leq |x|$ and $y \geq x^2$
36. $y \geq |x|$ and $y \leq 1$
37. $y \geq |x|$ and $x^2 + y^2 \leq 1$
38. $x^2 + \frac{y^2}{4} = 1$
39. $\frac{x^2}{4} + \frac{y^2}{25} \leq 1$

In Exercises 40–44 sketch the graph of each of the given equations. (Hint: Factor.)

40. $x^2 - 4y^2 = 0$ **41.** $x^2 - 5xy + 4y^2 = 0$

42. $x^2 - xy - 6y^2 = 0$ **43.** $xy + y - x - x^2 = 0$

44. $xy - y + x^2 - x = 0$

In Exercises 45–49, sketch the set of all points (x, y) for which

45. $(y - x)(x^2 + y^2 - 4) = 0$

46. $(x^2 + y^2 - 9)[(x - 1)^2 + (y + 1)^2 - 4] = 0$

47. $y \leq x$ and $x^2 + y^2 \leq 4$ **48.** $y \leq x$ and $x^2 + y^2 > 1$

49. $(x - 1)^2 + (y - 1)^2 \leq 4$ and $(x - 1)^2 + (y - 1)^2 \geq 2$

50. Find an equation for the set of all points (x, y) equidistant from the vertical and horizontal axes.

◊**51.** Find an equation for the set of points the sum of whose distances from $(0, c)$ and $(0, -c)$ is $2a$ (where a and c are positive real numbers).

◊**52.** Find an equation for the set of points the difference of whose distances from the two points $(2, 0)$ and $(-2, 0)$ is 3.

2-4 THE COMPLEX NUMBER SYSTEM

Passage from the real numbers to the complex numbers will be better understood if seen as a particular instance of the process of extending a number system. For our purposes here, a **number system** is a set of numbers that is closed under given operations of addition and multiplication; that is, the sums and products of numbers in the set are in the set. We are assuming that addition and multiplication have the associative, commutative, distributive, and identity properties [see Section 1-1].

An **extension** of a number system is itself a number system, and it extends the original system; that is, it includes the original numbers and leaves their sums and products unchanged. Also, the original identities 0 and 1 are the identities in the larger system, and the product of 0 and any number in the larger system is 0.

The *whole numbers* 0, 1, 2, 3, . . . , with the usual operations of addition and multiplication, constitute a simple and familiar number system. Note that subtraction is not always possible in this system; in particular, the equation $x + 1 = 0$ has no solution (i.e., there is *no whole number* x for which $x + 1 = 0$). We seek an extension of the system of whole numbers in which there is a solution—call it -1—to the equation $x + 1 = 0$. Toward this end, we start out with our original numbers 0, 1, 2, 3, . . . , and the new number -1, where $(-1) + 1 = 0$. We cannot, however, stop here. The requirement that the extension be a *system* forces the introduction of other new numbers, $(-1) + (-1) = -2$, $(-2) + (-1) = -3$, and so on, and also forces the usual definitions of sums and products. Thus, the extension generated by -1 is the system of *integers*. Similarly, the system of rational numbers is an extension of the system of integers, generated by the solutions of the equations $2x = 1$, $3x = 1$, $4x = 1$, and so on. Call these solutions $\frac{1}{2}, \frac{1}{3}, \frac{1}{4}, \ldots$.

It is instructive to look at the smaller extension generated by just one of these, $\frac{1}{2}$. It must include $\frac{1}{4} = (\frac{1}{2})^2$, $\frac{1}{8} = (\frac{1}{2})^3$, and so on; and hence all the numbers

$$n, \qquad \frac{n}{2} = n \cdot \frac{1}{2}, \qquad \frac{n}{4} = n \cdot \frac{1}{4}, \qquad \frac{n}{8} = n \cdot \frac{1}{8}$$

and so on, where n is any integer. The reader should verify that this set of numbers (which contains the integers and $\frac{1}{2}$) is closed under addition and multiplication and so constitutes the extension generated by $\frac{1}{2}$. These numbers are frequently employed in connection with measurement in inches. In fact, the marks on an ordinary ruler correspond to such numbers.

Although the real number system is an extension of the (system of) rational numbers, it is not generated by the solutions to algebraic equations. Rather, the real number system is characterized by being the (smallest) extension that has the property of completeness (see Section 4-6).

There are, however, many systems between the rational number and real number systems. For example, it can be shown that the equation $x^2 - 2 = 0$ has no rational number solution. Call the irrational number $\sqrt{2}$ a solution of this equation; then the extension of the rational number system generated by $\sqrt{2}$ must include all numbers $a + b\sqrt{2}$ where a and b are any rational numbers (since, by closure of multiplication, it must include $b\sqrt{2}$ for any rational number b, and hence, by closure of addition, it must include $a + b\sqrt{2}$). In fact, this set of numbers constitutes the extension, for it includes the rational numbers (take $b = 0$) and $\sqrt{2}$ (take $a = 0, b = 1$) and is closed under addition and multiplication. To see that it is closed, let a, b, c, and d be any rational numbers; then,

$$(a + b\sqrt{2}) + (c + d\sqrt{2}) = (a + c) + (b + d)\sqrt{2}$$

and

$$(a + b\sqrt{2})(c + d\sqrt{2}) = ac + ad\sqrt{2} + bc\sqrt{2} + 2bd$$
$$= (ac + 2bd) + (ad + bc)\sqrt{2}$$

The previous example closely parallels the extension we are primarily interested in. The **complex number system** is the extension of the real number system generated by a solution—call it i—of the equation $x^2 + 1 = 0$ (this equation has no real solution). Thus, $i^2 = -1$. The numbers $a + bi$ (where a and b are any real numbers) must all be complex numbers. It is easily checked* that

$$a + bi = a' + b'i \quad \text{if and only if} \quad a = a' \quad \text{and} \quad b = b'$$

This set of numbers is the entire system of complex numbers, since (1) it includes the real numbers (take $b = 0$), (2) it includes i (take $a = 0$, $b = 1$), and (3) it is closed under addition and multiplication. To verify (3), let a, b, c, and d be any real numbers and note that

$$(a + bi) + (c + di) = (a + c) + (b + d)i,$$

and

$$(a + bi)(c + di) = ac + adi + bci + bdi^2$$
$$= (ac - bd) + (ad + bc)i$$

Example 1

(a) $(2 - 3i) + (-5 - 7i) = -3 - 10i = -(3 + 10i)$
(b) $(2 - 3i) \cdot (-5 - 7i) = -10 + 15i - 14i + 21i^2$
$$= -10 - 21 + i$$
$$= -31 + i$$
(c) $(2 + 3i)(2 - 3i) = 4 - 9i^2 = 4 + 9 = 13$
(d) $i(2 - 3i)(-5 - 7i) = i(-31 + i)$
$$= -31i + i^2$$
$$= -1 - 31i$$

Example 2

$$i^3 = i \cdot i^2 = -i$$
$$i^4 = i^2 i^2 = (-1)(-1) = 1$$
$$i^5 = i^4 \cdot i = i,$$
$$i^{25} = i^{6 \cdot 4 + 1} = (i^4)^6 i = i$$
$$i^{43} = (i^4)^{10} i^3 = -i$$

If a and b are any real numbers, then $(-a) + (-b)i = -(a + bi)$ is an additive inverse of $a + bi$, since

$$(a + bi) + [(-a) + (-b)i] = 0 + 0i = 0$$

---

*If $a + bi = a' + b'i$, then adding $-a' + (-b')i$, we obtain $(a - a') + (b - b')i = 0$. If $b \neq b'$, then $i = -\dfrac{a - a'}{b - b'}$, which is impossible since i is not a real number. Therefore, $b = b'$, and hence $a - a' = 0$. The converse is immediate.

We define subtraction as follows:

$$(a + bi) - (c + di) = (a + bi) + [-(c + di)]$$
$$= (a - c) + (b - d)i$$

Example 3

$$-5i = (-5)i$$
$$-i = (-1)i$$
$$-(3 - 2i) = -3 + 2i$$
$$(-5 + 3i) - (3 - i) = -8 + 4i$$

Example 4

If $(2 - 3i) + z = -4 + i$, then z
$$= (-4 + i) - (2 - 3i) = -6 + 4i$$

Let a be any positive real number. In the complex number system the equation $x^2 + a = 0$ has two solutions; that is, the negative real number $-a$ has two square roots. These are $i\sqrt{a}$ and $-i\sqrt{a}$, and we define $\sqrt{-a} = i\sqrt{a}$. Note that this choice for $\sqrt{-a}$ is analogous to our choice of the positive square root for $\sqrt{a}$.

Example 5

$$\sqrt{-1} = i \qquad \sqrt{-4} = 2i \qquad \sqrt{-8} = \sqrt{8}i = 2\sqrt{2}i$$
$$\sqrt{-4}\sqrt{-9} = (2i)(3i) = 6i^2 = -6$$
$$\sqrt{(-4)(-9)} = \sqrt{36} = 6$$

□

There is a natural and useful identification of the complex numbers with the points of the coordinate plane (or equivalently, with the ordered pairs of real numbers) which extends the identification of real numbers with the points on the horizontal axis; namely, the complex number $a + bi$ is identified with the point (a, b) (see Figure 2-16).

In this context, the coordinate plane is called the *complex plane,* and the horizontal and vertical axes are called the *real* and *imaginary axes,* respectively. The addition and multiplication of complex numbers becomes

$$(a, b) + (c, d) = (a + c, b + d)$$
$$(a, b) \cdot (c, d) = (ac - bd, ad + bc)$$

Geometrically, the sum of the points (a, b) and (c, d) is the vertex opposite $(0, 0)$ of the parallelogram whose other vertices are (a, b) and (c, d). See Figure 2-17.

Although the multiplication of $a + bi$ by $c + di$ can also be

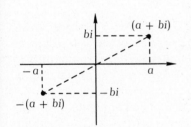

Figure 2-16

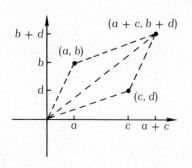

Figure 2-17

given a general geometric interpretation, we shall do so here only in two special cases: (1) when $d = 0$, that is, multiplication by a real number; and (2) when $c = 0$, $d = 1$, that is, multiplication by i.

In the first case, $c(a + bi) = ca + cbi$, or equivalently, $c(a, b) = (ca, cb)$. Geometrically, for the case that $c > 0$, this amounts to stretching the line segment from the origin to (a, b) by the factor c, whereas in the case $c < 0$, it amounts to stretching this line segment by the factor $|c|$ and then reflecting in the origin if $c < 0$. In Figure 2-18, we exhibit this for the case that $c > 0$. Notice in this figure that $-c(a + bi)$ is the reflection of $c(a + bi)$ in the origin.

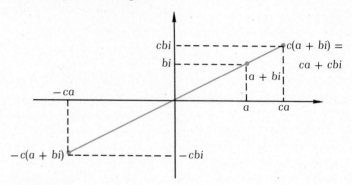

Figure 2-18

In the second case, $i(a + bi) = ai + bi^2 = -b + ai$, or equivalently, $i(a, b) = (-b, a)$. Geometrically, this amounts to rotating the line segment from the origin to (a, b) by 90° counterclockwise. See Figure 2-19.

Let $a + bi$ be any complex number. The **conjugate** of $a + bi$ is, by definition, $a - bi$, and is denoted by $\overline{a + bi}$. The **absolute value** of $a + bi$ is, by definition, $\sqrt{a^2 + b^2}$, and is denoted by $|a + bi|$. Letting $z = a + bi$, note that $z\overline{z} = (a + bi)(a - bi) = a^2 + b^2 = |z|^2$, and hence $|z| = \sqrt{z\overline{z}}$. Geometrically, $\overline{z}$ is the reflection of z in the real axis and $|z|$ is the distance from 0 to z. See Figure 2-20. This definition of absolute value is consistent with that given for real numbers.

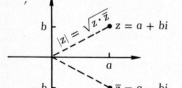

Figure 2-19

Figure 2-20

Example 6

$$\overline{-3 + 2i} = -3 - 2i$$
$$(-2 + i)(-2 - i) = 4 - i^2 = 4 + 1 = 5$$

and

$$|-2 + i| = \sqrt{4 + 1} = \sqrt{5}$$
$$|3 - 4i| = \sqrt{9 + 16} = 5$$

□

We can now easily show that any complex number $z \neq 0$ has a multiplicative inverse, given by

$$\frac{1}{z} = \frac{\bar{z}}{|z|^2}$$

where

$$\frac{\bar{z}}{|z|^2} = \frac{1}{|z|^2} \bar{z}$$

Simply observe that

$$z \frac{\bar{z}}{|z|^2} = \frac{z\bar{z}}{|z|^2} = \frac{|z|^2}{|z|^2} = 1$$

or, equivalently, that

$$(a + bi) \frac{a - bi}{a^2 + b^2} = \frac{a^2 + b^2}{a^2 + b^2} = 1$$

where a and b are any real numbers not both 0.

Example 7

Let $z = 4 - 3i$. Then

$$\frac{1}{z} = \frac{1}{4 - 3i} = \frac{4 + 3i}{(4 - 3i)(4 + 3i)} = \frac{4 + 3i}{25} = \frac{\bar{z}}{|z|^2}$$

Also

$$\frac{2 + i}{4 - 3i} = \frac{(2 + i)(4 + 3i)}{(4 - 3i)(4 + 3i)} = \frac{5 + (10)i}{25} = \frac{1 + 2i}{5}$$

or

$$\frac{1}{5} + \frac{2}{5} i$$

From the existence of multiplicative inverses, it follows that

$$z_1 z_2 = 0 \qquad \text{if and only if} \qquad z_1 = 0 \quad \text{or} \quad z_2 = 0$$

for if $z_1 z_2 = 0$ and $z_1 \neq 0$, then $\frac{1}{z_1}(z_1 z_2) = 0$; that is $z_2 = 0$. The converse is immediate.

Example 8

Find all complex numbers x for which $x^2 + 5 = 0$.

Solution

For any complex number x,

$$x^2 + 5 = x^2 - (-5) = (x - \sqrt{-5})(x + \sqrt{-5})$$
$$= (x - \sqrt{5}i)(x + \sqrt{5}i).$$

Hence, $x^2 + 5 = 0$ if and only if $x = \sqrt{5}i$ or $x = -\sqrt{5}i$.

Example 9

Find all complex numbers for which $x^2 + x + 1 = 0$.

Solution

For any complex number x,

$$x^2 + x + 1 = x^2 + x + (\tfrac{1}{2})^2 - (\tfrac{1}{2})^2 + 1$$
$$= (x + \tfrac{1}{2})^2 + \tfrac{3}{4}$$
$$= \left[\left(x + \frac{1}{2}\right) - \frac{\sqrt{3}}{2}i\right]\left[\left(x + \frac{1}{2}\right) + \frac{\sqrt{3}}{2}i\right]$$
$$= \left[x - \left(-\frac{1}{2} + \frac{\sqrt{3}}{2}i\right)\right]\left[x - \left(-\frac{1}{2} - \frac{\sqrt{3}}{2}i\right)\right]$$

Hence, $x^2 + x + 1 = 0$ if and only if $x = -\tfrac{1}{2} \pm (\sqrt{3}/2)i$.
□

The last example illustrates the fact that the derivation of the quadratic formula given in Chapter 1 holds even for $b^2 - 4ac < 0$ if we allow complex solutions. Hence, for any real numbers a, b, and c (with $a = 0$), the complex numbers x for which

$$ax^2 + bx + c = 0$$

are

$$x = \frac{-b \pm \sqrt{b^2 - 4ac}}{2a}$$

Example 10

Find all complex numbers x for which $x^2 + x + 1 = 0$.

Solution

Since $a = b = c = 1$, we have

$$x = \frac{-1 + \sqrt{1 - 4}}{2} = -\frac{1}{2} \pm \frac{\sqrt{3}}{2}i$$

Example 11

Find all complex numbers x for which $x^4 + x^2 - 1 = 0$.

Solution

$(x^2)^2 + x^2 - 1 = 0$ if and only if

$$x^2 = \frac{-1 \pm \sqrt{1^2 - 4(-1)}}{2} = \frac{-1 \pm \sqrt{5}}{2}$$

or, equivalently,

$$x = \pm \sqrt{\frac{-1 \pm \sqrt{5}}{2}}$$

That is, x is one of the four numbers

$$\sqrt{\frac{-1 + \sqrt{5}}{2}} \qquad -\sqrt{\frac{-1 + \sqrt{5}}{2}}$$

$$i\sqrt{\frac{1 + \sqrt{5}}{2}} \qquad -i\sqrt{\frac{1 + \sqrt{5}}{2}}$$

Example 12

Find all complex numbers x for which

$$\frac{x}{x^2 + 1} + \frac{x^2 + 1}{x} - 2 = 0$$

Solution

Let $z = x/(x^2 + 1)$. Observe that x is such a number if and only if

$$z + \frac{1}{z} - 2 = 0$$

$$z^2 - 2z + 1 = 0$$

$$(z - 1)^2 = 0$$

$$z = 1$$

$$\frac{x}{x^2 + 1} = 1$$

$$x^2 - x + 1 = 0$$

$$x = \frac{1 \pm \sqrt{3}i}{2}$$

Example 13

Find all real numbers x for which

$$\sqrt{2x + 5} + \sqrt{3x + 7} = \sqrt{x + 6}$$

Solution

If x is such a real number then

$$(\sqrt{2x + 5} + \sqrt{3x + 7})^2 = (\sqrt{x + 6})^2$$
$$2x + 5 + 2\sqrt{2x + 5} \cdot \sqrt{3x + 7} + 3x + 7 = x + 6$$

and so

$$\sqrt{2x + 5} \cdot \sqrt{3x + 7} = -(2x + 3)$$

Squaring again and simplifying, we obtain

$$2x^2 + 17x + 26 = 0$$

Hence, by the quadratic formula (or factoring), we obtain

$$x = \frac{-17 \pm \sqrt{17^2 - 4 \cdot 2 \cdot 26}}{2 \cdot 2} = \frac{-17 \pm \sqrt{81}}{4} = \frac{-17 \pm 9}{4}$$

Therefore, $x = -2$ or $x = -\frac{13}{2}$. We now check to see whether these numbers are actually solutions. For $x = -2$ we have

$$\sqrt{2 \cdot (-2) + 5} + \sqrt{3 \cdot (-2) + 7} = 1 + 1 = \sqrt{-2 + 6}$$

Therefore -2 *is* a solution.
    For $x = -\frac{13}{2}$ we have

$$\sqrt{2(-\tfrac{13}{2}) + 5} + \sqrt{3(-\tfrac{13}{2}) + 7} = \sqrt{-8} + \sqrt{-\tfrac{25}{2}}$$
$$= i\sqrt{8} + i\sqrt{\tfrac{25}{2}} = \tfrac{9}{2}i\sqrt{2} \neq \sqrt{-\tfrac{13}{2} + 6} = i\sqrt{\tfrac{1}{2}}$$

Therefore $-\frac{13}{2}$ is *not* a solution.

Exercises 2-4

In Exercises 1–7, compute the given sums.

1. $(2 + 3i) + (-2 - 3i)$
2. $(8 - i) + (3 + 2i)$
3. $i + (2 - i)$
4. $(\sqrt{2} + \sqrt{3}i) + (\sqrt{3} + \sqrt{2}i)$
5. $(-4 + 3i) - (2 - 5i)$
6. $3(-2 + i) + 2(-3 + i)$
7. $i(3 + 2i) + 2i(1 - i)$

In Exercises 8-11, find real numbers a and b for which

8. $(a, b) = (1, 3) + (2, 0)$

9. $(a, b) = (0, 4) - (5, 1)$

10. $(a, b) = (\frac{1}{2}, 4) - (\frac{3}{2}, \frac{1}{2})$

11. $(a, b) = (\sqrt{6}, 4) - (\sqrt{4}, 2)$

In Exercises 12-21, compute

12. $(1 + 3i)(1 - 3i)$

13. $(1 + 3i)(-1 + 3i)$

14. $(6 + \sqrt{2}i)(\sqrt{2} + 6i)$

15. $(\sqrt{2} + i)(-\sqrt{2} - i)$

16. $(2 + 3i)(-1 + 2i)$

17. $(1 + 2i)^2$

18. $\overline{3 - 5i}$

19. $\overline{-7i}$

20. $(2 + i)\overline{(2 + i)}$

21. $(2 + 3i)\overline{(2 + 3i)}$

In Exercises 22-27, find the reciprocal. In each case your answer should be a complex number $a + bi$ (where a and b are real numbers).

22. i

23. $-3i$

24. $2 + i$

25. $1 - 2i$

26. $2 + 3i$

27. $-3 - 5i$

In Exercises 28-33, find

28. $(-i)^4$

29. i^{101}

30. $(-ai)^4$, where a is a real number

31. $\sqrt{(-5)(-5)}$

32. $\dfrac{1}{i}$

33. $\left(\dfrac{1}{i}\right)^2$

34. Find all complex numbers z for which

$$(z + i)(z - 2 + i)(3z - 2i)(2z + 1 - i) = 0$$

In Exercises 35-43, find all complex numbers x for which

35. $x^2 + 2x + 4 = 0$

36. $x^4 + x^2 = 0$

37. $x^2 + \pi^2 = 0$

38. $x^2 - x + 1 = 0$

39. $x^2 + x + 1 = 0$

40. $5x^2 + 2x + 1 = 0$

41. $3x^2 - x + 1 = 0$

42. $x^3 - 1 = 0$

43. $x^3 - 8 = 0$

In Exercises 44-46, indicate whether the given numbers are real for every complex number z. Justify your answer.

44. $z + \bar{z}$

45. $z\bar{z}$

46. $|z|$

47. Prove that $\bar{z} = z$ if and only if z is real; that is, prove that if $\bar{z} = z$, then z is real, and conversely.

48. Prove that addition for complex numbers defined in this section is commutative and associative.

49. Verify that multiplication for complex numbers defined in this section is commutative, associative, and distributive with respect to addition.

50. Prove that the quadrilateral with vertices $(0, 0)$, (a, b), (c, d), and $(a + c, b + d)$ is a parallelogram. (*Hint:* Recall from plane geometry that a four-sided figure whose opposite sides have equal length is a parallelogram.)

51. If z and u are complex numbers, show that $\overline{zu} = \bar{z} \cdot \bar{u}$ and that $\left(\dfrac{z}{u}\right) = \dfrac{\bar{z}}{\bar{u}}$.

52. If z and u are complex numbers, show that $|zu| = |z| \cdot |u|$.

In Exercises 54–62, find all numbers x which satisfy the given condition.

53. $x + 1 = \sqrt{x + 7}$

54. $1 = -x + \sqrt{10 + 2x}$

55. $x + 2 = \sqrt{7x + 2}$

56. $\sqrt{x + 5} = \sqrt{2}x - \sqrt{2}$

57. $\sqrt{2x + 7} - \sqrt{x + 3} = 1$

58. $\sqrt{2x} - \sqrt{x - 2} = 2$

59. $\sqrt{6(x + 2)} - \sqrt{4x + 9} = 1$

60. $\sqrt{21x + 37} - \sqrt{19x + 24} = 1$

61. $\sqrt{3x - 2} + \sqrt{2x - 3} = \sqrt{3}\sqrt{x + 1}$

62. $\sqrt{2x - 2} - \sqrt{x - 15} = \sqrt{x - 3}$

*If $z = a + ib$, then a is defined as the **real part** of z, denoted by Re z; and b is called the **imaginary part** of z, denoted by Im z. In Exercises 63–66, show that*

63. $\text{Re } z \le |z|$ for all z **64.** $\text{Im } z \le |z|$ for all z

65. $|\text{Re } z| + |\text{Im } z| \ge |z|$ for all z

66. $\text{Re } z = z$ if and only if z is real

(*Hint:* Interpret these problems geometrically.)

Miscellaneous Problems

1. Find the distance between:
(a) $(r + s, s)$ and $(r - s, r)$
(b) $(\sqrt{r}, \sqrt{s})$ and $(-\sqrt{r}, -\sqrt{s})$

2. Find the point on the vertical axis equidistant from $(5, 7)$ and $(4, 11)$.

3. Find all numbers x for which the distance between $(2x - 1, (x - 1)^2)$ and $(x^2, 0)$ is $1/\sqrt{2}$.

4. Sketch the set of all complex numbers z for which (a) Re $z \leq 1$, (b) $|$Re $z| < 2$, (c) Im $z = 3$, (d) $|\text{Im}(z^2)| > 0$.

5. Sketch the set of all complex numbers z for which (a) $|z| \leq 1$, (b) $|z - 2| < 2$, (c) $|z + i| < 1$, (d) $|z - 1 + 2i| > 1$.

6. Show that $\overline{i + \overline{z}} = z - i$.

7. Show that $\dfrac{(4 + i)^2}{15 - 8i} = 1$.

8. Show that if $z = \dfrac{\sqrt{3} - i}{2}$, then $z^3 + i = 0$.

9. Show that $|z + w|^2 + |z - w|^2 = 2|z|^2 + 2|w|^2$ for any complex numbers z, w. This is called the *parallelogram law* for complex numbers. (Hint: $|u|^2 = u\bar{u}$ for any complex number u.)

10. Graph the set of all points (x, y) for which
(a) $|x| + |y| = 1$,
(b) $|x| + |y| \leq 1$.

11. Graph the set of all points (x, y) for which

$$|x - 5| + |y + 2| = 1.$$

12. Explain how the graph of Problem 11 can be obtained from the graph of Problem 10 by horizontal and vertical shifts.

13. Graph the set of all points (x, y) for which
(a) $|x| - |y| = 1$,
(b) $|x| - |y| \leq 1$.

14. Find an equation for the set of all points (x, y) the sum of whose distances from $(0, \sqrt{5})$ and $(0, -\sqrt{5})$ is 6. Sketch the graph of this equation.

15. For what number a is the equation

$$x^2 + y^2 + 2ax - 2y + a^2 - 3 = 0$$

a circle with center $(1, 1)$?

3 Functions and Their Graphs

3-1 INTRODUCTION

Many laws governing physical, biological, and social phenomena are stated in terms of interdependence among several quantities. In physics, for example, the *temperature* of a gas in a closed container determines the *pressure* exerted by the gas on the walls of the container. Similarly, the formula $e = mc^2$ shows how the energy, e, of an object is determined by its mass, m (with c, the speed of light, being constant).

In economics, supply of and demand for a product are related. In biology, the age of a once living object is related to the amount of radioactive carbon now present in the object.

The concept of *function* is abstracted from such situations and is one of the key concepts in all branches of mathematics. The reader may find it useful to think of a function (informally, of course) as a machine that produces, for a given *input*, precisely one *output*.

The experience of the preceding chapter is used here in a natural way. Associated with each function is a set of points called its *graph*. Graphs are one of the essential tools used in the study of functions.

A **function** is a special kind of correspondence (i.e., relation) between two sets of real numbers: that is, a function assigns *precisely one* real number to each number in a given set. This set is called the **domain** of the function.*

---

*Another definition of function in terms of ordered pairs is: A function from A to B is a set of ordered pairs (x, y), with x in A and y in B, with the property that if (x, y_1) and (x, y_2) are pairs in the relation, then $y_1 = y_2$.

Domain   Range

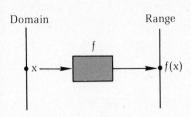

Figure 3-1

Let f be a function and x a number in the domain of f. Then $f(x)$ (read "f of x" or "f acting on x" or "the image of x under f") is the number that f assigns to x. The set of all numbers assigned by a function to the numbers in its domain is called the **range** of this function. Thus, if x is in the *domain* of f, then $f(x)$ is in the *range* of f. See Figure 3-1.

Some of the functions we shall discuss have sets called *intervals* as their domains. Specifically, if a and b are real numbers and $a < b$, then

1. the set of all real numbers x for which $a \le x \le b$ is called the **closed interval from a to b** and is denoted by $[a, b]$, and
2. the set of all real numbers x for which $a < x < b$ is called the **open interval from a to b** and is denoted by (a, b).

For example, the open interval $(0, 1)$ is the set of all real numbers x for which $0 < x < 1$; the closed interval $[-8, \pi]$ is the set of all real numbers x for which $-8 \le x \le \pi$; and the set of all real numbers x for which $|x| < 2$ is the open interval $(-2, 2)$.

Example 1

Let f be the function that assigns to each real number in the closed interval $[0, 2]$ its square: then $f(x) = x^2$ for each x in $[0, 2]$. The *domain* of f is the closed interval $[0, 2]$. Thus, for example,

$$f(0) = 0^2 = 0, \quad f(\tfrac{1}{2}) = (\tfrac{1}{2})^2 = \tfrac{1}{4}, \quad f(\tfrac{1}{3}) = \tfrac{1}{9}, \quad f(\sqrt{2}) = 2$$
$$f(f(\sqrt{2})) = f(2) = 4, \quad f(f(\tfrac{1}{2})) = f(\tfrac{1}{4}) = \tfrac{1}{16},$$

and so on. The range of f is the closed interval $[0, 4]$. In Figure 3-2 we have a *correspondence diagram* for f.

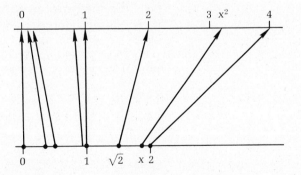

Figure 3-2

Example 2

Let h be the function defined by $h(x) = 3x + 1$ for any real number x. Then

$$h(0) = 3 \cdot 0 + 1 = 1, \qquad h(\tfrac{1}{2}) = 3(\tfrac{1}{2}) + 1 = \tfrac{5}{2},$$
$$h(\sqrt{3}) = 3\sqrt{3} + 1, \qquad h(-\tfrac{1}{3}) = 3(-\tfrac{1}{3}) + 1 = 0$$

and so on. A partial *correspondence diagram* for h is given in Figure 3-3. The domain of f is the set of all real numbers. What is the range?

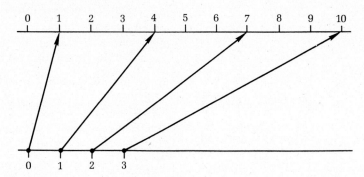

Figure 3-3

Example 3

Let g be defined by $g(x) = \sqrt{x}$ for $x \geq 0$. Then

$$g(0) = 0, \qquad g(\tfrac{1}{4}) = \tfrac{1}{2}, \qquad g(1) = 1, \qquad g(2) = \sqrt{2}$$

and so on. Note that, since negative numbers are not in the domain of g, it makes no sense, for example, to write $g(-\tfrac{1}{2})$.

Example 4

Let k be the function *defined on* [1, 3] (i.e., the domain of k is the closed interval [1, 3]) as follows:

$$k(x) = \begin{cases} x - 1 & \text{if } 1 \leq x \leq 2 \\ x + 2 & \text{if } 2 < x \leq 3 \end{cases}$$

Then

$$k(1) = 1 - 1 = 0, \qquad k(\tfrac{3}{2}) = \tfrac{3}{2} - 1 = \tfrac{1}{2}, \qquad k(2) = 2 - 1 = 1,$$
$$k(\tfrac{5}{2}) = \tfrac{5}{2} + 2 = \tfrac{9}{2}, \qquad k(3) = 3 + 2 = 5$$

What is the range of k?

Example 5

Let u be the function whose domain is the set of nonnegative integers and which is defined by $u(x) = 1^x + (-1)^x$ for every x in its domain. Then

$$u(0) = 2, \quad u(1) = 0, \quad u(2) = 2, \quad u(3) = 0, \quad u(4) = 2$$

and, in general,

$$u(x) = \begin{cases} 0 & \text{if } x \text{ is odd} \\ 2 & \text{if } x \text{ is even} \end{cases}$$

What does the correspondence diagram look like?

Exercises 3-1

In Exercises 1–11, let $f(x) = x^3 - x$ for any real number x. Find

1. $f(0)$
2. $f(1)$
3. $f(-1)$
4. $f(\sqrt{2})$
5. $f[f(2)]$
6. $f[f(-1)]$
7. $f[f(\sqrt{2})]$
8. $f(1 + \sqrt{2})$
9. $f[f(1 + \sqrt{2})]$
10. $f(\sqrt{2} + \sqrt{3})$
11. $f[f(\sqrt{2} + \sqrt{3})]$

In Exercises 12–19, let $g(x) = x^4 - x^2 + 1$ for any real number x. Compute

12. $g(0)$
13. $g(1)$
14. $g(\frac{1}{2})$
15. $g(\sqrt{3})$
16. $g[g(-1)]$
17. $g[g(\sqrt{2})]$
18. $g(1/\sqrt{2})$
19. $g[g(1/\sqrt{2})]$

In Exercises 20–25, let f and g be the functions given above. Compute

20. $g[f(0)]$
21. $g[f(1)]$
22. $g(f(\frac{1}{2}))$
23. $f[g(0)]$
24. $f[g(1)]$
25. $f[g(\frac{1}{2})]$

In Exercises 26–28, let $f(x) = \dfrac{2}{5 + x} + \dfrac{1}{1 + x^2}$, where x is any real number other than -5. Find

26. $f(0)$
27. $f(\frac{11}{17})$
28. $f(-\frac{1}{2})$

29. Let $f(x) = x^2 - 4$ for any real number x. Find all real numbers x for which $f(x) = 12$.

30. Let $f(x) = [1/(x - 1)] + [2/(x - 2)]$ for all real numbers x other than 1 and 2. Find all real numbers x for which $f(x) = 0$.

31. Let $f(x) = [1/(x^2 + x)] + 3/x$, $x \neq 0$, $x \neq -1$. Find all real numbers x for which $f(x) = 0$.

32. Let $f(x) = 3 + [2/(x + 2)]$, $x \neq -2$, and $g(x) = 4 + 5/(x + 1)$, $x \neq -1$. Find all real numbers x for which $[f(x)/g(x)] + 1 = 0$.

33. Let $f(x) = x^2$ for any real number x. Find all real numbers x for which $f(x + 1) = f(2x)$.

34. Let $f(x) = x^2 - 1$ for any real number x. Find all real numbers x for which $f[f(x)] = 0$.

35. What is the range of the function h of Example 2?

36. What is the range of the function g of Example 3?

37. Let $f(x) = 2x + 3$ for any number x in the closed interval $[0, 1]$. (a) What is the range of f? (b) Sketch a correspondence diagram for f.

38. Let $f(x) = x^2 + 3$, for any number x. (a) What is the range of f? (b) Sketch a correspondence diagram for f.

39. Let $f(x) = -x^2 + 2$, for any number x. (a) What is the range of f? (b) Sketch a correspondence diagram.

40. Let $f(x) = [1/(x - 1)] - 1$, $x \neq 1$. Find all numbers x for which $f(2x + 3) = 0$.

41. Let $f(x) = 3x - 1$, for any number x, and let h be a number not equal to 0. Find $[f(2 + h) - f(2)]/h$ and "simplify" your answer.

42. How many functions are there whose domain is the set $\{a, b\}$ and whose range is the set $\{1, 2\}$?

43. How many functions are there whose domain is the set $\{a, b, c\}$ and whose range is (a) either $\{1\}$ or $\{2\}$? (b) The set $\{1, 2\}$? Exhibit all these functions for both (a) and (b).

3-2 NOTATION. GRAPHS

The phrase "let f be the function that assigns to each number in the closed interval $[0, 2]$ its square," (see Example 1 of Section 3-1) may conveniently be replaced by any one of the following:

Let f be defined by

$$f(x) = x^2, \quad x \text{ in } [0, 2]$$

or

$$f: x \to x^2, \quad x \text{ in } [0, 2]$$

or simply

$$x \to x^2, \quad x \text{ in } [0, 2]$$

Furthermore, we shall adopt the following convention:

When a function is defined by a specific rule, or formula, and

the domain of the function is not specified, then it is understood that the domain is the set of all real numbers for which
the formula is meaningful, that is, for which the output is a
real number.*

Thus, according to this convention, the square root function
above is given simply by $x \rightarrow \sqrt{x}$, since $\sqrt{x}$ is a real number
if and only if $x \geq 0$; and the function h of Example 2, Section
1, may be given simply by $x \rightarrow 3x + 1$, since $3x + 1$ is a real
number if x is any real number.

Example 1

Let $f(x) = 1/(2x - 5)$. Then the domain of f is the set of all
numbers except $\frac{5}{2}$, since there is no division by 0.

Example 2

Let $f(x) = \sqrt{x - 1}$; then the domain of f is the set of all numbers ≥ 1, since $f(x)$ is a real number if and only if $x - 1 \geq 0$,
and $x - 1 \geq 0$ if and only if $x \geq 1$.
□

We shall now consider the graphs of certain basic functions.
By definition, given a coordinate plane,

the graph of a function f is the set of all points $(x, f(x))$ with
x in the domain of f.

In other words, it is *the set of all points (x, y) for which
$y = f(x)$; or equivalently, it is the graph of the equation
$y = f(x)$.*

Example 3

The graph of the **squaring function** $x \rightarrow x^2$ (or equivalently,
of the equation $y = x^2$) is given in Figure 3-4.

---

*Also, for convenience, we will frequently simply say "number" instead of
"real number."

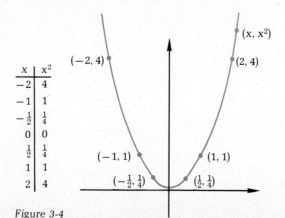

x	x^2
-2	4
-1	1
$-\frac{1}{2}$	$\frac{1}{4}$
0	0
$\frac{1}{2}$	$\frac{1}{4}$
1	1
2	4

Figure 3-4

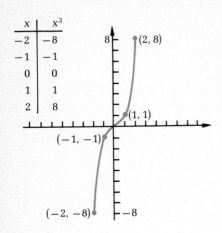

x	x^3
−2	−8
−1	−1
0	0
1	1
2	8

Figure 3-5

Example 4

The graph of the **cubing function** $x \to x^3$ is given in Figure 3-5.

Example 5

The graph of the **identity function** $I: x \to x$ (i.e., of the equation $y = x$) is given in Figure 3-6.

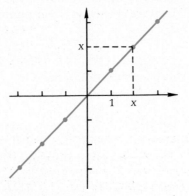

Figure 3-6

Example 6

The function that assigns to each number x the greatest integer that is ≤x is called the *greatest integer function* and is denoted by $x \to [x]$. Thus, $[\frac{5}{2}] = 2, [2] = 2, [-.3] = -1$, etc. The graphs of $x \to [x]$ and of some other functions are given in Figure 3-7. □

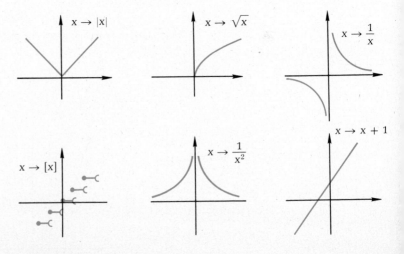

Figure 3-7

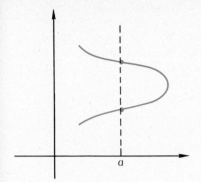

Figure 3-8

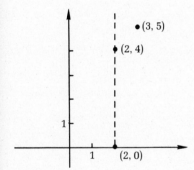

Figure 3-9

Observe that every vertical line intersects the graph of a function in one point at the most. (Why is this so?) Furthermore, any set of points having this property is the graph of one and only one function, namely, the function that assigns to each number that occurs as first coordinate the *unique* second coordinate occurring with it.

Example 7

The curve in Figure 3-8 is *not* the graph of a function, since the number a, for instance, occurs as the first coordinate of more than one point on the curve.

Example 8

The set of points $(2, 0)$, $(3, 5)$, $(4, 0)$ is the graph of the function $2 \to 0$, $3 \to 5$, $4 \to 0$.

Example 9

The set of points $(2, 0)$, $(3, 5)$, $(2, 4)$ is *not* the graph of a function, since the two points $(2, 0)$ and $(2, 4)$ have the same first coordinate. See Figure 3-9.
□

A **constant function** is a function whose range consists of precisely one number; in other words, if f is a constant function, then $f(x)$ is the same number for all x in the domain of f. If that number is c, for example, then the (constant) function $x \to c$ is frequently referred to as the *constant function c*.

Example 10

The graphs of the constant functions $x \to 1$, $x \to -1$, $x \to 3$, and $x \to 5$ are given in Figure 3-10.

Example 11

The graph of the **absolute value function** $x \to |x|$ is given in Figure 3-11. If $x \geq 0$, then $|x| = x$, and, hence, to the right of the vertical axis, the graph of the absolute value function is the same as the graph of I. If $x < 0$, then $|x| = -x$ (which is positive), and, thus, to the left of the vertical axis, the graph of the absolute value function is the reflection of the graph of I in the horizontal axis.

Figure 3-10

x	$|x|$
-3	3
-2	2
-1	1
0	0
1	1
2	2
3	3

$x \to |x|$

The absolute value function

Figure 3-11

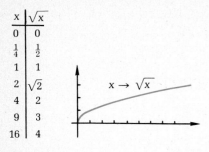

x	$\sqrt{x}$
0	0
$\frac{1}{4}$	$\frac{1}{2}$
1	1
2	$\sqrt{2}$
4	2
9	3
16	4

Figure 3-12

Example 12

The graph of the function f given by $f(x) = \sqrt{x}$ is shown in Figure 3-12. By the domain convention, the domain of f is the set of nonnegative real numbers. Also, $\sqrt{x} \geq 0$ for $x \geq 0$, and if x is large, so is $\sqrt{x}$.

Example 13

The graph of the function $x \to 1/\sqrt{x}$ is given in Figure 3-13. The domain of the function is the set of positive real numbers, and for any x in the domain, we have $1/\sqrt{x} > 0$. If x is near 0, then $1/\sqrt{x}$ is large, whereas if x is large, then $1/\sqrt{x}$ is near 0.

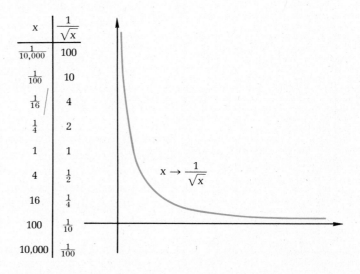

x	$\dfrac{1}{\sqrt{x}}$
$\frac{1}{10,000}$	100
$\frac{1}{100}$	10
$\frac{1}{16}$	4
$\frac{1}{4}$	2
1	1
4	$\frac{1}{2}$
16	$\frac{1}{4}$
100	$\frac{1}{10}$
10,000	$\frac{1}{100}$

Figure 3-13

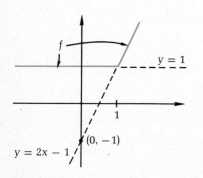

$y = 2x - 1$

Figure 3-14

Example 14

Let $f(x) = x + |x - 1|$. Sketch the graph of f.

Solution

Note that if $x \geq 1$, then $|x - 1| = x - 1$; whereas if $x < 1$, then $|x - 1| = 1 - x$. Thus, to sketch the graph of f, we consider the two cases $x \geq 1$ and $x < 1$. If $x \geq 1$, then $f(x) = x + (x - 1) = 2x - 1$; if $x < 1$, then $f(x) = x + (1 - x) = 1$. Also, $f(1) = 1$. The graph is sketched in Figure 3-14.

Example 15

Let $f(x) = |x| + |x - 1|$. Sketch the graph of f.

Solution

Let x be a real number. Then

if $x > 0$, then $|x| = x$
if $x < 0$, then $|x| = -x$
if $x \geq 1$, then $|x - 1| = x - 1$
if $x < 1$, then $|x - 1| = 1 - x$

Thus,

if $x < 0$, then $f(x) = -x + (1 - x) = -2x + 1$
if $0 \leq x \leq 1$, then $f(x) = x + (1 - x) = 1$
if $x > 1$, then $f(x) = x + (x - 1) = 2x - 1$

See Figure 3-15.

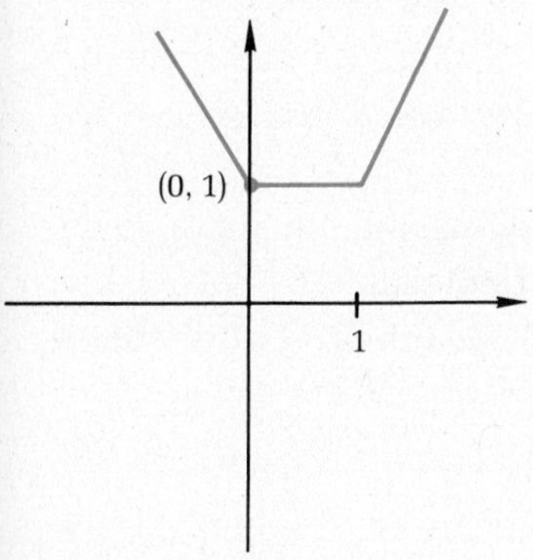

(0, 1)

1

Figure 3-15

Exercises 3-2

In Problems 1–16, sketch the graph of f

1. $f(x) = x + 3$ **2.** $f(x) = 2x$
3. $f(x) = -4x$ **4.** $f(x) = -x + 5$
5. $f(x) = -4$ **6.** $f(x) = 2$
7. $f(x) = -x^2 + 1$ **8.** $f(x) = x^3 + 1$
9. $f(x) = -x^3$ **10.** $f(x) = 5x^3$
11. $f(x) = 5x^3 + 1$ **12.** $f(x) = \sqrt{x} + 3$
13. $f(x) = -\sqrt{x}$ **14.** $f(x) = \dfrac{1}{x^3}$
15. $f(x) = \dfrac{1}{x - 1}$ **16.** $f(x) = \dfrac{1}{x - 2}$

Keeping the domain convention in mind, describe the domain of the functions given in Exercises 17–21.

17. $f(x) = \dfrac{1}{(x - 1)(x - 2)}$ **18.** $f(x) = \sqrt{3x + 4}$

19. $f(x) = \dfrac{1}{\sqrt{3x + 4}}$ **20.** $f(x) = x^2 + \dfrac{2}{\sqrt{x^2 + 1}}$

21. $f(x) = \dfrac{1}{\sqrt{x - 1}} + \dfrac{1}{\sqrt{x - 2}}$

In Exercises 22–27 give the domain and range and sketch the graph of each function.

22. $x \to \sqrt{x - 4}$ **23.** $x \to \sqrt{2x + 3}$
24. $x \to |x| + 7$ **25.** $x \to 2|x| + 7$
26. $x \to \sqrt{x^2 - 3}$ **27.** $x \to \sqrt{2x^2 - 5}$

In Exercises 28–36, graph the given functions.

28. $x \to [x] + 2$

29. $x \to [3x] + 2$

30. $x \to [3x + 2]$

31. $x \to x/x$

32. $x \to [x/|x|]$

33. $x \to 1/|x|$

34. $x \to x - [x]$

35. $x \to x + |x - 2|$

36. $x \to |x| + |x - 2|$

37. Is the following set of points the graph of a function? $\{(-1, -1), (0, -1), (1, -1), (2, -1)\}$ Explain.

38. Is $\{(0, 1), (2, 3), (4, 7), (5, 3), (2, 6)\}$ the graph of a function? Explain.

39. Is the set of all points (x, y) for which $y^4 = x$ the graph of a function?

In Exercises 40–46, sketch the graph of the given function.

40. $x \to |[x]|$

41. $x \to [|x|]$

42. $x \to \sqrt{|x|}$

43. $x \to |x| - x$

44. $x \to |x| + x$

45. $x \to \dfrac{x^2}{|x|}$

46. $x \to |x| - 1$

47. Let f and g be the greatest integer function and the absolute value function, respectively. Find all x such that (a) $f(x) = g(x)$, and (b) $f[g(x)] = g[f(x)]$.

3-3 OPERATIONS ON FUNCTIONS

Let f and g be any functions. We define the **sum**

$$f + g \text{ to be the function } x \to f(x) + g(x)$$

and the **product**

$$fg \text{ to be the function } x \to f(x)g(x)$$

The domain of $f + g$ and fg is the set of all numbers that belong to both the domain of f and the domain of g.

In Figure 3-16 the graphs of two functions f and g are given along with the graph of $f + g$.

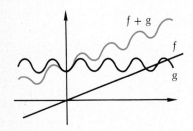

Figure 3-16

Example 1

Let f be the squaring function and g be the function $x \to x^3 + 1$. Then $f + g$ is the function $x \to x^2 + x^3 + 1$, and fg is the function $x \to x^2(x^3 + 1) = x^5 + x^2$.

□

If c is a real number and f is any function, the sum $f + c$ is the function $x \to f(x) + c$. The graph of $f + c$ is the graph of f *shifted* $|c|$ units—upward if $c > 0$ and downward if $c < 0$. See Figure 3-17.

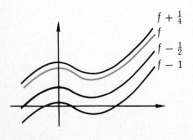

Figure 3-17

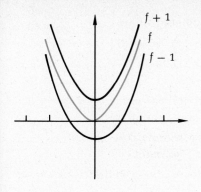

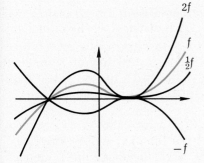

Figure 3-18

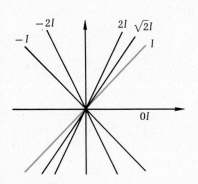

Figure 3-19

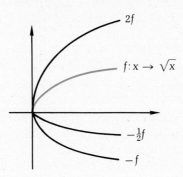

Figure 3-21

Example 2

Let f be the squaring function $x \to x^2$. The graph of f is shown in Figure 3-18, together with $f + c$ for various values of c.
□

By definition, if f and g are functions and c is a real number, then

$$cf \text{ is the function } x \to cf(x)$$
$$-f = (-1)f$$

and

$$f - g = f + (-g)$$

If $c \geq 0$, the graph of cf is the graph of f stretched vertically by a factor of c (which amounts to a contraction if $0 \leq c \leq 1$). For $c \leq 0$, the graph of cf is the graph of f stretched vertically by a factor of $|c|$ and *reflected in the horizontal axis*. See Figure 3-19.

Example 3

Let f be the square root function $x \to \sqrt{x}$. The graphs of f, $-f$, $2f$, and $-\frac{1}{2}f$ are given in Figure 3-20.

Figure 3-20

Example 4

The graphs of cI for various values of c are given in Figure 3-21. (Recall that I is the function $x \to x$.)
□

So far we have studied operations on functions that result in *vertical* changes: vertical shift, vertical stretch (or contraction), and reflection in the *horizontal* changes.

Again, let f be a function and c a real number. Then f_c is

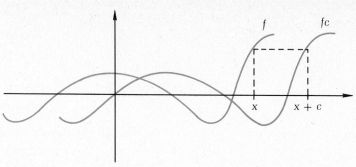

Figure 3-22

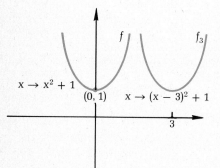

Figure 3-23

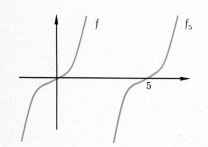

Figure 3-24

defined to be the function $x \to f(x - c)$ or, equivalently, $f_c(x) = f(x - c)$ for all x for which $x - c$ is in the domain of f. The function f_c is called a **horizontal shift,** or a **translate** of f, since its graph is the graph of f translated $|c|$ units—to the right if $c > 0$, and to the left if $c < 0$. This can be seen from the fact that $f_c(x + c) = f(x - c + c) = f(x)$ for all x in the domain of f; that is, the height of the graph of f_c above $x + c$ is exactly the same as the height of the graph of f above x. See Figure 3-22.

Example 5

Let f be the cubing function $x \to x^3$; then f_5 is the function $x \to (x - 5)^3$. See Figure 3-23.

Example 6

Let f be the function $x \to x^2 + 1$; then f_3 is the function $x \to (x - 3)^2 + 1$. See Figure 3-24.

□

Now suppose $c > 0$, and consider the function $x \to f(cx)$. The graph of this function is the graph of f *shrunk horizontally* by a factor of c if $c > 1$ (and expanded by a factor of $1/c$ if $c < 1$), since the height of the graph of $x \to f(cx)$ above x/c is the same as the height of the graph of f above x. Note also that the graph of $x \to f(-cx)$ is the graph of $x \to f(cx)$ reflected in the vertical axis. We shall illustrate this in Example 8. See Figures 3-25, 3-26, and 3-27.

Example 7

Let f be the absolute value function $x \to |x|$, g the function $x \to f(x/3) = |x/3|$, and h: $x \to f(3x) = |3x|$. The graphs of f, g, and h are given in Figure 3-26.

□

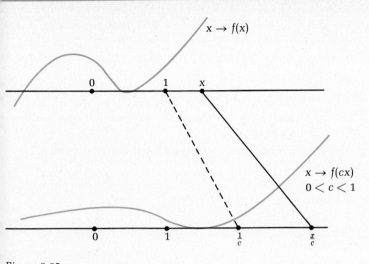

$x \to f(x)$

$x \to f(cx)$
$0 < c < 1$

Figure 3-25

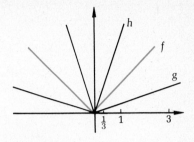

Figure 3-26

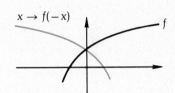

$x \to f(-x)$

Figure 3-27

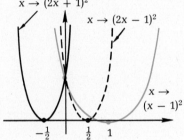

$x \to (2x + 1)^2$

$x \to (2x - 1)^2$

$x \to (x - 1)^2$

Figure 3-28

A special, yet frequently encountered, case occurs when the graph of a function f is merely reflected in the vertical axis: the resulting graph is the graph of $x \to f(-x)$. See Figure 3-27.

Example 8

If $f(x) = (x - 1)^2$, then $f(-2x) = (-2x - 1)^2 = (2x + 1)^2$. Hence, the graph of $x \to (2x + 1)^2$ is the graph of f shrunk horizontally by a factor of 2 and then reflected in the vertical axis. See Figure 3-28.

□

There are other ways in which new functions can be obtained from a given function. For example, if f is a given function then $x \to |f(x)|$ is a function whose graph can be obtained from the graph of f by reflecting about the horizontal axis that part of the graph of f which is below the horizontal axis. See Figure 3.29.

The reader is invited to consider the following question. Let $f(x) = x^2 - x$. What is the graph of each of the following functions?

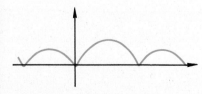

Figure 3-29

$$x \to f\left(\frac{1}{x}\right), \qquad x \to \frac{1}{f(x)}, \qquad x \to [f(x)],$$

$$x \to (f(x))^2, \qquad x \to \sqrt{f(x)}$$

Exercises 3-3

In Exercises 1–5, find the function whose graph is obtained from the graph of the cubing function by

1. a shift up of 1 unit

2. a vertical stretch by a factor of 3

3. a vertical stretch by a factor of -3 (i.e., a stretch by a factor of 3 followed by reflection)

4. a shift upward of 1 unit followed by a vertical stretch by a factor of 3

5. a vertical stretch by a factor of 3 followed by an upward shift of 1 unit

Let f be the cubing function. Graph f and the functions given in Exercises 6–8.

6. $x \to f(x - 1)$ **7.** $x \to f\left(\dfrac{x}{3}\right)$

8. $x \to f\left(\dfrac{x}{3} - 1\right)$

Suppose the graph of f is

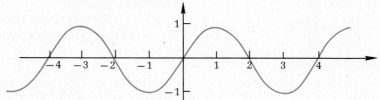

Exercises 9–11

In Exercises 9–11, sketch the graph of g, if for every real number x

9. $g(x) = f(x + 1)$ **10.** $g(x) = f\left(\dfrac{x}{2}\right)$

11. $g(x) = 2f(x)$

Suppose the graph of f is

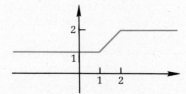

Exercises 12–14

In Exercises 12–14, sketch a graph of g, if for every real number x.

12. $g(x) = f(x) - 2$ **13.** $g(x) = f(x - 2)$

14. $g(x) = f(2x)$

15. Suppose the graph of f is

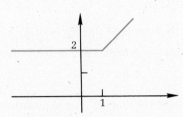

Exercise 15

If for every real number x, $g(x) = f(x - 1) - 2$, sketch the graph of g.

Let $f(x) = x^2 - 2$. Graph each of the functions in Exercises 16–18. Where do the graphs intersect the horizontal axis?

16. $x \rightarrow f(3x)$ **17.** $x \rightarrow f(x) - 4$

18. $x \rightarrow f(2x) - 1$

Let f be the squaring function. Graph the functions in Exercises 19–24.

19. $f + 1$ (i.e., $x \rightarrow x^2 + 1$) **20.** $2f$

21. $2f + 1$ **22.** $3f + 2$

23. $f_2 - 3$ **24.** $-2f_3 + \frac{1}{2}$

Let f be the function $x \rightarrow 1/x$. Graph the functions in Exercises 25–27.

25. $f + \frac{3}{2}$ $\left(\text{i.e., } x \rightarrow \frac{1}{x} + \frac{3}{2}\right)$ **26.** $\left(\frac{1}{2}\right)f$

27. $2f - 1$

Graph the functions in Exercises 28–30.

28. $3I$ **29.** $I - 2$

30. $3I + 5$

31. If $f(x) = \dfrac{1}{x - 1}$, and $g(x) = \dfrac{x}{x^2 - 1}$, find all x such that $(f + g)(x) = 0$.

32. If $f(x) = \dfrac{x + 1}{x + 3}$, and $g(x) = \dfrac{x^2 - 9}{1 - x^2}$, find all x for which $(fg)(x) = 0$.

33. If $f(x) = x + \dfrac{1}{x} + 2$, and $g(x) = \dfrac{x^2}{x^2 - 1}$, find all real numbers x for which $(fg)(x) = 0$.

34. True or false (justify): Let $f(x) = x^3$, and $g(x) = x + 1$; then (a) $fg = gf$; (b) $f + g = g + f$.

35. Let $f(x) = x + 1$. Find all real numbers x such that $f[f(x)] = 2$.

In Exercises 36–38, let $f(x) = x^2$. Find

36. f_3 [where $f_3(x) = f(x - 3)$ for every real number x]

37. $f_3 - 2$ **38.** $4f_3 - 2$

39. Let $f(x) = 2x^2 - x - 5$. Find g if $g(x) = f(x + 2)$ for every real number x. For what numbers x does $g(x) = 0$?

40. Let $f(x) = 3x^2 - 2$. Let g and h be given, respectively, by $g(x) = f(x + 2)$ and $h(x) = g(x - 1)$. Find all real numbers x for which $g(x) = h(x)$.

Miscellaneous Problems

1. Let $f(x) = 3x - 2$. Find all numbers x for which $f(x^2) = (f(x))^2$.

2. Let $f(x) = 2x + 1$. Compute $\dfrac{f(x + h) - f(x)}{h}$, where $h \neq 0$.

3. Let $f(x) = -5x + 2$.
(a) Find a function g for which $f(g(x)) = x$ for all x.
(b) Find a function h for which $f(h(x)) = x^2 - 1$.

4. Let $f(x) = 3x + 2$. Find all x for which $f(f(x)) = 3x$.

5. Let $f(x) = x^4 - 3x^2$. Which of the following holds for all numbers x?
(a) $f(-x) = f(x)$ (b) $f(-x) = -f(x)$

6. Repeat Problem 5 for the functions g and h given by $g(x) = x^3 + 3x$, $h(x) = \sqrt{x^2 + 1}$

7. Let $f(x) = x^2 + 1$ and $g(x) = x^3 + 2x^2 + x + 1$. Find the intersection of the graphs of f and g.

8. Let $f(x) = (x - 1)^2 + 2$ for $x \geq 1$
     $= 3x - 1$ for $x < 1$
Graph f.

9. Let $V(x) =$ the volume of an open rectangular box that is formed by cutting equal squares of side x inches from the corners of a 12-inch square piece of tin. Find $V(x)$ and graph V.

In Problems 10–15, suppose the graph of f is:

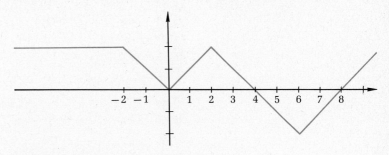

Problems 10–15

Sketch the graph of g where

10. $g(x) = |f(x)|$

11. $g(x) = [f(x)]$

◊**12.** $g(x) = f\left(\dfrac{1}{x}\right)$

13. $g(x) = \dfrac{1}{f(x)}$

14. $g(x) = (f(x))^2$

◊**15.** $g(x) = \sqrt{|f(x)|}$

16. Suppose f is a function whose graph crosses the horizontal axis at $-\frac{7}{6}$, $-\frac{3}{4}$, 2, 5, and $\frac{13}{2}$. Determine the points where the graph of the function $x \to -3f(x + \frac{1}{2})$ crosses the horizontal axis.

17. A charter airline flight requires 40 passengers and charges 100 dollars per passenger, but will reduce the fare by 50 cents per additional passenger. The plane can take 90 passengers. Let $f(x) =$ the fare when x passengers are taken. Find $f(x)$ and graph f.

4 Sequences

4-1 INTRODUCTION

The word "sequence" suggests a succession. Since a function whose domain is the set N of natural numbers can be *thought of as a succession of numbers,* we have the following definition:

A *sequence* is a function whose domain is the set N.*

Example 1

The function defined by $n \to 1/n$, n in N, is a sequence. It may be indicated by the succession $1, \frac{1}{2}, \frac{1}{3}, \frac{1}{4}, \frac{1}{5}$, and so on.
□

Given a sequence f, we shall frequently denote it by

$$f(1), f(2), f(3), \ldots, f(n), \ldots$$

or, more briefly, by

$$\langle f(n) \rangle$$

The sequence of Example 1 is thus given by

$$1, \frac{1}{2}, \frac{1}{3}, \ldots, \frac{1}{n}, \ldots$$

---

*Occasionally, it is convenient to modify this definition to include in the domain of a sequence all integers *from some integer on;* such a domain may be the set $0, 1, 2, 3, \ldots$, or $-3, -2, -1, 0, 1, 2, \ldots$, and so on. We shall not do this until Section 4-2.

or, more briefly, by $\left\langle \dfrac{1}{n} \right\rangle$.

For any n in N, $f(n)$ is called the **nth term** of the sequence f; in particular, $f(1)$ is the **first term,** $f(2)$ the **second term,** and so on.

Example 2

Consider the sequence whose first term is 1 with each term after the first being twice its predecessor.

The first few terms are 1, 2, 2^2, 2^3, 2^4, and, in fact, for any n in N the nth term is 2^{n-1}. Hence, this sequence is

$$1, 2, 2^2, \ldots, 2^{n-1}, \ldots$$

or, more briefly,

$$\langle 2^{n-1} \rangle$$

□

The notational practice for sequences differs somewhat from that for other functions: rather than denoting the nth term of a sequence f by $f(n)$, it is customary (but not essential) to denote it by a_n (or b_n, or c_n, etc.). Thus, for example, we may refer to a sequence

$$a_1, a_2, a_3, \ldots, a_n, \ldots \quad \text{or} \quad \langle a_n \rangle^*$$

Example 3

The sequence $\langle a_k \rangle$ for which $a_1 = 1$ and $a_{k+1} = 2 \cdot a_k$ for all k in N is precisely the sequence of Example 2. Here $a_k = 2^{k-1}$.

Example 4

The first five terms of the sequence $\langle b_m \rangle$ for which $b_m = 1/2^{m+1}$ are $\frac{1}{4}$, $\frac{1}{8}$, $\frac{1}{16}$, $\frac{1}{32}$, and $\frac{1}{64}$.

Example 5

The sequences $\langle a_n \rangle$ and $\langle b_n \rangle$ for which $a_n = (n+1)/(n+2)$ and $b_n = \sqrt{n}$ are

$$\frac{2}{3}, \frac{3}{4}, \frac{4}{5}, \ldots, \frac{n+1}{n+2}, \ldots \quad \text{and} \quad 1, \sqrt{2}, \sqrt{3}, \ldots, \sqrt{n}, \ldots$$

respectively.

---

*Note that the use of n is quite arbitrary. Clearly, we obtain the same sequence by describing "a_k for every k in N" or "a_i for every i in N."

Example 6

Find a_n if $a_1 = \frac{1}{2}$ and $a_{k+1} = 3 + a_k$ for all k in N (i.e., find the nth term of the sequence whose first term is $\frac{1}{2}$ with each term after the first being three more than its predecessor).

Solution

$$a_1 = \tfrac{1}{2}$$

$$a_2 = 3 + a_1 = 3 + \tfrac{1}{2}$$

$$a_3 = 3 + a_2 = \underbrace{3 + 3}_{2 \cdot 3} + \tfrac{1}{2}$$

$$a_4 = 3 + a_3 = \underbrace{3 + 3 + 3}_{3 \cdot 3} + \tfrac{1}{2}$$

$$a_5 = 3 + a_4 = \underbrace{3 + 3 + 3 + 3}_{4 \cdot 3} + \tfrac{1}{2}$$

.
.
.

$$a_{k+1} = 3 + a_k = \underbrace{3 + 3 + \cdots + 3}_{k \cdot 3} + \tfrac{1}{2}$$

Therefore, for any $n > 1$ in N, we have

$$a_n = 3 + a_{n-1} = \underbrace{3 + 3 + \cdots + 3}_{(n-1) \cdot 3} + \tfrac{1}{2}$$

$$= 3(n - 1) + \tfrac{1}{2} = 3n - 3 + \tfrac{1}{2} = 3n - \tfrac{5}{2}$$

Thus, $a_n = 3n - \frac{5}{2}$ for all n in N.

Exercises 4-1

In Exercises 1–6, find the first five terms for the given sequence.

1. $\langle 2^n + 1 \rangle$

2. $\langle n^2 + n \rangle$

3. $\left\langle \dfrac{1}{2^n} + \dfrac{1}{3^n} \right\rangle$

4. $\left\langle \dfrac{1 + (-1)^n}{2} \right\rangle$

5. $\left\langle \dfrac{(-1)^n \cdot n}{n + 1} \right\rangle$

6. $\langle (-1)^{n+1} + (-1)^n \rangle$

In Exercises 7–12, find a_n if

7. $a_1 = 7$ and $a_k = a_{k-1} - 2$

8. $a_1 = 5$ and $a_k = (1/2)a_{k-1}$

9. $a_1 = 3$ and $a_k = 5a_{k-1}$

10. $a_1 = 2$ and $a_k = (a_{k-1})^2$

11. $a_1 = 1$ and $a_k = \sqrt{a_{k-1}}$

12. $a_1 = 2$ and $a_k = \sqrt{a_{k-1}}$

In practice, if enough of the terms of a sequence are listed, the "law of formation" may be guessed at. In Exercises 13–18, the first five terms are given; in each case, find the nth term of a possible sequence indicated by the pattern.

13. $1, 4, 7, 10, 13, \ldots$

14. $2, 6, 18, 54, 162, \ldots$

15. $2, 4, 8, 16, 32, \ldots$

16. $3, 5, 9, 17, 33, \ldots$

17. $1, \frac{1}{4}, \frac{1}{9}, \frac{1}{16}, \frac{1}{25}, \ldots$

18. $\frac{1}{2}, -\frac{2}{20}, \frac{3}{38}, -\frac{4}{56}, \frac{5}{74}, \ldots$

19. Given a sequence $\langle a_n \rangle$ for which $a_1 = 1$, and a real number r, with the property that $a_n < r a_{n-1}$ for all $n > 1$, show that $a_n < r^{n-1}$.

20. If an initial amount P_0 of money is compounded annually at the interest rate of $r\%$, then each year the amount is multiplied by $\left(1 + \dfrac{r}{100}\right)$. Let P_n be the amount of money after n years. Show that

$$P_n = \left(1 + \frac{r}{100}\right)^n P_0$$

21. What will $1000 at 5% compounded annually amount to after 10 years?

4-2 THE SIGMA NOTATION

Suppose we wish to add, say, the first 103 terms of a sequence $\langle a_k \rangle$. A convenient notation for this sum uses the capital Greek letter *sigma*, Σ. Thus:

$$a_1 + a_2 + a_3 + \cdots + a_{103} = \sum_{k=1}^{103} a_k$$

This is read "the sum of the numbers a_k from $k = 1$ to $k = 103$." In general, *given a sequence* $\langle a_n \rangle$ *and natural numbers* n *and* m *with* $n \geq m$, $\displaystyle\sum_{k=m}^{n} a_k$ *is the sum of the numbers* a_k *from* $k = m$ *to* $k = n$; that is,

$$\sum_{k=m}^{n} a_k = a_m + a_{m+1} + \cdots + a_n$$

Remarks

1. As before, the use of k here is arbitrary, so that

$$\sum_{k=m}^{n} a_k = \sum_{i=m}^{n} a_i = \sum_{t=m}^{n} a_t \quad \text{and so on.}$$

2. If the domain of $\langle a_n \rangle$ is not N but, rather, the set of all integers greater than or equal to some given integer z, then, of course, m and n need not be in N, but must only be integers satisfying $n \geq m \geq z$. See the footnote on p. 84.

3. The case $m = n$ is interpreted as follows:

$$\sum_{k=m}^{m} a_k = a_m.$$

Example 1

(a) $\displaystyle\sum_{k=1}^{10} 3^k = 3 + 3^2 + 3^3 + \cdots + 3^{10}$

(b) $\displaystyle\sum_{t=3}^{15} \frac{1}{t + 5} = \frac{1}{3 + 5} + \frac{1}{4 + 5} + \frac{1}{5 + 5} + \cdots + \frac{1}{15 + 5}$

(c) $\displaystyle\sum_{i=-2}^{3} 3i = 3(-2) + 3(-1) + 3(0) + 3(1) + 3(2) + 3(3) = 9$

(d) $\displaystyle\sum_{j=5}^{5} b_j = b_5$

(e) For any nonzero real number x,

$$\sum_{k=8}^{20} x^k = x^8 + x^9 + x^{10} + \cdots + x^{19} + x^{20}$$

$$= x^8(1 + x + x^2 + \cdots + x^{11} + x^{12}) = x^8 \sum_{t=0}^{12} x^t$$

(f) $\displaystyle\sum_{u=3}^{8} u^2 = 3^2 + 4^2 + 5^2 + 6^2 + 7^2 + 8^2 = \sum_{i=3}^{4} i^2 + \sum_{j=5}^{8} j^2$

□

Examples 1(e) and 1(f) show the convenience of the sigma notation as well as some of its properties. We shall give several more examples of this. In what follows, c is a real number, $\langle a_k \rangle$ and $\langle b_k \rangle$ are sequences, and m and n are integers in the domains of both, with $n \geq m$.

The distributive property:

$$\sum_{k=m}^{n} ca_k = c \sum_{k=m}^{n} a_k$$

This is so because

$$\sum_{k=m}^{n} ca_k = ca_m + ca_{m+1} + \cdots + ca_n$$

$$= c(a_m + a_{m+1} + \cdots + a_n)$$

$$= c \sum_{k=m}^{n} a_k$$

The grouping property: If $m \leq p < n$, then

$$\sum_{k=m}^{n} a_k = \sum_{k=m}^{p} a_k + \sum_{k=p+1}^{n} a_k$$

This can be seen by observing that

$$\sum_{k=m}^{n} a_k = \underbrace{a_m + a_{m+1} + \cdots + a_p}_{\sum\limits_{k=m}^{p} a_k} + \underbrace{a_{p+1} + \cdots + a_n}_{\sum\limits_{k=p+1}^{n} a_k}$$

The additive property:

$$\sum_{k=m}^{n} (a_k + b_k) = \sum_{k=m}^{n} a_k + \sum_{k=m}^{n} b_k$$

To see this, note that

$$\sum_{k=m}^{n} (a_k + b_k) = (a_m + b_m) + (a_{m+1} + b_{m+1})$$

$$+ (a_{m+2} + b_{m+2}) + \cdots + (a_n + b_n)$$

$$= (a_m + a_{m+1} + a_{m+2} + \cdots + a_n)$$

$$+ (b_m + b_{m+1} + b_{m+2} + \cdots + b_n)$$

$$= \sum_{k=m}^{n} a_k + \sum_{k=m}^{n} b_k$$

Finally, in the case of the constant sequence $\langle c \rangle$ (i.e., the sequence $\langle a_k \rangle$ with $a_k = c$ for every k in the domain of $\langle a_k \rangle$) we have

$$\sum_{k=m}^{n} c = (n - m + 1)c$$

since

$$\sum_{k=m}^{n} c = \sum_{k=m}^{n} a_k = a_m + a_{m+1} + \cdots + a_n = \underbrace{c + c + \cdots + c}_{n-m+1 \text{ times}}$$

$$= (n - m + 1)c$$

In particular,

$$\sum_{k=1}^{n} c = nc$$

Example 2

(a) $\displaystyle\sum_{k=1}^{5} 5 \cdot 2^k = 5 \cdot 2^1 + 5 \cdot 2^2 + 5 \cdot 2^3 + 5 \cdot 2^4 + 5 \cdot 2^5$

$$= 5(2^1 + 2^2 + 2^3 + 2^4 + 2^5) = 5 \sum_{k=1}^{5} 2^k$$

(b) $\displaystyle\sum_{k=3}^{10} (5^k + k^5) = \sum_{k=3}^{10} 5^k + \sum_{k=3}^{10} k^5$

(c) $\displaystyle\sum_{k=1}^{80} 5 = 80 \cdot 5 = 400$

(d) $\displaystyle\sum_{k=10}^{20} (2 \cdot 3^k + 7 \cdot k^4) = 2 \sum_{k=10}^{20} 3^k + 7 \sum_{k=10}^{20} k^4$

(e) $\displaystyle\sum_{j=3}^{7} 8^j + \sum_{t=8}^{152} 8^t = \sum_{k=3}^{152} 8^k$

(f) $\displaystyle\sum_{k=1}^{17} (k - 1)^2 = 0^2 + 1^2 + 2^2 + \cdots + 16^2$

$$= \sum_{k=0}^{16} k^2 = \sum_{k=-1}^{15} (k + 1)^2$$

(g) For any sequence $\langle a_n \rangle$, any m in its domain, and any integer $p \geq m$, we have

$$\sum_{n=m}^{p} a_n = \sum_{n=m+k}^{p+k} a_{n-k}$$

for any nonnegative integer k. (Hint: Write it out.)

Exercises 4-2

In Exercises 1–7, compute.

1. $\displaystyle\sum_{k=2}^{5} \frac{1}{k-1}$

2. $\displaystyle\sum_{n=3}^{6} (-1)^n n$

3. $\displaystyle\sum_{n=1}^{4} \sqrt{n}$

4. $\displaystyle\sum_{k=4}^{8} (k-1)k$

5. $\displaystyle\sum_{n=1}^{100} (-1)^n$

6. $\displaystyle\sum_{n=1}^{100} (-1)^{n+1}$

7. $\displaystyle\sum_{n=1}^{101} (-1)^{n+1}$

8. Let $\langle a_n \rangle$ be a sequence. If $\displaystyle\sum_{k=10}^{20} a_k = 30$ and $\displaystyle\sum_{k=16}^{20} a_k = 18$, find $\displaystyle\sum_{k=10}^{15} a_k$.

9. If $\langle a_n \rangle$ and $\langle b_n \rangle$ are sequences for which $\displaystyle\sum_{k=1}^{100} a_k = 28$ and $\displaystyle\sum_{k=1}^{100} b_k = 43$, find $\displaystyle\sum_{k=1}^{100} (4a_k - 5b_k)$.

10. Let $\langle a_n \rangle$ and $\langle b_n \rangle$ be sequences. If $\displaystyle\sum_{k=1}^{100} a_k^2 = 50$, $\displaystyle\sum_{k=1}^{100} b_k^2 = 30$, and $\displaystyle\sum_{k=1}^{100} a_k b_k = 40$, find $\displaystyle\sum_{k=1}^{100} (a_k + b_k)^2$.

11. If $\langle a_n \rangle$ is a sequence for which $\displaystyle\sum_{k=1}^{10} a_k = 12$, find $\displaystyle\sum_{k=1}^{10} (-3a_k + 8)$ and $\displaystyle\sum_{k=1}^{10} \left(\frac{a_k}{4} - 10 \right)$.

12. Find $\displaystyle\sum_{k=10}^{100} a_k$ if $\displaystyle\sum_{k=10}^{100} (-3a_k + 8) = 602$.

13. Find $\displaystyle\sum_{k=1}^{100} b_k$ if $\displaystyle\sum_{k=1}^{100} (4b_k - 5) = 16$.

14. If $\langle a_n \rangle$ is a sequence for which $a_k = a_{k-1}$ for every k, and $\sum\limits_{i=1}^{50} a_i = 100$, find a_n.

Indicate whether the equalities in Exercises 15–18 are true or false.

15. $\sum\limits_{k=3}^{18} 2^{k+5} = \sum\limits_{k=1}^{16} 2^{k+7}$

16. $\left(\sum\limits_{k=1}^{3} k \right)^2 = \sum\limits_{k=1}^{3} k^2$

17. $\sum\limits_{k=1}^{10} \sqrt{k} = \sqrt{\sum\limits_{k=1}^{10} k}$

18. $\sum\limits_{k=20}^{30} \frac{1}{k^2} = \sum\limits_{k=8}^{18} \frac{1}{(k+12)^2}$

19. Let $\langle a_n \rangle$ be a sequence. Show that $\sum\limits_{k=2}^{n} (a_k - a_{k-1})$

$= a_n - a_1$.

20. Compute $\sum\limits_{k=1}^{1000} \frac{1}{k(k+1)}$. (*Hint:* $\frac{1}{k(k+1)} = \frac{1}{k} - \frac{1}{k+1}$ for any natural number k.)

4-3 ARITHMETIC PROGRESSION

A sequence in which the *difference* between each term and its successor is always the same is called an **arithmetic progression.** For example, 5, 8, 11, 14, and 17 are the first five terms of an arithmetic progression, because $8 - 5 = 11 - 8 = 14 - 11 = 17 - 14 = 3$.

An arithmetic progression is fully determined by its first term and the **common difference** (which is 3 in the example just given). Suppose a is the *first term* and d the *common difference* of the arithmetic progression $\langle a_n \rangle$. Then we have

$$
\begin{aligned}
a_1 &= a & &= a \\
a_2 &= a_1 + d = a + d & &= a + d \\
a_3 &= a_2 + d = a + d + d & &= a + 2d \\
a_4 &= a_3 + d = a + d + d + d & &= a + 3d
\end{aligned}
$$

.
.
.

$$
a_n = a_{n-1} + d = a + \underbrace{d + \cdots + d + d}_{n-1 \text{ times}} = a + (n-1)d
$$

In other words, for any n in N, the nth term a_n of an arithmetic progression whose first term is a and whose common difference is d, is

$$
a_n = a + (n-1)d
$$

Example 1

Find the nth term of the arithmetic progression whose first term is $\frac{3}{4}$ and whose common difference is $\frac{1}{2}$.

Solution

The nth term is $\frac{3}{4} + (n - 1)\frac{1}{2} = \frac{1}{2}n + \frac{1}{4}$, for any n in N.

Example 2

If $\langle a_n \rangle$ is an arithmetic progression with $a_{10} = -13$ and $a_4 = -1$, find a_n for any n in N and give the first few terms.

Solution

We have

$$a_{10} = a_1 + 9d = -13$$

and

$$a_4 = a_1 + 3d = -1$$

Solving this system, we find that $6d = -12$, so that $d = -2$, and hence $a_1 = -1 - 3d = -1 + 6 = 5$. Hence, $a_n = 5 - 2(n - 1) = -2n + 7$ for any n in N and, explicitly, $\langle a_n \rangle$ is the sequence

$$5, 3, 1, \ldots, -2n + 7, \ldots$$

Example 3

Find the ninety-fourth term of the arithmetic progression whose fourth term is $3\sqrt{3} + 1$ and whose forty-seventh term is $46\sqrt{3} + 44$.

Solution

Here we have

$$a_4 = a_1 + 3d = 3\sqrt{3} + 1$$

and

$$a_{47} = a_1 + 46d = 46\sqrt{3} + 44.$$

Hence, subtracting,

$$43d = 43\sqrt{3} + 43, \qquad d = \sqrt{3} + 1$$

and so

$$a_1 = 3\sqrt{3} + 1 - 3d = 3\sqrt{3} + 1 - 3\sqrt{3} - 3 = -2.$$

Therefore, $a_n = -2 + (n - 1)(\sqrt{3} + 1)$ for any n in N and, in particular, $a_{94} = -2 + 93(\sqrt{3} + 1) = 93\sqrt{3} + 91$.

□

Certain problems involve the addition of several terms of an arithmetic progression. We shall now develop a general method for such addition. We begin by finding *the sum of the first n natural numbers* or, equivalently, of the first n terms of the arithmetic progression whose first term is 1 and whose common difference is 1.

We have

$$\sum_{k=1}^{n} k = 1 + 2 + 3 + \cdots + (n-1) + n$$

and hence also

$$\sum_{k=1}^{n} k = n + (n-1) + (n-2) + \cdots + 2 + 1.$$

Adding, we obtain

$$2 \sum_{k=1}^{n} k$$

$$= \underbrace{(n+1) + (n+1) + (n+1) + \cdots + (n+1) + (n+1)}_{n \text{ times}}$$

$$= n(n+1)$$

and so

$$\sum_{k=1}^{n} k = \tfrac{1}{2}n(n+1) \quad \text{for any } n \text{ in } N.$$

Example 4

Find $\sum_{k=1}^{n} k$ for $n = 42$ and $n = 1000$.

Solution

$$1 + 2 + 3 + \cdots + 42 = \sum_{k=1}^{42} k = \tfrac{42}{2}(43) = 903$$

and

$$\sum_{k=1}^{1000} k = \frac{1000(1001)}{2} = 500{,}500$$

In general, *the sum of the first n terms of an arithmetic progression with first term a and common difference d is* $\frac{1}{2}n[2a + (n - 1)d]$; *that is*

$$\sum_{k=1}^{n} [a + (k - 1)d] = \tfrac{1}{2}n[2a + (n - 1)d]$$

For

$$\sum_{k=1}^{n} [a + (k - 1)d] = \sum_{k=1}^{n} a + d \sum_{k=1}^{n} (k - 1)$$

$$= na + d\left(\sum_{k=1}^{n} k - \sum_{k=1}^{n} 1\right)$$

$$= na + d[\tfrac{1}{2}n(n + 1) - n]$$
$$= \tfrac{1}{2}n(2a) + d[\tfrac{1}{2}n^2 - \tfrac{1}{2}n]$$
$$= \tfrac{1}{2}n[2a + d(n - 1)]$$

Remark   Since $2a + d(n - 1) = a + [a + d(n - 1)] = a_1 + a_n$, we see that

$$\sum_{k=1}^{n} [a + (k - 1)d] = n\left(\frac{a_1 + a_n}{2}\right)$$

that is, *the sum of the first n terms of an arithmetic progression is n times the average of the first and nth terms.*

Example 5

Find the sum of the first twenty terms of the arithmetic progression whose first term is 3 and whose common difference is 5.

Solution

Here $d = 5$, $a_1 = 3$, and so $a_{20} = 3 + 19 \cdot 5 = 98$. Hence, the sum is

$$20 \cdot \frac{98 + 3}{2} = 10 \cdot 101 = 1010$$

or

$$\frac{20}{2}(2 \cdot 3 + 19 \cdot 5) = 10(6 + 95) = 1010$$

Exercises 4-3

In Exercises 1–8, find the sum of each of the following arithmetic progressions.

1. $2 + 5 + 8 + 11 + 14 + 17 + 20$

2. $4 + 10 + 16 + 22 + 28 + 34 + 40$

3. $2 + 7 + 12 + \cdots + 52$

4. $3 + 10 + 17 + \cdots + 143$

5. $\displaystyle\sum_{k=1}^{20} (1 + 3k)$

6. $\displaystyle\sum_{k=1}^{50} (-3 + 2k)$

7. $\displaystyle\sum_{j=1}^{100} (4j + 1)$

8. $\displaystyle\sum_{i=1}^{100} (2 - 3i)$

9. Find the sum of the first twenty terms of the arithmetic progression whose first term is 2 and whose common difference is 4.

10. Find the sum of the first hundred terms of the arithmetic progression whose first term is 3 and whose common difference is 7.

In Exercises 11–16, $\langle a_n \rangle$ is an arithmetic progression whose common difference is d.

11. If $a_1 = 3$ and $d = 2$, find a_{33}.

12. If $a_1 = \pi$ and $d = 3\pi$, find a_{51}.

13. If $a_1 = -2$ and $d = -4$, find a_{10}.

14. If $a_1 = -\sqrt{2}$ and $d = -2\sqrt{2}$, find a_{83}.

15. If $a_{13} = 34$ and $d = 2$, find a_{100}.

16. If $a_{56} = -282$ and $d = -5$, find a_4.

In Exercises 17–21, $\langle a_n \rangle$ is an arithmetic progression.

17. If $a_1 = 3$ and $a_5 = 19$, find a_{10}.

18. If $a_1 = -2$ and $a_{20} = 55$, find a_{34}.

19. If $a_1 = 3$ and $a_{16} = -72$, find a_{26}.

20. If $a_1 = \sqrt{2}$ and $a_5 = 13\sqrt{2}$, find a_{10}.

21. If $a_1 = \frac{1}{2}$ and $a_5 = \frac{7}{2}$, find a_{17}.

In Exercises 22–26, $\langle a_n \rangle$ is an arithmetic sequence. Find a_n if

22. $a_{10} = -69$ and $a_{21} = -157$

23. $a_{100} = 202$ and $a_{200} = 402$

24. $a_{15} = -50$ and $a_{40} = -125$

25. $a_{20} = 23\sqrt{2}$ and $a_{50} = 53\sqrt{2}$

26. $a_{10} = -\frac{29}{4}$ and $a_{30} = -\frac{91}{4}$

27. Let a and b be real numbers. Show that the sequence $\langle a + bn \rangle$, is an arithmetic progression (in particular, find the first term and the common difference).

28. Give the first term and common difference for the arithmetic progressions $\langle 5n - 2 \rangle$ and $\langle -6 - 17n \rangle$.

29. Compute $\displaystyle\sum_{k=1}^{35} [(k + 1)^2 - k^2]$.

30. If $\langle a_n \rangle$ is an arithmetic progression for which $a_3 = 17$ and $a_{10} = 52$, find $\displaystyle\sum_{k=1}^{20} a_k$.

31. Find
$$1 + 2(1 + \tfrac{1}{2}) + 3(1 + \tfrac{1}{3}) + 4(1 + \tfrac{1}{4}) + \cdots + 100(1 + \tfrac{1}{100})$$
(*Hint:* Expand.)

32. Show that the sum of the first n *even* natural numbers is $n^2 + n$.

33. Show that the sum of the first n *odd* natural numbers is n^2.

34. Verify that the sum of the first n even natural numbers added to the sum of the first n odd natural numbers is the sum of the first $2n$ natural numbers.

35. Let n be a natural number. Show that

$$\sum_{k=1}^{n} k = \sum_{k=1}^{n} (n - k + 1)$$

36. Show that

$$\sum_{k=1}^{n} k + \sum_{k=1}^{n} (n - k + 1) = n(n + 1)$$

37. Use Exercises 35 and 36 to show that

$$\sum_{k=1}^{n} k = \frac{n(n + 1)}{2}$$

In Exercises 38–41, find the common difference d of the arithmetic progression $\langle a_n \rangle$, given that

38. $\displaystyle\sum_{k=1}^{14} a_k = 210$ $a_1 = 2$ **39.** $\displaystyle\sum_{k=4}^{12} a_k = 72$ $a_1 = 1$

40. $\displaystyle\sum_{k=1}^{15} a_k = 60$ $a_{15} = 18$

41. $\displaystyle\sum_{k=1}^{n} a_k = n(5n - 6)$ $a_1 = -1$

42. Find real numbers a and b for which

$$\sum_{k=1}^{5} (ak + b) = 55 \quad \text{and} \quad \sum_{k=1}^{7} (ak + b) = 98.$$

Given that $\displaystyle\sum_{k=1}^{n} k^2 = \frac{n(n + 1)(2n + 1)}{6}$, find the sums in Exercises 43 and 44.

43. $\displaystyle\sum_{k=1}^{n} (k + 1)^2$

44. $1 \cdot 2 + 2 \cdot 3 + 3 \cdot 4 + \cdots + n(n + 1)$
(Hint: $k(k + 1) = k^2 + k$.)

4-4 GEOMETRIC PROGRESSION

Whereas arithmetic progressions are characterized by a common difference between successive terms, geometric progressions are characterized by a common ratio; that is, a sequence $\langle a_n \rangle$ is a **geometric progression** if for some number r, called the common ratio, we have

$$a_{n+1} = r \cdot a_n$$

for all n in N.

Example 1

Consider the sequence whose first term is 5 and each of whose other terms is twice its predecessor. Then the first five terms of this sequence are 5, $5 \cdot 2$, $5 \cdot 2^2$, $5 \cdot 2^3$, and $5 \cdot 2^4$. This sequence is a geometric progression with common ratio 2.

□

The nth term a_n of a geometric progression whose first term is a and whose common ratio is a nonzero number r is given by

$$a_n = ar^{n-1}$$

To see this, suppose $\langle a_n \rangle$ is this geometric progression. Then we have

$$
\begin{aligned}
a_1 &= a & &= a \\
a_2 &= a_1 \cdot r & &= a \cdot r \\
a_3 &= a_2 \cdot r = a_1 \cdot r \cdot r & &= a \cdot r^2 \\
a_4 &= a_3 \cdot r = a_1 \cdot \underbrace{r \cdot r \cdot r}_{3 \text{ times}} & &= a \cdot r^3
\end{aligned}
$$

$\cdot$

$\cdot$

$\cdot$

and

$$a_n = a_{n-1} \cdot r = a_1 \cdot \underbrace{r \cdot r \cdot r \cdots r}_{n-1 \text{ times}} = a \cdot r^{n-1} \quad \text{for any } n \text{ in N}$$

Example 2

The nth term of a geometric progression whose first term is 3 and whose common ratio is $\frac{1}{2}$ is $3 \cdot (\frac{1}{2})^{n-1}$. Thus, the sequence is given explicitly by

$$3, \frac{3}{2}, \frac{3}{4}, \ldots, 3 \cdot (\tfrac{1}{2})^{n-1}, \ldots$$

The sixth term is $3 \cdot (\frac{1}{2})^5 = \frac{3}{32}$.

Example 3

Find the nth term of the geometric progression whose fourth term is $\frac{3}{4}$ and whose ninth term is $-\frac{3}{128}$.

Solution

Let a and r be the first term and the common ratio, respectively. Then we have $a \cdot r^{4-1} = ar^3 = \frac{3}{4}$ and $ar^8 = -\frac{3}{128}$. Hence,

$$\frac{ar^8}{ar^3} = r^5 = -\frac{3}{128} \frac{4}{3} = -\frac{1}{32}$$

so that $r = \sqrt[5]{-\frac{1}{32}} = -\frac{1}{2}$. We also have $ar^3 = a(-\frac{1}{2})^3 = \frac{3}{4}$ or, equivalently, $a = \frac{3}{4}(-8) = -6$. Hence, the nth term is

$$ar^{n-1} = (-6)(-\tfrac{1}{2})^{n-1}$$

That is, the sequence is given by

$$-6, 3, -\frac{3}{2}, \ldots, -6(-\tfrac{1}{2})^{n-1}, \ldots$$

$\square$

As in the case of arithmetic progressions, it is frequently desirable to find the sum of a number of terms of a geometric progression. The result we shall now prove is that *the sum of the first n terms of a geometric progression with first term a and common ratio r, with $r \neq 1$, is $a(1 - r^n)/(1 - r)$*; that is,

$$\sum_{k=1}^{n} ar^{k-1} = a + ar + ar^2 + \cdots + ar^{n-1} = a\left(\frac{1 - r^n}{1 - r}\right)$$

To see this, note that

$$(1 - r)(1 + r + r^2 + \cdots + r^{n-1})$$
$$= (1 + r + r^2 + \cdots + r^{n-1})$$
$$- (r + r^2 + \cdots + r^n) = 1 - r^n$$

or, using the sigma notation,

$$(1 - r)\sum_{k=1}^{n} r^{k-1} = \sum_{k=1}^{n} r^{k-1} - r\sum_{k=1}^{n} r^{k-1} = \sum_{k=0}^{n-1} r^k - \sum_{k=1}^{n} r^k$$
$$= 1 - r^n$$

Multiplying by a and dividing by $1 - r$, we obtain the indicated result.

Example 4

If a and x are real numbers and $x \neq 0$, then, by the preceding result, we have

$$1 - \left(\frac{a}{x}\right)^n = \left(1 - \frac{a}{x}\right)\sum_{k=1}^{n}\left(\frac{a}{x}\right)^{k-1}$$

and, hence, multiplying by x^n, we have

$$x^n\left(1 - \frac{a^n}{x^n}\right) = x\left(1 - \frac{a}{x}\right)x^{n-1}\sum_{k=1}^{n}\frac{a^{k-1}}{x^{k-1}}$$

or, equivalently,

$$x^n - a^n = (x - a)\sum_{k=1}^{n} a^{k-1}x^{n-k}$$
$$= (x - a)(x^{n-1} + ax^{n-2} + \cdots + a^{n-2}x + a^{n-1})$$

Note that for $x = 0$ this result is self-evident.

Example 5

$\frac{1}{2} + \frac{1}{2^2} + \frac{1}{2^3} + \cdots + \frac{1}{2^{10}}$ is the sum of the first ten terms

of the geometric progression with first term $\frac{1}{2}$ and ratio $\frac{1}{2}$.

Hence,

$$\frac{1}{2} + \frac{1}{2^2} + \frac{1}{2^3} + \cdots + \frac{1}{2^{10}} = \frac{1}{2} \cdot \frac{1 - \dfrac{1}{2^{10}}}{1 - \dfrac{1}{2}} = 1 - \frac{1}{2^{10}}$$

Example 6

$$\frac{3}{8} + \frac{3}{8^3} + \frac{3}{8^5} + \frac{3}{8^7} + \frac{3}{8^9} = \sum_{k=1}^{5} \frac{3}{8}\left(\frac{1}{8^2}\right)^{k-1}$$

$$= \frac{3}{8} \cdot \frac{1 - \left(\dfrac{1}{8^2}\right)^5}{1 - \left(\dfrac{1}{8}\right)^2}$$

$$= \frac{3}{8} \cdot \frac{1 - \dfrac{1}{8^{10}}}{\dfrac{63}{64}}$$

$$= \frac{64}{63} \cdot \frac{3}{8}\left(1 - \frac{1}{8^{10}}\right)$$

$$= \frac{8}{21}\left(1 - \frac{1}{8^{10}}\right)$$

Example 7

Compute $3 - 6 + 12 - 24 + 48 - 96 + 192 - 384$.

Solution

This result is given by

$$3\left(\frac{1 - (-2)^8}{1 - (-2)}\right) = 1 - 2^8 = -255$$

Exercises 4-4

In Exercises 1–8, find the sum of the given geometric progression.

   1. $2 + 6 + 18 + 54 + 162 + 486$

   2. $3 + 12 + 48 + 192 + 768 + 3072 + 12288$

3. $2 - 6 + 18 - 54 + 162 - 486$

4. $3 - 12 + 48 - 192 + 768 - 3072 + 12288$

5. $\dfrac{1}{3} + \dfrac{1}{3^2} + \dfrac{1}{3^3} + \cdots + \dfrac{1}{3^7}$

6. $\dfrac{2}{5^2} + \dfrac{2}{5^4} + \dfrac{2}{5^6} + \dfrac{2}{5^8} + \cdots + \dfrac{2}{5^{20}}$

7. $\dfrac{1}{3} - \dfrac{1}{3^2} + \dfrac{1}{3^3} - \dfrac{1}{3^4} + \cdots + \dfrac{1}{3^{11}}$

8. $\dfrac{2}{7^2} - \dfrac{2}{7^5} + \dfrac{2}{7^8} - \dfrac{2}{7^{11}} + \cdots - \dfrac{2}{7^{29}}$

In Exercises 9–16, $\langle a_n \rangle$ is a geometric progression whose common ratio is r.

9. If $a_1 = 3$ and $r = 3$, find a_{13}.

10. If $a_1 = 1$ and $r = 2$, find a_{16}.

11. If $a_1 = 100$ and $r = -\frac{1}{10}$, find a_6.

12. If $a_1 = \sqrt{2}$ and $r = \sqrt{3}$, find a_{23}.

13. If $a_1 = \frac{1}{2}$ and $r = 4$, find a_{11}.

14.* If $a_1 = i$ and $r = -1$, find a_{64}.

15. If $a_1 = \frac{1}{3}$ and $r = -i$, find a_{501} and a_{504}.

16. If $a_1 = 1 + i$ and $r = 1 - i$, find a_3.

In Exercises 17–20, give the next three terms and the nth term of each of the geometric progressions.

17. $1, 3, 9, \ldots$ **18.** $27, -9, 3, \ldots$

19. $-3, +3, -3, \ldots$

20. $1, 1 + r, (1 + r)^2, \ldots$ (where r is a real number)

In Exercises 21–28, $\langle a_n \rangle$ is a geometric progression.

21. If $a_1 = \frac{1}{4}$ and $a_6 = 256$, find a_n.

22. If $a_2 = 12$ and $a_7 = \frac{128}{81}$, find a_{11}.

23. If $a_2 = 6$ and $a_5 = -48$, find a_n.

24. If $a_3 = \sqrt{2}$ and $a_6 = -4$, find a_n.

25. If $a_3 = \frac{25}{4}$, $a_7 = \frac{4}{25}$, and r is positive, find a_6.

26. If $a_4 = \frac{2048}{729}$, $a_{14} = \frac{2}{3}$, and r is negative, find a_{15}.

27. If $a_{13} = \sqrt{2}$ and $a_{23} = 32\sqrt{2}$, find a_{21}.

28. If $a_4 = c^4$ and $a_{10} = c^7$, find a_n. (Here c is a positive real number.)

---

*Note that the definition of sequence can be extended to complex numbers.

In Exercises 29–33, compute the sum of the specified number of terms of the given geometric progression.

29. $2, 4, 8, \ldots;$ first 13 terms

30. $1, 0.4, 0.16, \ldots;$ first 8 terms

31. $\frac{2}{3}, 1, \frac{3}{2}; \ldots;$ first 6 terms

32. $1, \frac{2}{3}, \frac{4}{9}, \ldots;$ first n terms, for any n

33. $\frac{1}{125}, -\frac{1}{25}, \frac{1}{5}, \ldots;$ first 5 terms

34. If $\langle a_n \rangle$ is a geometric progression for which $a_3 = 1$ and $a_6 = -27$, find $\displaystyle\sum_{k=1}^{15} a_k$.

35. Let $\langle a_n \rangle$ be a geometric progression whose first term and common ratio are a and r, respectively. Show that

$$\sum_{k=m}^{n} a_k = \frac{a_m(1 - r^{n-m+1})}{1 - r} = \frac{a(r^{m-1} - r^n)}{1 - r} \quad \text{where } n \geq m$$

36. Find a common ratio for the geometric progression $\langle a_n \rangle$ if $\displaystyle\sum_{k=1}^{3} a_k = 13$ and $a_1 = 1$. How many solutions are there?

37. Find the first term for the geometric progression $\langle a_n \rangle$ if we know that $\displaystyle\sum_{k=1}^{5} a_k = \frac{4}{9}$ and $r = \frac{1}{3}$.

38. Let $\langle a_n \rangle$ be a geometric progression. Given $a_1 = 3$ and $r = \frac{2}{5}$, find the integer m for which $\displaystyle\sum_{k=1}^{m} a_k = \frac{609}{125}$.

39. For any geometric progression, show that the products formed by multiplying each term by any fixed real number also form a geometric progression. Find the common ratio of $\langle ca_n \rangle$ if the common ratio of $\langle a_n \rangle$ is r.

40. An automobile purchased for $2500 depreciates in value 10% every year. Find its value at the end of 4 years.

41. The population of a certain town is 10,000. If it increases 10% every year, what will be the population at the end of 5 years?

42. A tank holds 64 gallons of wine. Suppose that 16 gallons are removed and the tank is refilled by replacing wine with water. Then an additional 16 gallons of this mixture are removed and the tank is again filled with water. This operation is repeated until 6 batches have been removed. How much of the original wine remains in the tank?

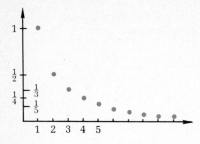

Figure 4-1

4-5 CONVERGENCE. INFINITE SERIES

Consider the sequence $1, \dfrac{1}{2}, \dfrac{1}{3}, \ldots, \dfrac{1}{n}, \ldots$ and observe that the larger n is, the smaller is the distance between $1/n$ and 0; in fact, $1/n$ is as close to 0 as desired for n large enough. We say, therefore, that the sequence $\langle 1/n \rangle$ converges to 0. See Figure 4-1.

Similarly, the sequence $\left\langle 1 + \dfrac{1}{2^n} \right\rangle$ converges to 1, because the distance between $1 + \dfrac{1}{2^n}$ and 1 can be made as small as we please by taking n large enough (because the distance between $1 + \dfrac{1}{2^n}$ and 1 is $\left| 1 + \dfrac{1}{2^n} - 1 \right| = \dfrac{1}{2^n}$, which is as close to 0 as desired for n large enough).

In general, we say that *a sequence $\langle a_n \rangle$ converges to the number a if the distance $|a_n - a|$ between a_n and a can be made as small as desired by taking n large enough;** or, equivalently, all of the numbers $a_n, a_{n+1}, a_{n+2}, \ldots$, fall within a preassigned distance of a if we choose n large enough.

If $\langle a_n \rangle$ converges to a, then a is called the **limit of** $\langle a_n \rangle$. Thus, we write

$$\lim_{n \to \infty} a_n = a$$

(read "the limit of a_n as n approaches infinity† is a") or, simply,

$$a_n \to a$$

(read "a_n converges to a").

Example 1

The examples at the beginning of this section can be restated as follows:

(a) $\displaystyle\lim_{n \to \infty} \dfrac{1}{n} = 0$

(b) $\displaystyle\lim_{n \to \infty} \left(1 + \dfrac{1}{2^n} \right) = 1$ or, equivalently,

$$\lim_{n \to \infty} \left(1 + \dfrac{1}{2^n} - 1 \right) = \lim_{n \to \infty} \dfrac{1}{2^n} = 0$$

---

*More precisely, "$\langle a_n \rangle$ converges to a" means that if we are presented with any positive number, say ϵ, then for this number ϵ we can find a natural number, say N, so that for any $n \geq N$ we have $|a_n - a| < \epsilon$ (i.e., the distance between a_n and a is less than ϵ).
†*Warning:* the word "infinity" and the symbol ∞ are used merely for notational convenience. Neither denotes a number.

Remark For any sequence $\langle a_n \rangle$,

$$\lim_{n \to \infty} a_n = a \quad \text{if and only if} \quad \lim_{n \to \infty} |a_n - a| = 0$$

A sequence is said to **converge**, or to be **convergent**, if it has a limit. Otherwise, it is said to **diverge**, or to be **divergent**. (It is sometimes possible to determine that a sequence converges without being able to compute the limit.)

Example 2

Show that the sequence $\left\langle \dfrac{2n + 5}{n} \right\rangle$ converges, and find its limit.

Solution

Observe that for any natural number n,

$$\frac{2n + 5}{n} = \frac{2n}{n} + \frac{5}{n} = 2 + \frac{5}{n}$$

Since $5/n$ approaches 0 as n increases, we see that $2 + 5/n$ approaches 2. Consequently, $\left\langle \dfrac{2n + 5}{n} \right\rangle$ converges, and the limit is 2. That is, $\lim\limits_{n \to \infty} \dfrac{2n + 5}{n} = 2$.

Example 3

Find $\lim\limits_{n \to \infty} \dfrac{3n^2 + 10}{2n^2 + n}$.

Solution

$$\lim_{n \to \infty} \frac{3n^2 + 10}{2n^2 + n} = \lim_{n \to \infty} \frac{n^2 \left(3 + \dfrac{10}{n^2} \right)}{n^2 \left(2 + \dfrac{1}{n} \right)}$$

$$= \lim_{n \to \infty} \frac{3 + \dfrac{10}{n^2}}{2 + \dfrac{1}{n}} = \frac{3}{2}$$

This result follows since both $10/n^2$ and $1/n$ approach 0 as n increases.*

---

*It is possible to prove that the limit of a sum, product, quotient, and nth root of sequences is equal to the sum, product, quotient, and nth root of the corresponding limits, whenever all these limits exist (for the quotient, the limit in the denominator must not be 0).

Example 4

Find $\lim_{n \to \infty} (\sqrt{n^2 + n} - n)$.

Solution

A technique to compute this limit involves "rationalizing" as follows:

$$\sqrt{n^2 + n} - n = \frac{(\sqrt{n^2 + n} - n)(\sqrt{n^2 + n} + n)}{\sqrt{n^2 + n} + n}$$

$$= \frac{(n^2 + n) - n^2}{\sqrt{n^2 + n} + n}$$

$$= \frac{n}{\sqrt{n^2 + n} + n}$$

$$= \frac{n}{n\left(\dfrac{\sqrt{n^2 + n}}{n} + 1\right)}$$

$$= \frac{1}{\dfrac{\sqrt{n^2 + n}}{\sqrt{n^2}} + 1}$$

$$= \frac{1}{\sqrt{1 + \dfrac{1}{n}} + 1}$$

Hence,

$$\lim_{n \to \infty} (\sqrt{n^2 + n} - n) = \lim_{n \to \infty} \frac{1}{\sqrt{1 + \dfrac{1}{n}} + 1} = \frac{1}{2}$$

□

Using the concept of *convergence*, we can define the "sum of an infinite series." We have already seen in our work on geometric progressions that, for example,

$$\sum_{k=1}^{n} \frac{1}{2^{k-1}} = 1 + \frac{1}{2} + \frac{1}{2^2} + \cdots + \frac{1}{2^{n-1}} + \cdots$$

$$= \frac{1 - (\frac{1}{2})^n}{1 - \frac{1}{2}} = 2 - \frac{1}{2^{n-1}}$$

Notice that $2 - (1/2^{n-1})$ approaches 2 as $n \to \infty$. Thus,

$$\lim_{n \to \infty} \sum_{k=1}^{n} \frac{1}{2^{k-1}} = \lim_{n \to \infty} \left(2 - \frac{1}{2^{n-1}}\right) = 2$$

or, equivalently, the sequence $\langle s_n \rangle$, given by

$$s_1 = 1 \qquad\qquad\qquad\qquad = 2 - 1$$

$$s_2 = 1 + \frac{1}{2} \qquad\qquad\qquad = 2 - \frac{1}{2}$$

$$s_3 = 1 + \frac{1}{2} + \frac{1}{2^2} \qquad\qquad = 2 - \frac{1}{2^2}$$

.
.
.

$$s_n = 1 + \frac{1}{2} + \frac{1}{2^2} + \cdots + \frac{1}{2^{n-1}} = 2 - \frac{1}{2^{n-1}}$$

.
.
.

converges to 2. This is frequently indicated by

$$\sum_{k=1}^{\infty} \frac{1}{2^{k-1}} = 2 \quad \text{or} \quad 1 + \frac{1}{2} + \frac{1}{2^2} + \cdots + \frac{1}{2^{n-1}} + \cdots = 2$$

$$\text{or} \quad 1 + \frac{1}{2} + \frac{1}{2^2} + \cdots = 2$$

Thus, the sequence $\left\langle \dfrac{1}{2^{n-1}} \right\rangle$ gives rise to the sequence $\langle s_n \rangle$ *of partial sums*, which converges to 2.

In general, given any sequence $\langle a_n \rangle$, the **sequence** $\langle s_n \rangle$ **of partial sums** is given by

$$s_1 = a_1 \qquad\qquad\qquad = \sum_{k=1}^{1} a_k$$

$$s_2 = a_1 + a_2 \qquad\qquad = \sum_{k=1}^{2} a_k$$

$$s_3 = a_1 + a_2 + a_3 \qquad = \sum_{k=1}^{3} a_k$$

.
.
.

$$s_n = a_1 + a_2 + a_3 + \cdots + a_n = \sum_{k=1}^{n} a_k$$

.
.
.

The sequence of partial sums is also called the **infinite series generated by** $\langle a_n \rangle$.*

Now, if, as in the preceding example, $\langle s_n \rangle$ has a limit, say a, that is, if $\lim\limits_{n \to \infty} s_n = \lim\limits_{n \to \infty} \sum\limits_{k=1}^{n} a_k = a$, then a is called the **sum of the infinite series,** and we write

$$\sum_{k=1}^{\infty} a_k = a \quad \text{or} \quad a_1 + a_2 + a_3 + \cdots + a_n + \cdots = a$$

$$\text{or} \quad a_1 + a_2 + a_3 + \cdots = a$$

The following theorem relates infinite series to geometric progressions.

Theorem 1

If a and r are real numbers and $|r| < 1$, then

$$\sum_{k=1}^{\infty} ar^{k-1} = a + ar + ar^2 + \cdots + ar^{n-1} + \cdots = \frac{a}{1-r}$$

In order to understand this, recall that

$$a + ar + \cdots + ar^{n-1} = a\left(\frac{1-r^n}{1-r}\right)$$

and since $|r| < 1$, it follows that $r^n \to 0$.† Hence, the sequence of partial sums converges to $a/(1-r)$. In particular, if $|r| < 1$, then

$$\sum_{k=1}^{\infty} r^{k-1} = 1 + r + r^2 + \cdots + r^{n-1} + \cdots = \frac{1}{1-r}$$

Remark   If $|r| \geq 1$, then $\left\langle \sum\limits_{k=1}^{n} r^{k-1} \right\rangle$ diverges. (Why?)

Example 5

Compute $\dfrac{7}{3} + \dfrac{7}{3^2} + \dfrac{7}{3^3} + \cdots + \dfrac{7}{3^n} + \cdots$

---

*Common ways of denoting this infinite series are

$$\Sigma a_n, \qquad a_1 + a_2 + a_3 + \cdots + a_n + \cdots, \quad \text{and} \quad a_1 + a_2 + a_3 \cdots$$

†For a proof of this, see R. Courant's *Integral and Differential Calculus,* I (Interscience Publishers, 1947), 32.

Solution

$$\frac{7}{3} + \frac{7}{3^2} + \frac{7}{3^3} + \cdots + \frac{7}{3^n} + \cdots$$

$$= \frac{7}{3}\left(1 + \frac{1}{3} + \frac{1}{3^2} + \cdots + \frac{1}{3^{n-1}} + \cdots\right)$$

$$= \frac{7}{3} \cdot \frac{1}{1 - \frac{1}{3}}$$

$$= \frac{7}{2}$$

Here we used the fact that the absolute value of the common ratio, $\frac{1}{3}$, is <1.

Example 6

A ball is dropped from a height of 16 feet, and each time it bounces it rises to $\frac{3}{4}$ its previous height. How many feet does the ball travel?

Solution

On the first rebound, it rises $\frac{3}{4} \cdot 16 = 12$ feet, and then falls 12 feet. On the second rebound, it rises $\frac{3}{4} \cdot \frac{3}{4} \cdot 16$ feet, and then falls the same distance. Continuing this way, we see that the ball travels a total distance of

$$16 + 2 \cdot \frac{3}{4} \cdot 16 + 2 \cdot \frac{3}{4} \cdot \frac{3}{4} \cdot 16 + 2 \cdot \frac{3}{4} \cdot \frac{3}{4} \cdot \frac{3}{4} \cdot 16 + \cdots$$

$$= 16 + 32 \cdot \frac{3}{4}[1 + \frac{3}{4} + (\frac{3}{4})^2 + \cdots]$$

$$= 16 + 24\left(\frac{1}{1 - \frac{3}{4}}\right)$$

$$= 112 \text{ (feet)}$$

□

An important application of the concepts of infinite series is to repeating decimals. We shall illustrate this by example.

Example 7

Note that the number

$$.232323 = .23 + .0023 + .000023$$

$$= \frac{23}{10^2} + \frac{23}{10^4} + \frac{23}{10^6}$$

is a six-place approximation of the number $.\overline{23}$, where the bar over .23 indicates that 23 repeats indefinitely, in the following sense:

$$.\overline{23} = .23 + .0023 + .00\ 00\ 23 + .00\ 00\ 00\ 23 + \cdots$$

$$= \frac{23}{10^2} + \frac{23}{10^4} + \frac{23}{10^6} + \frac{23}{10^8} + \cdots$$

To compute $.\overline{23}$, we regard it as the sum of an infinite series:

$$.\overline{23} = \frac{23}{10^2} + \frac{23}{10^4} + \frac{23}{10^6} + \cdots$$

$$= \frac{23}{10^2}\left(1 + \frac{1}{10^2} + \frac{1}{10^4} + \cdots\right)$$

$$= \frac{23}{100}\frac{1}{1 - \frac{1}{100}}$$

$$= \frac{23}{99}$$

Example 8

(a) $.\overline{123} = \dfrac{123}{10^3} + \dfrac{123}{10^6} + \dfrac{123}{10^9} + \cdots$

$$= \frac{123}{10^3}\left(1 + \frac{1}{10^3} + \frac{1}{10^6} + \cdots\right)$$

$$= \frac{123}{1000}\frac{1}{1 - \frac{1}{1000}}$$

$$= \frac{123}{999}$$

(b) $.\overline{1234} = \dfrac{.1234}{1 - .0001}$

$$= \frac{1234}{9999}$$

(c) $5.1\overline{2} = 5.1 + \dfrac{1}{10}(.\overline{2})$

$$= \frac{51}{10} + \frac{1}{10}\left(\frac{.2}{1 - .1}\right)$$

$$= \frac{51}{10} + \frac{1}{10} \cdot \frac{2}{9}$$

$$= \frac{9 \cdot 51 + 2}{90}$$

$$= \frac{461}{90}$$

Exercises 4-5

In Exercises 1–10, state whether or not the given sequence converges. If the sequence does converge, find the limit.

1. $1, \dfrac{3}{2}, \dfrac{7}{4}, \dfrac{15}{8}, \ldots, \dfrac{2^n - 1}{2^{n-1}}, \ldots$

2. $1, 2, 3, 4, \ldots, n, \ldots$

3. $1, -\dfrac{1}{2}, \dfrac{1}{4}, -\dfrac{1}{8}, \ldots, (-1)^{n-1} \dfrac{1}{2^{n-1}}, \ldots$

4. $2, -\dfrac{3}{2}, \dfrac{4}{3}, -\dfrac{5}{4}, \dfrac{6}{5}, \ldots, (-1)^{n+1} \dfrac{(n+1)}{n}, \ldots$

5. $5, \dfrac{7}{2}, 3, \ldots, \dfrac{2n+3}{n}, \ldots$

6. $-1, \dfrac{1}{2}, -\dfrac{1}{3}, \ldots, (-1)^n \dfrac{1}{n}, \ldots$

7. $\dfrac{5}{32}, \dfrac{2}{32}, -\dfrac{1}{32}, \ldots, \dfrac{8}{32} - \dfrac{3n}{32}, \ldots$

8. $1, -1, 1, -1, \ldots, (-1)^{n+1}, \ldots$

9. $\dfrac{8}{5}, \dfrac{34}{25}, \dfrac{152}{125}, \ldots, 1 + (\tfrac{3}{5})^n, \ldots$

10. $4, 2\sqrt{2}, 2, \sqrt{2}, \ldots, 4(\sqrt{2})^{-n+1}, \ldots$

In Exercises 11–16, compute the limit of the given sequence, if it has one.

11. $\left\langle \dfrac{6n - 17}{n} \right\rangle$

12. $\left\langle \dfrac{n^3 + n}{n^4 + 1} \right\rangle$

13. $\left\langle \dfrac{n^2 + n}{3n - 1} \right\rangle$

14. $\left\langle \dfrac{3n^2 + 2n - 5}{10 - n - 2n^2} \right\rangle$

15. $\left\langle \dfrac{2k^2 - 1}{k(k + 1)} \right\rangle$

16. $\left\langle \dfrac{(2n + 1)(n - 1)}{(4n + 1)(2n - 3)} \right\rangle$

In Exercises 17–26, find the sum of the geometric progression, if there is one.

17. $1 - \tfrac{1}{5} + \tfrac{1}{25} - \tfrac{1}{125} + \cdots$

18. $\tfrac{1}{2} + \tfrac{1}{4} + \tfrac{1}{8} + \cdots$

19. $\displaystyle\sum_{n=0}^{\infty} \dfrac{2^n}{3^{n-1}}$

20. $\displaystyle\sum_{n=0}^{\infty} \dfrac{1}{2^n} (-1)^n$

21. $\displaystyle\sum_{n=0}^{\infty} \dfrac{3^{n+1}}{4^{n-1}}$

22. $1 + 0.8 + 0.64 + \cdots$

23. $\tfrac{1}{81} - \tfrac{1}{54} + \tfrac{1}{36} - \tfrac{1}{24} + \cdots$

24. $\tfrac{3}{2} + 1 + \tfrac{2}{3} + \cdots$

25. $2 - 2 + 2 - 2 + \cdots$

26. $\displaystyle\sum_{n=1}^{\infty} \left(\dfrac{3}{8}\right)^n$

In Exercises 27–36, compute as in Example 7.

27. $2.\overline{4}$

28. $0.\overline{8}$

29. $3.\overline{7}$

30. $2.\overline{45}$

31. $0.\overline{31}$

32. $8.\overline{51}$

33. $0.1\overline{42}$

34. $1.32\overline{132}$

35. $6.\overline{415}$

36. $0.1\overline{288}$

37. Let x be a real number with $|x| < \frac{3}{2}$. Find $\sum\limits_{n=1}^{\infty} (\frac{2}{3}x)^{n-1}$.

38. Compute $\dfrac{8}{7} + \dfrac{8}{7^2} + \dfrac{8}{7^3} + \cdots + \dfrac{8}{7^n} + \cdots$.

39. Compute $\dfrac{2}{5} + \dfrac{2}{5 \cdot 7} + \dfrac{2}{5 \cdot 7^2} + \cdots + \dfrac{2}{5 \cdot 7^{n-1}} + \cdots$.

40. The sum of a geometric progression is $\frac{9}{2}$ and the first term is 2. What is the common ratio?

41. The length of the side of a square is 4 inches. A second square is inscribed by connecting the midpoints of the sides of the first square, a third by connecting the midpoints of the sides of the second, and so on. Find the sum of the areas of the squares so formed, including the first.

42. Find the sum of the perimeters of all squares described in Exercise 41.

◊43. Compute $\lim\limits_{n \to \infty} (\sqrt{n + 5n^{1/2}} - \sqrt{n})$.

4-6 THE COMPLETENESS PROPERTY OF THE REAL NUMBER SYSTEM

There is an important property of the real number system that is basic to a more careful study of that system. In fact, this property is so basic that it is sometimes used as part of the definition of the real number system. Its understanding requires the following two definitions:

Definition 1

A sequence $\langle a_n \rangle$ with the property that

$$a_1 \leq a_2 \leq a_3 \leq \cdots \leq a_n \leq a_{n+1} \leq \cdots$$

is called an **increasing sequence.** Similarly, if

$$a_1 \geq a_2 \geq a_3 \geq \cdots \geq a_n \geq a_{n+1} \geq \cdots$$

then $\langle a_n \rangle$ is said to be a **decreasing sequence.**

Example 1

(a) The sequence $\langle n \rangle$ is an increasing sequence, since $1 <$ $2 < 3 < \cdots < n < n + 1 < \cdots$.
(b) The sequence $\langle 1/n \rangle$ is a decreasing sequence, since $1 >$ $\frac{1}{2} > \frac{1}{3} > \cdots$.
(c) The sequence $\langle (n - 1)/n \rangle$ is an increasing sequence, since $(n - 1)/n = 1 - (1/n)$ for every n and $\langle 1/n \rangle$ decreases.
(d) The sequence $\langle -2n \rangle$ is a decreasing sequence, since $-2 > -4 > -6 > \cdots$.

□

Definition 2

If, for some real number b, the sequence $\langle a_n \rangle$ has the property that

$$a_n \leq b \qquad \text{for all } n \text{ in N}$$

then the sequence $\langle a_n \rangle$ is said to be **bounded above (by b)**, and b is called an **upper bound** for $\langle a_n \rangle$. Similarly, if

$$a_n \geq b \qquad \text{for all } n \text{ in N},$$

then the sequence $\langle a_n \rangle$ is **bounded below (by b)**, and b is called a **lower bound** for $\langle a_n \rangle$.

Example 2

(a) The sequence $\langle (n - 1)/n \rangle$ is bounded above by 1 (or, in fact, by any number greater than 1) since, for any n in N, $n - 1 < n$; hence $(n - 1)/n < 1$. It is bounded below by 0 (or, in fact, by any number smaller than 0).
(b) The sequence $\langle (-1)^n/n \rangle = \langle -1, \frac{1}{2}, -\frac{1}{3}, \frac{1}{4}, \ldots \rangle$ is bounded above by 1, and bounded below by -1. Notice that this sequence is neither increasing nor decreasing (it is also said to "oscillate").
(c) The sequence $\langle 1/n \rangle$ is bounded above by 1 and below by 0.
(d) The sequence $\langle n \rangle$ is bounded below by 1 and is not bounded above, and the sequence $\langle -2n \rangle$ is bounded above by -2 but is not bounded below.

□

We can now state the basic property of real numbers proposed in the beginning of this section.

The completeness property of the real number system: Every sequence of real numbers which is bounded above and increasing (or bounded below and decreasing) converges.

Example 3

From the two previous examples we see that

(a) $\langle n/(n + 1)\rangle$ is increasing and bounded above. Hence, $\langle n/(n + 1)\rangle$ converges. In fact,

$$\lim_{n \to \infty} \frac{n}{n + 1} = \lim_{n \to \infty} \frac{n}{n[1 + (1/n)]} = \lim_{n \to \infty} \frac{1}{1 + (1/n)} = 1$$

See Figure 4-2.

(b) The sequence $\langle 1/n \rangle$ is decreasing and bounded below and, hence, by the completeness property converges.

Example 4

Let $a_n = \dfrac{1 \cdot 3 \cdot 5 \cdot 7 \cdots (2n - 1)}{2 \cdot 4 \cdot 6 \cdots (2n)}$, n in N. Show that $\langle a_n \rangle$ is decreasing and bounded below.

Solution

To understand more fully the sequence being investigated, we shall write out a few of the early terms. Notice that

$$a_1 = \frac{1}{2}$$

$$a_2 = \frac{1 \cdot 3}{2 \cdot 4} = \frac{1}{2} \cdot \frac{3}{4} = a_1 \cdot \frac{3}{4} = \frac{3}{8}$$

$$a_3 = \frac{1 \cdot 3 \cdot 5}{2 \cdot 4 \cdot 6} \cdot \frac{1 \cdot 3}{2 \cdot 4} \cdot \frac{5}{6} = a_2 \cdot \frac{5}{6} = \frac{5}{16}$$

$$a_4 = a_3 \cdot \frac{7}{8}.$$

Thus, we see that, in general,

$$a_{n+1} = \frac{2n + 1}{2n + 2} \cdot a_n$$

and since, for any n in N, $2n + 1 < 2n + 2$, we see that $(2n + 1)/(2n + 2) < 1$. Hence,

$$\frac{a_{n+1}}{a_n} = \frac{2n + 1}{2n + 2} < 1$$

that is, $a_{n+1} < a_n$, and $\langle a_n \rangle$ is decreasing. Next, note that $a_n > 0$ for any n in N, so that $\langle a_n \rangle$ is bounded below. From the completeness property, the sequence $\langle a_n \rangle$ converges; however, this fact alone does not tell us what the limit is.

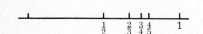

Figure 4-2

Example 5

The sequence $\frac{14}{10}, \frac{141}{100}, \frac{1414}{1000}, \ldots,$ of rational numbers obtained in the computation of the decimal expansion of $\sqrt{2}$ is increasing and bounded above, and hence converges. In fact, it converges to $\sqrt{2}$.

Remark It is useful to note for purposes of computation (and this fact has already been used in some of the preceding examples) that to show that a sequence $\langle a_n \rangle$ of *positive* real numbers is increasing (i.e., $a_n < a_{n+1}$, n in N), it is enough to show that for any n in N, $a_{n+1}/a_n > 1$. Similarly, a sequence $\langle a_n \rangle$ of *positive* real numbers is decreasing if and only if $a_{n+1}/a_n < 1$ for every n in N.

Exercises 4-6

Which of the following statements are true? Justify.

1. Any arithmetic progression is an increasing sequence if its common difference is positive.

2. Any geometric progression is an increasing sequence if its common ratio is positive.

3. Any geometric progression whose common ratio is negative is a decreasing sequence.

4. No oscillating sequence has a limit.

In Exercises 5–9, find a lower and upper bound for the given sequence. (If there is no bound, explain.)

5. $\langle n^2 \rangle$

6. $\left\langle (-1)^n \dfrac{n}{n+1} \right\rangle$

7. $\left\langle \dfrac{1}{2^n} \right\rangle$

8. $\left\langle \sqrt{n + \dfrac{1}{n}} \right\rangle$

9. $\langle a^n \rangle \; 0 \le a \le 1$

Show that the sequences in Exercises 10–13 are increasing sequences.

10. $\langle 2^n \rangle$

11. $\left\langle \dfrac{n+1}{n+4} \right\rangle$

12. $\left\langle \dfrac{n^2 + n + 1}{n+1} \right\rangle$

13. $\left\langle \dfrac{n^2 + 2n}{n+2} \right\rangle$

14. Show that the sequence $\langle 4 - [(n+2)/n] \rangle$ is increasing, bounded above, and hence convergent. What is the limit?

15. Show that the sequence $\langle 3 + [(n+5)/n] \rangle$ is decreasing, bounded below, and hence converges. What is the limit?

16. Show that the sequence $\langle n/2^n \rangle$ is decreasing, bounded below, and hence converges.

17. Show that for $n \ge 4$, the sequence $\langle (3n - 7)/2^n \rangle$ is decreasing, bounded below, and hence converges.

4-7 MATHEMATICAL INDUCTION

Consider the sequence of dominos indicated in Figure 4-3(a) and suppose that (1) the first domino falls [see Figure 4-3(b)], and (2) if any domino falls, then the next one (to the right) falls [see Figure 4-3(c)]. *Conclusion:* each domino will fall.

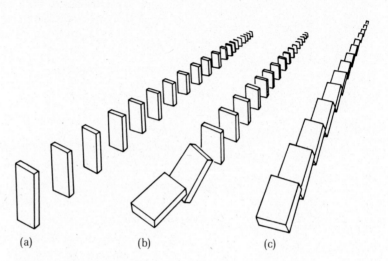

(a)   (b)   (c)

Figure 4-3

This kind of reasoning, when used in mathematics, is called a proof by **induction.**

The following are examples of theorems that are actually sequences of statements, and the proof of these theorems consists of the following reasoning:

(1) show that the first statement is true;
(2) show that if any statement in the sequence is true, then the next statement is true, thus reaching the conclusion that
(3) every statement in the sequence is true.

Example 1

Prove the theorem

$$1 + 2 + 3 + \cdots + n = \frac{n(n + 1)}{2}$$

for every positive integer n.

Solution

Notice that the theorem asserts that each statement in the following sequence of statements is true:

Statement 1

$$1 = \frac{1(1 + 1)}{2}$$

Statement 2

$$1 + 2 = \frac{2(2 + 1)}{2}$$

Statement 3

$$1 + 2 + 3 = \frac{3(3 + 1)}{2}$$

$$\cdot$$
$$\cdot$$
$$\cdot$$

Statement k

$$1 + 2 + 3 + \cdots + k = \frac{k(k + 1)}{2}$$

$$\cdot$$
$$\cdot$$
$$\cdot$$

Statement k + 1

$$1 + 2 + 3 + \cdots + k + (k + 1) = \frac{(k + 1)(k + 2)}{2}$$

To demonstrate that each statement is true, we apply the procedure outlined above:

1. The first statement is true since $1 = \dfrac{1(1 + 1)}{2}$.

2. Suppose the kth statement is true; that is, suppose

$$1 + 2 + 3 + \cdots + k = \frac{k(k + 1)}{2}$$

We want to show that *if* the kth statement is true, then the next (i.e., $k + 1$st) statement is also true. But the $k + 1$st statement is

$$1 + 2 + 3 + \cdots + k + (k + 1) = \frac{(k + 1)(k + 2)}{2}$$

Thus, the problem reduces to:

assume

(a) $1 + 2 + 3 + \cdots + k = \dfrac{k(k + 1)}{2}$

and *show*

(b) $1 + 2 + 3 + \cdots + k + (k + 1) = \dfrac{(k + 1)(k + 2)}{2}$

From (a), we can conclude that

$$1 + 2 + 3 + \cdots + k + (k + 1) = \frac{k(k + 1)}{2} + (k + 1)$$

$$= (k + 1)\left(\frac{k}{2} + 1\right)$$

$$= \frac{(k + 1)(k + 2)}{2}$$

which completes the proof.

In the next example, we use the concept of *divisibility* of integers. The statement "the integer a is **divisible by** the integer b" means that the quotient a/b is an integer; that is, that there is an integer p for which $a = bp$. Thus, for example, 6 is divisible by 3 (since $\frac{6}{3} = 2$ or, equivalently, $6 = 3 \cdot 2$).

Example 2

Show that $2^{(2^n)} - 1$ is divisible by 3 for any positive integer n.

Solution

Consider the sequence of statements whose kth statement is "$2^{2^k} - 1$ is divisible by 3." We want to show that each statement in this sequence is true.

1. The first statement is true, since $2^{2^1} - 1 = 3$.
2. Assume that the kth statement is true; that is, suppose $2^{2^k} - 1$ is in fact divisible by 3; that is, that $2^{(2^k)} - 1 = 3p$ for some positive integer p. We wish to show that the $k + 1$st statement "$2^{2^{k+1}} - 1$ is divisible by 3" is true. But note that

$$2^{2^{k+1}} - 1 = 2^{2^k \cdot 2} - 1$$
$$= (2^{2^k})^2 - 1$$
$$= (2^{2^k} - 1)(2^{2^k} + 1)$$

Since $2^{2^k} - 1 = 3p$, it follows that

$$2^{2^{k+1}} - 1 = 3p \cdot (2^{(2^k)} + 1)$$
$$= 3 \cdot (\text{a positive integer})$$

Thus, the $k + 1$st statement is true, and hence, the statement "$2^{(2^n)} - 1$ is divisible by 3 for any positive integer n" is also true.

Example 3

(Bernoulli's Inequality) Show that if n is a positive integer and x is a real number with $x \geq -1$, then

$$(1 + x)^n \geq 1 + nx$$

Solution

1. We note that $(1 + x)^1 = 1 + 1 \cdot x$ (i.e., the statement holds for $n = 1$).
2. Suppose $(1 + x)^k \geq 1 + kx$ for a positive integer k. Then

$$
\begin{aligned}
(1 + x)^{k+1} &= (1 + x)^k(1 + x) \\
&\geq (1 + kx)(1 + x) && \text{since } 1 + x \geq 0 \\
&= 1 + (k + 1)x + kx^2 \\
&\geq 1 + (k + 1)x && \text{since } kx^2 \geq 0
\end{aligned}
$$

which proves the theorem.

Exercises 4-7

Use mathematical induction to prove each of the following exercises. In each case, n is a positive integer.

1. $\displaystyle\sum_{k=1}^{n} k^2 = \frac{n(n + 1)(2n + 1)}{6}$

2. $1^2 + 3^2 + 5^2 + \cdots + (2n - 1)^2 = \dfrac{n(4n^2 - 1)}{3}$

3. $\displaystyle\sum_{k=1}^{n} k^3 = \left[\frac{n(n + 1)}{2}\right]^2$

4. $1^3 + 3^3 + 5^3 + \cdots + (2n - 1)^3 = n^2(2n^2 - 1)$

5. $\dfrac{1}{1 \cdot 2} + \dfrac{1}{2 \cdot 3} + \dfrac{1}{3 \cdot 4} + \cdots \dfrac{1}{n(n + 1)} = \dfrac{n}{n + 1}$

6. $\displaystyle\sum_{k=1}^{n} 2^k = 2(2^n - 1)$

7. $\displaystyle\sum_{k=1}^{n} x^{k-1} = \dfrac{1 - x^n}{1 - x} \qquad (x \neq 1)$

8. $\frac{1}{3}(n^3 + 2n)$ is a positive integer

9. $\frac{1}{3}(n^3 + 6n^2 + 2n)$ is a positive integer

10. $a^n - b^n$ is divisible by $a - b$ (where a and b are any numbers, with $a \neq b$). (*Hint:* $a^{k+1} - b^{k+1} = a^{k+1} - ab^k + ab^k - b^{k+1}$.)

11. $x^{2n} - y^{2n}$ is divisible by $x + y$ (where x and y are numbers, and $x \neq y$).

12. Prove that the sum of the interior angles of a polygon of n sides is $(n - 2)\pi$ (radians).

13. Consider a sequence of statements whose kth statement is

$$1 + 3 + 5 + \cdots + (2k - 1) = k^2 + \sqrt{2}$$

Show that if you *assume* the kth statement is true, then the next statement would also be true. What can be said about the theorem

$$1 + 3 + 5 + \cdots + (2n - 1) = n^2 + \sqrt{2}$$

for any positive integer n?

14. Use mathematical induction to prove the *Binomial Theorem**: For any numbers a, b, and any positive integer n,

$$(a + b)^n = a^n + na^{n-1}b + \frac{n(n - 1)}{1 \cdot 2} a^{n-2}b^2$$

$$+ \frac{n(n - 1)(n - 2)}{1 \cdot 2 \cdot 3} a^{n-3}b^3 + \cdots + b^n$$

Miscellaneous Problems

1. Find all numbers a for which the numbers $2a + 1$, $3a - 2$, $a + 10$ are consecutive terms in an arithmetic progression.

2. Show that there is *no* number a for which $3a + 1$, $2a + 8$, $a - 5$ are consecutive terms of an arithmetic progression.

3. Find all numbers a for which $a - 2$, $a - 6$, and $2a + 3$ are consecutive terms of a geometric progression.

4. Compute:

(a) $\displaystyle\sum_{k=1}^{100} [(k + 1)^3 - k^3]$

(c) $\displaystyle\sum_{k=0}^{\infty} (-\frac{2}{3})^k$

(b) $\displaystyle\sum_{k=0}^{\infty} (\frac{2}{3})^k$

(d) $\displaystyle\sum_{k=2}^{1000} (-1)^k$

---

*Another proof of this theorem is given in Section 13-4.

5. *Definition: n!* is the product of the first *n* positive integers. For example, $4! = 1 \cdot 2 \cdot 3 \cdot 4 = 24$.

Show that the sequence $\left\langle \dfrac{2^n}{n!} \right\rangle$ is decreasing and bounded below and hence converges.

6. Let $a_n = \dfrac{10^n}{n!}$. (See problem 5 for definition of *n!*) Find a number *N* such that $a_{n+1} < a_n$ for $n > N$. It follows that the sequence $\left\langle \dfrac{10^n}{n!} \right\rangle$ converges. Why?

7. Use mathematical induction to show that

$$\frac{1}{1 \cdot 3} + \frac{1}{3 \cdot 5} + \frac{1}{5 \cdot 7} + \cdots + \frac{1}{(2n-1)(2n+1)} = \frac{n}{2n+1}$$

for every positive integer *n*.

8. Let P_n be the statement

$$1 + 3 + 5 + \cdots (2n - 1) = n^2 + 4.$$

Show that if P_k is true then P_{k+1} is true for every positive integer *k*. Nevertheless, P_n is not true for every positive integer *n*. Explain.

9. Show by mathematical induction that $\frac{1}{2}(n^4 + 5n)$ is a positive integer for every positive integer *n*.

10. A balloon contains 1 cubic foot of air. Air is pumped into the balloon so that at each stroke the amount of air added to the balloon is $\frac{1}{10}$ of the amount in the balloon. Will the pump ever increase the amount of air in the balloon to 2 cubic feet?

11. A ball is dropped from a height of 10 feet. Each time the ball bounces, it rises three-fifths of the distance it has just fallen. What is the "total" distance traveled by the ball?

12. Suppose the market price asked in the sale of *n* gallons of oil is p_n dollars per gallon, where

$$p_n = \frac{6n^2 + 7}{5n^2 + 1}$$

Show that the asking price stabilizes with increasing demand; that is, the price levels off for arbitrarily large demand.

13. Compute $\displaystyle\lim_{n \to \infty} \frac{2n}{n + 7\sqrt{n}}$.

$\diamond$**14.** Compute $\displaystyle\lim_{n \to \infty} \frac{3\sqrt{n} + \sqrt[4]{n}}{2\sqrt{n} - 10\sqrt[4]{n}}$.

$\diamond$**15.** Compute $\displaystyle\lim_{n \to \infty} n(\sqrt{n^2 + 1} - n)$.

$\diamond$**16.** Compute $\displaystyle\lim_{n \to \infty} (\sqrt{n + \sqrt{n}} - \sqrt{n - \sqrt{n}})$.

2

POLYNOMIAL AND RATIONAL FUNCTIONS

Chapter 5 deals with *polynomial functions*, which, roughly speaking, are those functions that can be built up from the identity function $I: x \rightarrow x$, and from the constant functions, by means of *addition* and *multiplication*. They constitute the simplest class of functions, and much of their significance lies in their use in approximating other, more complicated, functions.

For example, it is shown in the calculus that if x is a small enough number, then $x - \frac{1}{6}x^3 + \frac{1}{120}x^5$ is a very close approximation to sin x (i.e., the trigonometric function *sine* may be approximated by a polynomial function).

If we now add the operation of *division*, we obtain the class of *rational functions*. In other words, rational functions are quotients of polynomial functions. They are considered in Chapter 6.

5 Polynomial Functions

5-1 INTRODUCTION

Let $a_0, a_1, a_2, \ldots, a_n$ be real numbers with $a_n \neq 0$; then the function f given by

$$f(x) = a_0 + a_1x + a_2x^2 + \cdots + a_nx^n$$

is called a *polynomial function*. The zero constant function is also called a polynomial function.

The reader has already encountered polynomial functions. For example, the *constant* functions, the *identity* function, the *squaring* function, and the *cubing* function are all polynomial functions.

Example 1

(a) $x \to 5$, with $a_0 = 5$
(b) $x \to x$, with $a_0 = 0$, $a_1 = 1$
(c) $x \to x^2$, with $a_0 = 0$, $a_1 = 0$, $a_2 = 1$
(d) $x \to x^3$, with $a_0 = 0$, $a_1 = 0$, $a_2 = 0$, $a_3 = 1$

Remark It will be shown in Section 5-6 that the polynomial function $f: x \to a_0 + a_1x + \cdots + a_nx^n$ determine the numbers $a_0, \ldots, a_n$; that is, if

$$f(x) = a_0 + a_1x + \cdots + a_nx^n = b_0 + b_1x + \cdots + b_mx^m$$

for all real numbers x, then $m = n$ and $a_0 = b_0$, $a_1 = b_1, \ldots,$

$a_n = b_n$. When this fact is used it is called "equating coefficients."

The integer n is called the **degree** of f; a_n, the **leading coefficient** (of f); a_0, the **constant term**; a_1, the coefficient of x; a_2, the coefficient of x^2; and so on. To say that c is a **zero** (or **root**) of f means that $f(c) = 0$.

Remark No degree is assigned to the zero function.

Example 2

Let $f(x) = x^3 - 2x^2 - 15x$. Find the zeros of f.

Solution

Since $f(x) = x(x^2 - 2x - 15) = x(x - 5)(x + 3)$ we see that

$$f(0) = 0,$$
$$f(5) = 0,$$
$$f(-3) = 0$$

and thus 0, 5, -3 are the *zeros* or *roots* of f.

Example 3

(a) Let $f(x) = x^3$. The degree of f is 3; its leading coefficient is 1; its constant term is 0, as are the coefficients of x and x^2. The only root of f is 0.
(b) Let $f(x) = 3x^7 - \sqrt{2}x^5 + \pi x - 17$. The degree of f is 7; the leading coefficient is 3; the constant term is -17; the coefficients of x^2, x^3, x^4, and x^6 are 0; and the coefficients of x and x^5 are π and $-\sqrt{2}$, respectively.
(c) If f is a nonzero constant function, then the degree of f is 0, and its constant term is also its leading coefficient. Note that f will have no zeros.

Exercises 5-1

1. Prove that if f and g are each polynomial functions, then $f + g$ and $f - g$ are polynomial functions.

2. If m and n are nonnegative integers, and if f and g are polynomial functions, with (degree of f) $= m$ and (degree of g) $= n$, what can be said about (degree of $f + g$)? Be sure to consider the cases $m = n$ and $m < n$.

3. Give an example of polynomial functions f and g, each of degree 6, for which $f + g$ has degree 3.

4. Let $f(x) = x^2 + 2$ and $g(x) = x^3 - 1$. Find the degree and coefficients of each of the following functions: $2f + g$, $f - g$, and fg. Let $h(x) = f[g(x)]$. What is the degree of h?

5. Let $f(x) = x^3 - 2x^2 + 1$ and $g(x) = 2x^2 - 1$. Find the degree and coefficients of each of the following polynomial functions: $f + g$, $2f - 3g$, and $\frac{1}{2}f \cdot g$.

6. Let $f(x) = 2x^2 + 5x - 10$. Define g by $g(x) = f(x - 3)$. Find the coefficients of g.

7. Let $f(x) = 3x^2 - 2x + 5$. Define g by $g(x) = f(2x + 3)$. Find the coefficients of g.

8. Let $f(x) = -x^3 + 2x - 4$. Define g by $g(x) = f(x + 2)$. Find the coefficients of g.

9. Let f and g be quadratic and cubic functions, respectively. Let $h(x) = f[g(x)]$. What is the degree of h? Generalize your result to the case that f and g have degrees m and n, respectively (where, of course, m and n are nonnegative integers).

10. Let f, g, and h be nonzero polynomial functions. Show that (degree of fgh) = (degree of f) + (degree of g) + (degree of h).

5-2 LINES AND FIRST DEGREE POLYNOMIAL FUNCTIONS

In this section, we shall investigate polynomial functions of degree 1 and 0 and their graphs, which turn out to be straight lines.

Associated with each nonvertical line L in the coordinate plane is a measure of its steepness, called the *slope,* and defined as follows: If (x_1, y_1) and (x_2, y_2) are any distinct points on L,

then the slope of L is equal to $\dfrac{(y_2 - y_1)}{(x_2 - x_1)}$

It is customary to use the notation

$$\Delta x = x_2 - x_1$$

and

$$\Delta y = y_2 - y_1$$

Thus, the slope of L can be written as $\Delta y/\Delta x$.

The slope does not depend on the particular points used (see the similar triangles in Figure 5-1), and as the next examples illustrate, lines with positive slope rise (going from left to right) and lines with negative slope fall (going from left to right). Although no slope is assigned to vertical lines, it is useful to note that almost-vertical lines have large slope (positive or negative). Observe also that two lines are parallel if and only if they have the same slope.

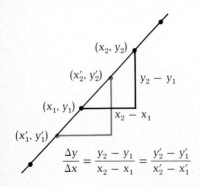

$$\frac{\Delta y}{\Delta x} = \frac{y_2 - y_1}{x_2 - x_1} = \frac{y_2' - y_1'}{x_2' - x_1'}$$

Figure 5-1

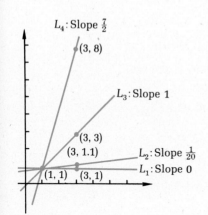

Figure 5-2

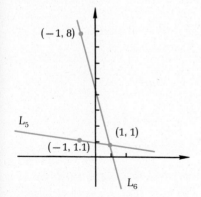

Figure 5-3

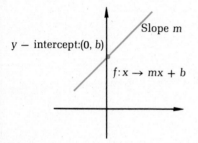

Figure 5-4

Example 1

Sketch and find the slope of each of the following lines:
(a) L_1, the line through $(1, 1)$ and $(3, 1)$
(b) L_2, the line through $(1, 1)$ and $(3, 1.1)$
(c) L_3, the line through $(1, 1)$ and $(3, 3)$
(d) L_4, the line through $(1, 1)$ and $(3, 8)$

Solution

We compute $\Delta y/\Delta x$ for each of the given lines.

(a) Slope of $L_1 = \dfrac{1 - 1}{3 - 1} = 0$ (note that L_1 is horizontal)

(b) Slope of $L_2 = \dfrac{1.1 - 1}{3 - 1} = \dfrac{.1}{2} = \dfrac{1}{20}$

(c) Slope of $L_3 = \dfrac{3 - 1}{3 - 1} = 1$

(d) Slope of $L_4 = \dfrac{8 - 1}{3 - 1} = \dfrac{7}{2}$

See Figure 5-2.

Example 2

Sketch and find the slope of each of the following lines:
(a) L_5, the line through $(-1, 1.1)$ and $(1, 1)$
(b) L_6, the line through $(-1, 8)$ and $(1, 1)$

Solution

(a) Slope of $L_5 = \dfrac{1 - 1.1}{1 - (-1)} = \dfrac{-.1}{2} = -\dfrac{1}{20}$

(b) Slope of $L_6 = \dfrac{1 - 8}{1 - (-1)} = -\dfrac{7}{2}$

See Figure 5-3.

□

The following theorem shows that nonvertical lines are graphs of polynomial functions of degree 1 (or 0).

Theorem 1

Let m and b be any numbers, and let f be the function whose graph is the (nonvertical) line L that passes through $(0, b)$ and has slope m. Then, for any number x,

$$f(x) = mx + b$$

(see Figure 5-4).

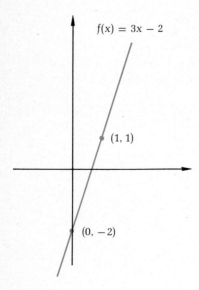

$f(x) = 3x - 2$

(1, 1)

(0, −2)

Figure 5-5

Proof Let $(x, f(x))$ be any point on L distinct from $(0, b)$. Since L has slope m,

$$m = \frac{f(x) - b}{x - 0}$$

or, equivalently,

$$f(x) = mx + b$$

Example 3

Find the function f whose graph is the line through $(0, 3)$ with slope $\frac{1}{4}$.

Solution

By Theorem 1, $f(x) = \frac{1}{4}x + 3$ for all numbers x.

Example 4

Let $f(x) = 3x - 2$. Sketch the graph of f.

Solution

From Theorem 1 we know that the graph is a line [with slope 3 and passing through $(0, -2)$]. Hence, to sketch it only two points are needed. Now $(0, -2)$ is one such point and $(1, 1)$ is another [since $f(1) = 3 \cdot 1 - 2 = 1$]. See Figure 5-5.

Theorem 2

Let L be the (nonvertical) line with slope m that passes through (x_1, y_1). Then an equation for L is

$$y - y_1 = m(x - x_1)$$

Proof From Theorem 1 we know that, for some number b, L is the graph of the function f given by $f(x) = mx + b$. Since (x_1, y_1) is on L, we have $y_1 = f(x_1) = mx_1 + b$ and hence $b = y_1 - mx_1$. Now *any* point (x, y) is on L if and only if $y = f(x)$, since L is the graph of f. That is,

$$y = mx + b = mx + y_1 - mx_1$$

or, equivalently,

$$y - y_1 = m(x - x_1)$$

Example 5

Find an equation for the line through $(-1, 3)$ with slope $\frac{1}{3}$.

Solution

This line is the graph of the equation

$$y - 3 = \tfrac{1}{3}(x + 1)$$

or, equivalently,

$$y = \tfrac{1}{3}x + \tfrac{10}{3} \qquad (\text{or } 3y - x = 10)$$

The line can be easily sketched if we use the fact that $(-1, 3)$ and $(0, \tfrac{10}{3})$ are on the line.

Example 6

Find an equation of the line through $(-2, 1)$ and $(3, 5)$.

Solution

The slope is

$$\frac{5 - 1}{3 - (-2)} = \frac{4}{5}$$

Using the point $(-2, 1)$, we obtain the equation

$$y - 1 = \tfrac{4}{5}(x + 2)$$

or, equivalently,

$$y = \tfrac{4}{5}x + \tfrac{13}{5}$$

[If $(3, 5)$ is used, we obtain the equivalent equation $y - 5 = \tfrac{4}{5}(x - 3)$.]

□

The following theorem gives a condition for two lines to be perpendicular. The proof of this theorem is given at the end of this section.

Theorem 3

Two lines are perpendicular if and only if the product of their slopes is -1.

Remark The theorem does not apply to horizontal or vertical lines. Recall that if a line is vertical its slope is undefined, and that if a line is horizontal its slope is zero.

Example 7

Let L_1 be the graph of $x \rightarrow 3x + 5$. Let L_2 be the graph of $x \rightarrow -\tfrac{1}{3}x + 7$. The product of the slopes is $3 \cdot (-\tfrac{1}{3}) = -1$. Hence, L_1 and L_2 are perpendicular. (See Figure 5-6.)

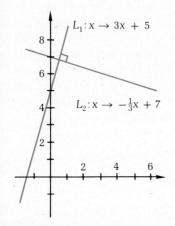

$L_1 : x \rightarrow 3x + 5$

$L_2 : x \rightarrow -\tfrac{1}{3}x + 7$

Figure 5-6

Example 8

Find an equation for the perpendicular bisector of the segment joining $(1, 3)$ and $(-5, 7)$. (See problem 17, Section 2-3.)

Solution

The midpoint of the segment is

$$\left(\frac{-5 + 1}{2}, \frac{7 + 3}{2}\right) = (-2, 5)$$

and its slope is

$$\frac{\Delta y}{\Delta x} = \frac{7 - 3}{-5 - 1} = -\frac{4}{6} = -\frac{2}{3}.$$

Hence an equation for the perpendicular bisector is

$$y - 5 = \tfrac{3}{2}(x - (-2))$$

that is,

$$y = \tfrac{3}{2}x + 8$$

□

We shall now verify this result on perpendicular lines. It is possible to do this by the following geometrical considerations. First of all, since parallel lines have the same slopes, we might just as well suppose that L_1 and L_2 pass through the origin. If they did not, we could replace L_1 and L_2, respectively, by lines L_1' and L_2' that pass through the origin, with $L_1 \| L_1'$ and $L_2 \| L_2'$; observe that $L_1 \perp L_2$ if and only if $L_1' \perp L_2'$.

Suppose m_1 and m_2 are, respectively, the slopes of L_1 and L_2, with $L_1 \perp L_2$. Refer to Figure 5-7. Since L_1 has slope equal to m_1, we see that $(1, m_1)$ is a point on L_1. Note that $L_1 \perp L_2$ is equivalent to the fact that a 90° rotation of L_1 through the origin carries L_1 into L_2, and at the same time this rotation carries triangle A in Figure 5-7 into triangle B. Under this 90° rotation, $(1, 0)$ is carried into $(0, 1)$, and $(1, m_1)$ is carried into $(-m_1, 1)$. Summarizing, $L_1 \perp L_2$ if and only if $(1, m_1)$ on L_1 is carried onto $(-m_1, 1)$ on L_2. Since $(-m_1, 1)$ and $(0, 0)$ are on L_2, we see that m_2, the slope of L_2, is given by

$$m_2 = \frac{0 - 1}{0 - (-m_1)} = -\frac{1}{m_1},$$

which is the desired result.

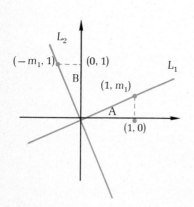

Figure 5-7

Exercises 5-2

In Exercises 1–7, find the linear function whose graph passes through the two given points.

1. $(0, 0)$ and $(2, 5)$ **2.** $(1, 0)$ and $(-1, 0)$

3. $(1, 2)$ and $(2, 1)$ **4.** $(-1, -4)$ and $(3, -5)$

5. $(\frac{1}{2}, -\frac{3}{4})$ and $(-\frac{1}{5}, \frac{5}{8})$ **6.** $(-\frac{1}{2}, \frac{1}{7})$ and $(-\frac{1}{5}, \frac{1}{3})$

7. $(0, 0)$ and (a, b), where a and b are any numbers with $a \neq 0$.

In Exercises 8–14, sketch a graph for each of the functions.

8. $x \rightarrow x - 5$ **9.** $x \rightarrow x + 1$

10. $x \rightarrow -x + 3$ **11.** $x \rightarrow 2x + 3$

12. $x \rightarrow -5x + 1$ **13.** $x \rightarrow \frac{1}{2}x - \frac{1}{2}$

14. $x \rightarrow -\frac{1}{2}x + 7$

In Exercises 15–24, sketch the set of all points (x, y) for which

15. $y \geq -x + 1$ **16.** $y > -x + 1$

17. $y < 2x - 3$ **18.** $y < x - 5$

19. $3x + 2y \leq 1$ **20.** $2x - y > 3$

21. $y \leq 2x + 1$ and $y > -x$ **22.** $y > 2x$ and $x \geq 0$

23. $x + y \leq 1$ and $y - x > 1$

24. $y > 3x - 1$ and $y \leq 3x + 1$

In Exercises 25 and 26, let $f(x) = 5x + 7$.

25. Find the point where the graph of f crosses the horizontal axis.

26. Find the point where the graph crosses the vertical axis.

27. Find the linear function whose graph has slope 5 and passes through the point $(2, 7)$.

28. Let $f(x) = -3x + \frac{1}{2}$. Find the linear function whose graph passes through $(-1, 4)$ and is parallel to the graph of f.

29. Let $f(x) = (-\frac{3}{5})x + 1$. Find the linear function whose graph passes through $(\frac{1}{5}, -\frac{1}{10})$ and is parallel to the graph of f.

In Exercises 30–33, decide whether or not the lines determined by the given pairs of points are parallel, perpendicular, or neither.

30. $(\frac{1}{3}, 2)$, $(1, 4)$, and $(-1, -4)$, $(2, 5)$

31. $(8, -7)$, $(-7, 8)$, and $(10, -7)$, $(-4, 6)$

32. $(2, -3)$, $(-1, 3)$ and $(1, -\frac{1}{2})$, $(-4, -3)$

33. $(1, 1)$, $(\frac{3}{2}, \frac{4}{3})$ and $(-1, 2)$, $(\frac{2}{3}, -\frac{1}{2})$

34. Determine an equation of the line passing through the point $(1, -2)$ and perpendicular to the line given by $2x - 3y + 1 = 0$.

35. Find the real number a for which the lines given by

$$-4x + 3y = 7$$

and

$$ax - 2y = 4$$

will be perpendicular to each other.

In Exercises 36 and 37, show that the given four points are the vertices of a parallelogram. (Hint: Plot the points and show that the appropriate lines are parallel.)

36. $(-2, 8)$, $(1, -4)$, $(-3, 6)$, and $(0, -6)$

37. $(1, 9)$, $(4, 0)$, $(0, 6)$, and $(3, -3)$

In Exercises 38–41, show that the three given points are collinear. (Hint: Consider the slopes.)

38. $(1, -1)$, $(-1, -7)$, and $(2, 2)$

39. $(1, 5)$, $(-1, -3)$, and $(-2, -7)$

40. $(2, -1)$, $(3, -3)$, and $(4, -5)$

41. $(1, \frac{1}{4})$, $(2, \frac{3}{4})$, and $(-2, -\frac{5}{4})$

In Exercises 42–45, find the number a for which the three given points are collinear.

42. $(a, 13)$, $(-2, 7)$, and $(1, -2)$

43. $(-3, 1)$, $(a, -1)$, and $(4, -13)$

44. $(0, 1)$, $(-\frac{1}{2}, a)$, and $(1, 3)$

45. $(1, \frac{1}{2})$, $(0, a)$, and $(\frac{1}{2}, \frac{1}{3})$

46. Let f be a linear function with $f(1) = 2$ and $f(3) = 4$, find $f(5)$ and $f(x)$, where x is any number.

◊**47.** Let $f(x) = 3x - 7$. Find the linear function g for which $f(g) = I$ (i.e. for which $f[g(x)] = x$ for any number x).

◊**48.** Repeat Exercise 43 for the function f given by $f(x) = (-\frac{3}{4})x + \frac{1}{5}$.

5-3 QUADRATIC FUNCTIONS

A **quadratic function** is a polynomial function of degree 2. In other words, f is a quadratic function if and only if there are numbers a, b, and c, with $a \neq 0$, for which

$$f(x) = ax^2 + bx + c \qquad \text{for all real numbers } x$$

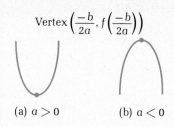

Vertex $\left(\dfrac{-b}{2a}, f\left(\dfrac{-b}{2a}\right)\right)$

(a) $a > 0$ (b) $a < 0$

Figure 5-8

We shall obtain the following information about the graph of f:

1. The graph of f has one of the shapes indicated in Figure 5-8 and is called a **parabola.**
2. For $a > 0$, the graph of f has a lowest point, whereas for $a < 0$, the graph has a highest point. In either case, this point is called the **vertex** of the parabola.
3. The first coordinate of the vertex is $-b/2a$. Thus, for $a > 0$, we see that $f(x)$ is *minimum* for $x = -b/2a$, whereas for $a < 0$, $f(x)$ is *maximum* for $x = -b/2a$.
4. The graph of f intersects the horizontal axis in:
(a) two points, namely

$$\left(\frac{-b \pm \sqrt{b^2 - 4ac}}{2a}, 0\right) \qquad \text{if } b^2 - 4ac > 0$$

(b) one point, namely

$$\left(-\frac{b}{2a}, 0\right) \qquad \text{if } b^2 - 4ac = 0$$

(c) no points, if $b^2 - 4ac < 0$

We shall now verify these four statements. First, completing the square, we see that for all numbers x,

$$f(x) = ax^2 + bx + c = a\left(x^2 + \frac{b}{a}x\right) + c$$

$$= a\left(x^2 + \frac{b}{a}x + \frac{b^2}{4a^2}\right) + c - \frac{b^2}{4a}$$

$$= a\left(x + \frac{b}{2a}\right)^2 + c - \frac{b^2}{4a}$$

$$= a\left[x - \left(-\frac{b}{2a}\right)\right]^2 + c - \frac{b^2}{4a}$$

To verify (1), observe that the graph of f can be obtained from the graph of the squaring function, $s: x \rightarrow x^2$, as follows:

1. A stretch (or shrink) of the graph of s by the factor $|a|$, followed by reflection in the horizontal axis if $a < 0$, to obtain the graph of $x \rightarrow ax^2$.
2. A horizontal shift of the graph of $x \rightarrow ax^2$ by $|b/2a|$ units, to the right if $b/2a < 0$ and to the left if $b/2a > 0$, to obtain the graph of $x \rightarrow a\left(x + \frac{b}{2a}\right)^2$.

3. A vertical shift of the graph of $x \to a\left(x + \dfrac{b}{2a}\right)^2$ by $\left|\dfrac{4ac - b^2}{4a}\right|$ units, upward if $\dfrac{4ac - b^2}{4a} > 0$ and downward if $\dfrac{4ac - b^2}{4a} < 0$, to obtain the graph of

$$x \to a\left(x + \frac{b}{2a}\right)^2 + \frac{4ac - b^2}{4a}$$

Combining steps 2 and 3, we see that the graph of $x \to ax^2$ is shifted so that the vertex is moved from $(0, 0)$ to $\left(-\dfrac{b}{2a}, c - \dfrac{b^2}{4a}\right)$. Hence, the first coordinate of the vertex is $-b/2a$.

We can obtain the points of intersection of the graph of f and the horizontal axis by observing that the graph of f intersects the horizontal axis at x if and only if $f(x) = 0$, or, equivalently, $ax^2 + bx + c = 0$. Using the quadratic formula (see Section 1-10), we obtain (4).

Example 1

Graph the quadratic function $f \colon x \to 2x^2 + 4x - 1$.

Solution

In this case $a = 2$, $b = 4$, and $c = -1$. Hence,

$$-\frac{b}{2a} = -\frac{4}{4} = -1$$

and so the minimum of f is $f(-1) = -3$. Thus, the vertex is at $(-1, -3)$. The zeros of f are

$$\frac{-4 \pm \sqrt{16 + 8}}{4} = \frac{-4 \pm \sqrt{24}}{4} = \frac{-2 \pm \sqrt{6}}{2}$$

that is, $\dfrac{-2 - \sqrt{6}}{2}$ and $\dfrac{-2 + \sqrt{6}}{2}$. The graph is given in Figure 5-9.

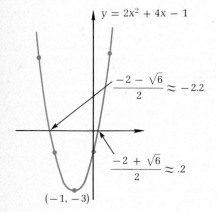

$y = 2x^2 + 4x - 1$

$\dfrac{-2 - \sqrt{6}}{2} \approx -2.2$

$\dfrac{-2 + \sqrt{6}}{2} \approx .2$

$(-1, -3)$

Figure 5-9

Example 2

Let $f(x) = 4x^2 - 4x + 1$. To find the points of intersection of the horizontal axis and the graph of f, we must find all numbers x for which $4x^2 - 4x + 1 = (2x - 1)^2 = 0$. Since the only solution is $\frac{1}{2}$, there is exactly one point of intersection, $(\frac{1}{2}, 0)$. Here, $a = 4$ and $b = -4$, so $-b/2a = \frac{1}{2}$. Since $f(\frac{1}{2}) = 0$, the lowest point on the graph is also $(\frac{1}{2}, 0)$. See Figure 5-10.

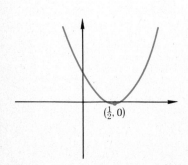

$(\frac{1}{2}, 0)$

Figure 5-10

Example 3

If an object is shot straight up with velocity v feet per second, then (ignoring air resistance) its height t seconds later is $-\frac{1}{2}gt^2 + vt$, where $g = 32$ is the acceleration due to gravity. What is the greatest height attained by the object? When does it strike the ground?

Solution

Letting $a = -\frac{1}{2}g$ and $b = v$, we see that $-\frac{1}{2}gt^2 + vt$ is maximum for

$$t = -\frac{b}{2a} = -\frac{v}{2(-\frac{1}{2}g)} = \frac{v}{g}$$

The maximum height is therefore

$$-\frac{1}{2}g\left(\frac{v}{g}\right)^2 + v\frac{v}{g} = -\frac{1}{2}\frac{v^2}{g} + \frac{v^2}{g} = \frac{v^2}{2g} \text{ feet}$$

The object strikes the ground when the height is 0 again, that is, at t seconds where $t > 0$ and $-\frac{1}{2}gt^2 + vt = 0$. Solving, we find $t = 2v/g$ (see Figure 5-11). If, for instance, a stone is hurled straight up at 100 feet per second, then it takes $\frac{100}{32} \approx 3$ seconds for the stone to reach its maximum height and 3 more seconds to come down. The stone hits the ground 6 seconds later, and the maximum height is $(100)^2/64 = 156$ feet.

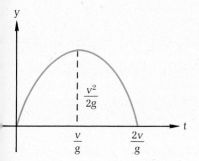

Figure 5-11

Example 4

Find the dimensions of the rectangle of maximum area for a given perimeter.

Solution

Note that whatever the given perimeter, we can assign it the unit of length, and thus we can assume that the perimeter is 1 unit. Let the base and height be x and y units, respectively. Then $2x + 2y = 1$, so

$$y = \frac{1 - 2x}{2} = \frac{1}{2} - x$$

Letting the area be A square units, we have

$$A = xy = x(\tfrac{1}{2} - x) = -x^2 + \tfrac{1}{2}x$$

which is maximum for $x = -\frac{1}{2}/2(-1) = \frac{1}{4}$ (here $a = -1$ and $b = \frac{1}{2}$). For $x = \frac{1}{4}$ we have $y = \frac{1}{4}$, so the rectangle of maximum area for a given perimeter is a square.

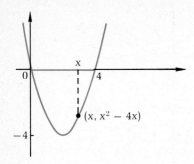

Figure 5-12

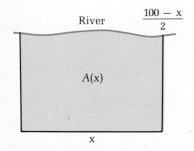

Figure 5-13

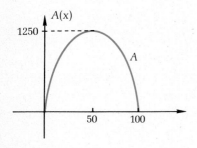

Figure 5-14

Example 5

Find all numbers x for which $x^2 - 4x < 0$.

Solution

First, the zeros of $x \rightarrow x^2 - 4x$ are 0 and 4. Since the leading coefficient is greater than zero, the graph opens upward, and hence the points on the graph below the horizontal axis are those whose first coordinate is between the zeros; that is, $x^2 - 4x < 0$ if and only if $0 < x < 4$. See Figure 5-12.

Example 6

Find the numbers k for which the graph of $x \rightarrow x^2 - kx + k$ is entirely above the horizontal axis.

Solution

Here $a = 1$, $b = -k$, and $c = k$. Since $a > 0$, the graph opens upward and hence is entirely above the horizontal axis if and only if it does not intersect the horizontal axis. This is equivalent to $b^2 - 4ac < 0$, which holds if and only if $0 < k < 4$ (by Example 5).

Example 7

A farmer wants to fence in a rectangular pasture, one side of which is bounded by a river. What is the largest area that can be enclosed if he has 100 feet of fence available?

Solution

Let x be the length of the side opposite the river, and let $A(x)$ be the area. Then $A(x) = x \left(\dfrac{100 - x}{2} \right)$. See Figure 5-13.

Hence $A(x) = -\frac{1}{2}x^2 + 50x$. So A is a quadratic function with $a = -\frac{1}{2}$, $b = 50$. Hence, the maximum is obtained when $x = \dfrac{-b}{2a} = 50$. Therefore the largest area that can be enclosed is $A(50) = \dfrac{50(100 - 50)}{2} = 1250$ square feet. See Figure 5-14.

Example 8

A Norman window has the shape of a rectangle surmounted by a semicircle. See Figure 5-15. Given that the perimeter is 12 feet, find the dimensions that maximize the area.

Solution

Taking the perimeter to be 1 unit ($= 12$ feet), let the base and height of the rectangle be x units and y units, respectively, and

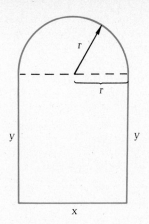

Figure 5-15

let the radius of the semicircle be r units. Then $x = 2r$ and

$$1 = x + 2y + \pi r = 2y + (\pi + 2)r$$

so

$$y = \frac{1 - (\pi + 2)r}{2}$$

Letting the area be A square units, we have

$$A = xy + \frac{\pi r^2}{2} = 2r\left(\frac{1 - (\pi + 2)r}{2}\right) + \frac{\pi r^2}{2}$$

$$= r - (\pi + 2)r^2 + \frac{\pi}{2}r^2 = r - \left(\pi + 2 - \frac{\pi}{2}\right)r^2$$

$$= r - \left(\frac{\pi}{2} + 2\right)r^2$$

which is maximum for

$$r = -\frac{1}{2[-(\pi/2 + 2)]} = \frac{1}{\pi + 4}$$

In this case,

$$x = 2r = \frac{2}{\pi + 4}$$

and

$$y = \frac{1 - (\pi + 2)r}{2} = \frac{1}{\pi + 4}$$

Hence, the base is $2/(\pi + 4)$ units $= 24/(\pi + 4)$ feet, and the height is $12/(\pi + 4)$ feet. Note that the base is twice the height.

Exercises 5-3

In Exercises 1–10, graph f, where

1. $f(x) = x^2$
2. $f(x) = 2x^2$
3. $f(x) = -3x^2$
4. $f(x) = 2x^2 + 5$
5. $f(x) = -3x^2 + 7$
6. $f(x) = x^2 + 3x - 1$
7. $f(x) = x^2 + 3x + 2$
8. $f(x) = -x^2 + x + 1$
9. $f(x) = 2x^2 + 4x - 3$
10. $f(x) = 2x^2 - 6x + 7$

From the graph of $x \rightarrow x^2$, obtain by appropriate shifts and stretches the graph of f in Exercises 11–14.

11. $f(x) = (x - 1)^2 + 3$
12. $f(x) = (x + 2)^2 + 5$
13. $f(x) = 3(x - 1)^2 + 4$
14. $f(x) = -2(x + 2)^2 + 7$

In Exercises 15–22, sketch the set of points (x, y) for which

15. $y \leq x^2$

16. $y \leq x^2 + 1$

17. $y < -x^2$

18. $y > -x^2 + 3$

19. $y < x^2 + 7x + 12$

20. $y \geq -x^2 - x + 12$

21. $y \geq x^2 - 2$ and $y \leq -2x^2 + 3$

22. $y \geq 2x^2 - 1$ and $y \leq -2x^2 + 1$

In Exercises 23–34, find all numbers x for which

23. $x^2 + x - 6 > 0$

24. $2x^2 - 5x - 3 \leq 0$

25. $x^2 - \pi^2 < 0$

26. $4x^2 + 12x + 9 > 0$

27. $6 - x - x^2 \geq 0$

28. $4 - x^2 > 0$

29. $x^2 + x < 0$

30. $2x^2 - 5x \leq 0$

31. $x^2 + 3 \geq -x$

32. $6 + x \leq x^2$

33. $x^2 - 2x > 3$

34. $2x^2 + \sqrt{2}x - 1 < 0$

35. Find the numbers k for which the graph of the quadratic function $x \to x^2 + kx + 4$ is entirely above the horizontal axis. For what numbers k does the graph touch the horizontal axis exactly once?

36. Find the numbers k for which the graph of the quadratic function $x \to -x^2 - 3x + 2k$ is entirely below the horizontal axis.

37. Find the numbers k for which the graph of the quadratic function $x \to x^2 + 2kx - k$ (a) intersects the horizontal axis twice, and (b) is entirely above the horizontal axis.

38. Show that there is no number k for which the graph of $x \to x^2 + (k + 1)x + k$ is entirely above the horizontal axis.

In Exercises 39–41, let a, b, and c be any numbers with $a \neq 0$, and let $f(x) = ax^2 + bx + c$. Show that if r_1 and r_2 are the zeros of f, that is, $f(r_1) = 0$ and $f(r_2) = 0$, then

39. $r_1 + r_2 = -\dfrac{b}{a}$

40. $r_1 \cdot r_2 = \dfrac{c}{a}$

41. The first coordinate of the vertex of f is $\frac{1}{2}(r_1 + r_2)$.

42. Find two numbers whose sum is 50 and whose product is as large as possible.

◇**43.** At midnight ship B was 90 miles due south of ship A. Ship A sailed east at 15 miles per hour, and ship B sailed north at 20 miles per hour. At what time were they closest together?

◇**44.** For what number is $-5x^4 + 10x^2 + 100$ maximum, and what is the maximum?

◇**45.** Let $f(x) = 2x$. Find the coordinates of the point or points on the graph of f which are closest to $(9, 0)$.

◇**46.** A horizontal gutter is made from a piece of tin 8 inches wide by turning up equal strips along the edges into vertical positions. How many inches should be turned up at each side to yield the maximum carrying capacity?

◇**47.** A piece of wire of length l is cut into two parts, one of which is bent into the shape of an equilateral triangle, and the other into a semicircle, not including the line connecting the endpoints. How should the wire be cut so that the sum of the enclosed areas is as large as possible? As small as possible?

◇**48.** A piece of wire of length l is to be cut into two pieces, one to be bent to form a square, and the other to form an equilateral triangle. How should the wire be cut so that the sum of the areas is maximum? Minimum?

◇**49.** Find the shortest distance between the point $(1, 3)$ and the straight line given by $y = 2x + 3$.

◇**50.** Let $f(x) = \sqrt{4 - x^2}$, for all numbers x such that $-2 \leq x \leq 2$. Find the area of the rectangle of maximum area that can be inscribed in the semicircle formed by the graph of f and the interval $[-2, 2]$ on the horizontal axis. (*Hint:* Maximize the square of the area.)

5-4 THE REMAINDER AND FACTOR THEOREMS

The concept of divisibility for polynomial functions is quite analogous to that for the integers. Recall that if m and n are integers, then the statement "m divides n" (or "m is a factor of n" or "n is divisible by m") means that $n = mq$ for some integer q. Similarly, if f and g are polynomial functions, then we say that g **divides** f (or g is a **factor** of f, or f is **divisible** by g) if and only if

$$f = g \cdot h$$

for some polynomial function h.

Example 1

Let $g(x) = x - 3$ and $f(x) = x^4 - 81$; then g divides f since, for all numbers x,

$$x^4 - 81 = x^4 - 3^4 = (x - 3)(x^3 + 3x^2 + 9x + 27)$$

that is, $f = g \cdot h$, where $h(x) = x^3 + 3x^2 + 9x + 27$.

Example 2

Let $g(x) = x + 2$ and $f(x) = x^3 - x^2 - 6x$; then g divides f since, for all numbers x,

$$x^3 - x^2 - 6x = (x + 2)(x^2 - 3x)$$

that is, $f = g \cdot h$, where $h(x) = x^2 - 3x$.

□

The following theorem is basic for the investigation of polynomial functions.

Theorem 1

The Remainder Theorem.* Let f be a polynomial function (of degree $n \geq 1$) and let c be any real number. Then there is polynomial function g (of degree $n - 1$) and a real number r such that

$$f(x) = (x - c)g(x) + r \qquad \text{for all real numbers } x$$

Remark Notice in the Remainder Theorem that $r = f(c)$ since

$$f(c) = (c - c)g(c) + r = 0 + r = r$$

Hence, for all numbers x,

$$f(x) = (x - c)g(x) + f(c)$$

Proof Let $f(x) = a_n x^n + a_{n-1}x^{n-1} + \cdots + a_1 x + a_0$ (the numbers $a_0, \ldots, a_n$ are, of course, given). Let

$$g(x) = b_{n-1}x^{n-1} + \cdots + b_1 x + b_0$$

To find $b_0, \ldots, b_{n-1}$ and r, observe that for all numbers x

$$(x - c)g(x) + r = b_{n-1}x^n + (b_{n-2} - cb_{n-1})x^{n-1} + \cdots$$
$$+ (b_0 - cb_1)x + r - cb_0$$

Equating coefficients, we have

$$b_{n-1} = a_n$$
$$b_{n-2} = a_{n-1} + cb_{n-1}$$
$$b_{n-3} = a_{n-2} + cb_{n-2}$$
$$\cdot$$
$$\cdot$$
$$\cdot$$
$$b_0 = a_1 + cb_1$$
$$r = a_0 + cb_0$$

---

*The name of the Remainder Theorem comes from the fact that r is called, by analogy with the integers, the **remainder** (that is left after f is divided by another function). For example, since $29 = 9 \cdot 3 + 2$, 2 is the remainder after 29 is divided by 9. Similarly, since, for all numbers x, $x^3 = (x - 3)(x^2 + 3x + 9) + 27$, 27 is the remainder.

Thus, $b_0, b_1, \ldots, b_{n-1}$ can be successively computed: first b_{n-1}, then b_{n-2}, and so on. Since $r = f(c)$, the last number computed is $f(c)$. We shall see that this provides an efficient way of computing $f(c)$ for any number c.

The above pattern enables us to use the following convenient scheme, called **synthetic division,** for computing g and $f(c)$:

$\underline{c}$	a_n		a_{n-1}	a_{n-2}	a_{n-3}	$\cdots$	a_3	a_2	a_1	a_0	f
			$+$	$+$	$+$		$+$	$+$	$+$	$+$	
			$c \cdot b_{n-1}$	$c \cdot b_{n-2}$	$c \cdot b_{n-3}$		$c \cdot b_3$	$c \cdot b_2$	$c \cdot b_1$	$c \cdot b_0$	
$b_{n-1} = a_n$			$= b_{n-2}$	$= b_{n-3}$	$= b_{n-4}$	$\cdots$	$= b_2$	$= b_1$	$= b_0$	$= f(c)$	g

Example 3

If $f(x) = 4x^5 + 3x^4 + 6x - 13$, find g and r, where $f(x) = (x - 2) \cdot g(x) + r$.

Solution

Here $a_5 = 4$, $a_4 = 3$, $a_3 = a_2 = 0$, $a_1 = 6$, and $a_0 = -13$. We thus have the following scheme:

2	4	3	0	0	6	-13
		$2 \cdot 4$	$2 \cdot 11$	$2 \cdot 22$	$2 \cdot 44$	$2 \cdot 94$
	4	11	22	44	94	175

Hence,

$$g(x) = 4x^4 + 11x^3 + 22x^2 + 44x + 94,$$

and

$$r = f(2) = 175.$$

Check: $f(2) = 4(2)^5 + 3(2)^4 + 6(2) - 13 = 4 \cdot 32 + 3 \cdot 16 + 12 - 13 = 128 + 48 - 1 = 175$, and $(x - 2)(4x^4 + 11x^3 + 22x^2 + 44x + 94) + 175 = 4x^5 + 11x^4 + 22x^3 + 44x^2 + 94x - 8x^4 - 22x^3 - 44x^2 - 88x - 188 + 175 = 4x^5 + 3x^4 + 6x - 13 = f(x)$.

Compare Example 3 with the familiar division process:

$$
\begin{array}{r}
4x^4 + 11x^3 + 22x^2 + 44x\ + 94 \\
x - 2\ \overline{)\,4x^5 +\ \ 3x^4 +\ \ 0x^3 +\ \ 0x^2 +\ \ 6x\ -\ \ 13} \\
4x^5 -\ \ 8x^4 \\
\hline
11x^4 \\
11x^4 - 22x^3 \\
\hline
22x^3 \\
22x^3 - 44x^2 \\
\hline
44x^2 +\ \ 6x \\
44x^2 - 88x \\
\hline
94x\ -\ \ 13 \\
94x\ -\ 188 \\
\hline
175
\end{array}
$$

If c is a zero of f, then $r = f(c) = 0$, and we have, from the Remainder Theorem:

Theorem 2

The Factor Theorem. If f is a polynomial function, then a real number c is a zero of f [i.e., $f(c) = 0$] if and only if $x \rightarrow x - c$ is a factor of f.

Example 4

Let $f(x) = x^5 + x^4 + x^3 + x^2 + x + 1$. Show that $x \rightarrow x + 1$ is a factor of f.

Solution

$$
\begin{array}{r|rrrrr|r}
-1 & 1 & 1 & 1 & 1 & 1 & 1 \\
   &   & -1 & 0 & -1 & 0 & -1 \\
\hline
   & 1 & 0 & 1 & 0 & 1 & 0
\end{array}
$$

The remainder is 0, and so $x \rightarrow x + 1$ *is* a factor. The other factor is the function g, given by $g(x) = x^4 + x^2 + 1$.

Example 5

Let $f(x) = 2x^3 + 4x^2 - x - 2$. Decide whether $x \rightarrow x - 3$ and $x \rightarrow x + 2$ are factors of f.

Solution

Because of the simplicity of f, instead of using synthetic division, we may compute directly:

$$f(3) = 2(3)^3 + 4(3)^2 - 3 - 2 = 85 \neq 0$$

and

$$f(-2) = 2(-8) + 4(4) + 2 - 2 = 0$$

Hence, $x \to x - 3$ is not a factor of f and $x \to x + 2$ is.

Example 6

Let $f(x) = 3x^4 - 4x^3 - x^2 + 3$. Find $f(-3)$.

Solution

Omitting the middle step, we have

$$\underline{-3|} \quad \begin{array}{ccccc|c} 3 & -4 & -1 & 0 & & 3 \\ & 3 & -13 & 38 & -114 & 345 \end{array}$$

Hence, $f(-3) = 345$.

Exercises 5-4

For each of the functions f and w in Exercises 1-8, find the function g and the number r for which $f(x) = w(x) \cdot g(x) + r$.

1. $f(x) = x^3 - 100x^2 - 100x + 1$, $\quad w(x) = x + 1$

2. $f(x) = x^4 - 10x^2 - 10x - 1$, $\quad w(x) = x + 1$

3. $f(x) = x^4 - 10x^2 - 10x - 1$, $\quad w(x) = x - 1$

4. $f(x) = 3x^4 - 8x^2 + 5x - 7$, $\quad w(x) = x - 1$

5. $f(x) = x^5 + 2$, $\quad w(x) = x - 2$

6. $f(x) = 2x^4 + 12x^3 + 13x^2 - 75$, $\quad w(x) = x + 5$

7. $f(x) = -x^6 + 2x^4 + x^2 + 3$, $\quad w(x) = x - 2$

8. $f(x) = x^7 - 1$, $\quad w(x) = x - 1$

In Exercises 9-16, complete as in the following example. If $x \neq 2$, then

$$\frac{x^3 + 3x^2 - x - 15}{x - 2} = ?$$

Solution: By synthetic division, we find that for any number x,

$$x^3 + 3x^2 - x - 15 = (x - 2)(x^2 + 5x + 9) + 3$$

Thus, for $x \neq 2$,

$$\frac{x^3 + 3x^2 - x - 15}{x - 2} = x^2 + 5x + 9 + \frac{3}{x - 2}$$

9. If $x \neq -2$, then $\dfrac{2x^3 - x^2 + 2x - 18}{x + 2} = ?$

10. If $x \neq 2$, then $\dfrac{x^4 - 3x^3 + 2x^2 - 1}{x - 2} = ?$

11. If $x \neq 3$, then $\dfrac{x^4 + 2x^2 - 3x + 5}{x - 3} = ?$

12. If $x \neq -1$, then $\dfrac{2x^3 + x - 5}{x + 1} = ?$

13. If $x \neq 5$, then $\dfrac{2x^4 - x + 6}{x - 5} = ?$

14. If $x \neq \dfrac{1}{2}$, then $\dfrac{2x^4 - x^2 + 1}{x - \frac{1}{2}} = ?$

15. If $x \neq 1$, then $\dfrac{x^5 - 1}{x - 1} = ?$

16. If $x \neq 1$, then $\dfrac{x^6 - 1}{x - 1} = ?$

Let $f(x) = x^2 - x - 5$. Find the polynomial function g such that for all real numbers x

17. $f(x) = (x - 4)g(x) + f(4)$
18. $f(x) = (x - \frac{1}{2})g(x) + f(\frac{1}{2})$

Let $f(x) = x^3 - x^2 + x - 1$. Find a polynomial function g such that for all real numbers x

19. $f(x) - f(2) = (x - 2)g(x)$
20. $f(x) - f(-2) = (x + 2)g(x)$

21. Let $f(x) = 3x^3 - 5x^2 + 3x - 10$. Find the polynomial function g such that for all real numbers x

$$f(x) - f(2) = (x - 2)g(x)$$

Then find all the zeros of f.

22. Let $f(x) = 6x^3 - 3x^2 - 4x - \frac{1}{2}$. Find the polynomial function g such that for all real numbers x

$$f(x) - f(-\tfrac{1}{2}) = (x + \tfrac{1}{2})g(x)$$

Then find all the zeros of f.

23. Let $f(x) = 2x^3 - x^2 - 36x - 45$. Let $u(x) = x + 3$. Show that u is a factor of f. Find all the zeros of f.

24. Let $f(x) = 2x^3 - 5x^2 + 14x + 39$. Let $u(x) = x + \frac{3}{2}$. Show that u is a factor of f. Find all the zeros of f.

25. Let $f(x) = 2x^4 + 3x^3 + 2x^2 + 5x + 3$, and let $u(x) = 2x + 3$. Show that u is a factor of f. (*Hint:* u is a factor of f if and only if $\frac{1}{2}u$ is a factor of f.)

26. Let $f(x) = 3x^4 - x^3 + 15x^2 + 4x - 3$, and let $u(x) = 3x - 1$. Show that u is a factor of f. (*Hint:* u is a factor of f if and only if $\frac{1}{3}u$ is a factor of f.)

27. Let n be a positive odd integer, and let a be any real number. Let $f(x) = x^n + a^n$. Show that $x \to x + a$ is a factor of f.

28. Let $f(x) = 6x^{100} - 5x^{73} + 4x^{52} + 3x^{17} + 2$. Find the remainder r, where $f(x) = (x + 1)g(x) + r$ for all numbers x and the appropriate function g (you do not have to find g).

29. Let k be a number, and let $f(x) = 2x^3 - kx^2 + x - k$. Find that k for which function $x \to x + 1$ is a factor of f.

30. Let k be a number, and let $f(x) = k^2x^4 - kx^2 - 6$. Find k if the function $x \to x - 1$ is a factor of f.

31. Let $f(x) = x^5 - 2x^4 + x^2 - 3$. Use synthetic division to compute $f(x)$ for every *integer* x in the interval $[-2, 2]$.

5-5 ZEROS OF POLYNOMIAL FUNCTIONS

We know from the Factor Theorem that a polynomial function has a linear factor for each zero. It is natural then to consider *all* the linear factors and a resulting factorization, which will then reveal all the zeros.

Theorem 1

Let f be a polynomial function. Either f has no linear factors, or there are real numbers $c_1, \ldots, c_m$ not necessarily distinct, and a polynomial function g having no zeros such that

$$f(x) = (x - c_1) \cdots (x - c_m)g(x)^*$$

Proof If f has a linear factor (which we can take to have leading coefficient 1), then $f(x) = (x - c_1)f_1(x)$ for some real number c_1 and polynomial function f_1. Similarly, if f_1 has a linear factor, then $f_1(x) = (x - c_2)f_2(x)$ for some c_2 and f_2 and, hence, $f(x) = (x - c_1)(x - c_2)f_2(x)$. If we continue in this manner, we realize that the process must terminate, since the degree of the last factor is reduced by 1 at each stage. Thus we finally obtain

$$f(x) = (x - c_1)(x - c_2) \cdots (x - c_m)f_m(x)$$

---

*It can be shown that this factorization is unique (except for order, of course).

where $f_m(x)$ has no linear factors. It follows from the Factor Theorem that $f_m(x)$ has no zeros, for if it did it would have a linear factor.

As a consequence of Theorem 1 we have the following important result.

Theorem 2

A polynomial function of degree n has at most n zeros.

Proof   Referring to the factorization of Theorem 1, we observe that if x is not one of $c_1, \ldots, c_m$, then $f(x) \neq 0$, since the $m + 1$ factors of $f(x)$ (in Theorem 1) are not 0. From the fact that $n = $ degree $f = m + $ degree g, it follows that $m \leq$ degree f.

Example 1

Let $f(x) = x^4 + x^3 - x - 1$. Find c_1, c_2, and g such that $f(x) = (x - c_1)(x - c_2)g(x)$, where $g(x) \neq 0$ for any real number x.

Solution
$$f(x) = x^4 + x^3 - x - 1 = x^3(x + 1) - (x + 1)$$
$$= (x + 1)(x^3 - 1) = (x + 1)(x - 1)(x^2 + x + 1)$$

for every real number x. Hence, $c_1 = -1$, $c_2 = 1$, and $g(x) = x^2 + x + 1$. Note that $x^2 + x + 1 \neq 0$ for any real number x. (Why?)

Remark   The number of times a factor $x \to x - c$ occurs in the factorization of f of Theorem 1 is the **multiplicity** of the zero c.

Example 2

Let $f(x) = (x - 1)^3(x - \sqrt{2})^4(x - \pi)^7(x^4 + 5)$. Then the zeros of f are 1, $\sqrt{2}$, and π, and their multiplicities are, respectively, 3, 4, and 7.

Exercises 5-5

In Exercises 1–25, find the zeros of the given functions and their multiplicities.

1. $f(x) = (x + 3)^5(x - 3)(x + \pi)^4(x^2 - 9)$
2. $f(x) = x^2(3x - 2)^5(-x - 1)^5(x^2 - \frac{2}{3}x)(x^2 + 5)$
3. $f(x) = (\frac{1}{2}x + \frac{1}{3})^4(6x + \frac{1}{5})^7(x^6 + 2x^2 + 5)$
4. $f(x) = x^4 - 2x^2 + 1$
5. $f(x) = x^4 + 21x^2 - 100$
6. $f(x) = x^4 + 2x^2 - 15$
7. $f(x) = 9x^4 - 1$
8. $f(x) = 4x^4 - 35x^2 - 9$
9. $f(x) = 4x^4 + 16x^2 + 7$
10. $f(x) = x^9 - 2x^6 + x^3$
11. $f(x) = x^3 - 3x^2 + 3x - 1$

12. $f(x) = x^8 - 16$ **13.** $f(x) = x^6 - 8x^4 + 16x^2$

14. $f(x) = x^3 + 2x^2 - 16x - 32$

15. $f(x) = x^3 - 2x^2 - x + 2$

16. $f(x) = x^3 + 4x^2 - 4x - 16$

17. $f(x) = x^3 - 3x^2 - 9x + 27$

18. $f(x) = x^3 - 5x^2 + x - 5$

19. $f(x) = x^7 + \pi x^6 + x + \pi$

20. $f(x) = x^4 + 4x^3 - x - 4$

21. $f(x) = x^4 + 2x^3 + 8x + 16$

22. $f(x) = x^4 + 2x^3 - 8x - 16$

23. $f(x) = x^6 - 4x^4 - 16x^2 + 64$

24. $f(x) = x^6 - \sqrt{2}x^4 - 2x^2 + 2\sqrt{2}$

25. $f(x) = x^5 - 3x^3 - 27x^2 + 81$

26. Let $f(x) = x^4 + 4x^3 + 2x^2 - 4x - 3$. Use synthetic division to show that -1 is a zero of multiplicity 2 of f, and then find all the zeros of f and their multiplicities.

27. Let $f(x) = x^4 + 2x^3 - 3x^2 - 4x + 4$. Show that 1 is a zero of multiplicity 2 of f. Then find all the zeros of f and their multiplicities.

28. Let $f(x) = x^4 - 2x^3 + 2x - 1$. Use synthetic division to show that 1 is a zero of multiplicity 3, and then find the remaining zero of f.

29. Let $f(x) = 6x^4 + 7x^3 - 27x^2 - 28x + 12$, and let $u(x) = x^2 - 4$. Show that u is a factor of f. Then find all the zeros of f.

30. Let $f(x) = 2x^4 + 5x^3 - 11x^2 - 20x + 12$, and let $u(x) = x^2 + x - 6$. Show that u is a factor of f. Then find all the zeros of f.

31. Suppose the zeros of a cubic function f are -2, 1, and 2, and that $f(0) = 8$. Find $f(x)$ for any number x.

32. Suppose the zeros of a cubic function f are 3 (with multiplicity 2) and -3. Suppose further that $f(0) = 1$. Find $f(x)$ for any number x.

33. Let f be a polynomial function of degree 7, and let its zeros be 3 of multiplicity 2, and -1 of multiplicity 5; let $f(0) = 18$. Find f.

34. Let f be a polynomial function of degree 12. What is the greatest number of times the graph of f can cross or touch the horizontal axis?

35. Let f be a polynomial function of degree 5. Suppose the graph of f crosses the horizontal axis only at -1, 0, and 2, and that it touches the horizontal axis only at 1. What are the zeros of f, and what are their multiplicities?

5-6 THE FUNDAMENTAL THEOREM OF ALGEBRA. COMPLEX ZEROS

So far, by the very definition of polynomial functions, the domain of any polynomial function is the real line, while its range is some subset of the real line. Some interesting results may be obtained if the domain and range of such functions are extended to the complex plane; that is, given *complex* numbers $a_n, a_{n-1}, \ldots, a_0$, with $a_n \neq 0$, we shall consider the function f defined by

$$f(z) = a_n z^n + a_{n-1} z^{n-1} + \cdots + a_1 z + a_0, \qquad z \text{ complex}.$$

We shall call such functions **complex polynomial functions of degree n.**[*]

Example 1

Let the complex polynomial functions f, g, and h be defined by

$$f(z) = z^2 - 2z + 1$$
$$g(z) = z^2 - (1 + 2i)z + (i - 1)$$

and

$$h(z) = z^2 - 2z + 2$$

Note that f and h have real coefficients, whereas g has some complex coefficients that are not real numbers. (Keep in mind that every real number is a complex number.)

□

The complex number a is called a **complex zero** of f if $f(a) = 0$. Thus, for the functions in Example 1, 1 is a complex zero of f (of multiplicity 2); i and $1 + i$ are complex zeros of g (as may be verified by a direct computation); and $1 + i$ and $\overline{1 + i} = 1 - i$ are the complex zeros of h (as can be seen from the quadratic formula).[†]

We can now state the *Fundamental Theorem of Algebra*. It complements the results of Theorem 1 of the previous section and, although its proof is beyond the scope of this book, is stated here because of its great importance.

---

[*]As in the case of "real" polynomial functions, it can be shown that the coefficients $a_n, a_{n-1}, \ldots, a_0$ are uniquely determined, as is the integer n (which we again call the *degree of f*).
[†]See Section 2-4.

Theorem 1

The Fundamental Theorem of Algebra. Every complex polynomial function of degree n, with $n \geq 1$, has a complex zero.

Example 2

For all complex numbers z,
(a) if $f(z) = z^2 + 1$, then $f(z) = (z + i)(z - i)$
(b) if $f(z) = z^3 + 2z$, then $f(z) = z(z^2 + 2)$
$$= (z - 0)(z + \sqrt{2i})(z - \sqrt{2i})$$

(c) if $f(z) = z^3 - 1$, then

$$f(z) = (z - 1)\left[z - \left(-\frac{1}{2} + \frac{\sqrt{3}}{2}i\right)\right]\left[z - \left(-\frac{1}{2} - \frac{\sqrt{3}}{2}i\right)\right]$$

$\square$

As a consequence of the Fundamental Theorem of Algebra, we have the following theorem.

Theorem 2

Let f be any complex polynomial function, of degree n and with leading coefficient a. Then

$$f(z) = (z - c_1)(z - c_2) \cdots (z - c_n)a$$

where $c_1, \ldots, c_n$ are the (not necessarily distinct) zeros of f.

Proof Theorem 1 of Section 5-5 and its proof apply here unchanged for complex polynomial functions. Hence

$$f(z) = (z - c_1) \cdots (z - c_m)g(z)$$

where g has no zeros. By the Fundamental Theorem of Algebra, g must be a complex constant function. It follows that the constant must be a and that $m = n$.

Notice now that, for example, the function h given in Example 1 has real coefficients, and that $1 + i$ and $\overline{1 + i}$ are the complex zeros of h. This is an example of the following interesting result.

Theorem 3

If z is a zero of a complex polynomial function with real coefficients, then so is $\bar{z}$, its conjugate.

Recall first the following facts concerning conjugates: If z_1 and z_2 are complex numbers, then $\overline{z_1 + z_2} = \bar{z}_1 + \bar{z}_2$ and $\overline{z_1 z_2} = \bar{z}_1 \bar{z}_2$; that is, the conjugate of a sum (or product) is the sum (or product) of conjugates. Furthermore, if z is a real number, then $z = \bar{z}$. Some of these results have appeared in previous problem sets (see the exercises at the end of Section 2-4). The reader is asked to verify them again.

Proof Suppose that $f(z) = \displaystyle\sum_{k=0}^{n} a_k z^k$, z complex, where a_0, $a_1, \ldots, a_n$ are real numbers and $a_n \neq 0$. Suppose that z is a complex zero of f. Thus, $f(z) = 0$ and, hence, $\overline{f(z)} = \bar{0} = 0$. We have, then,

$$0 = \overline{f(z)}$$

$$= \overline{\sum_{k=0}^{n} a_k z^k}$$

$$= \sum_{k=0}^{n} \overline{a_k z^k} \qquad \text{(because the conjugate of a sum is the sum of conjugates)}$$

$$= \sum_{k=0}^{n} \bar{a}_k (\bar{z})^k \qquad \text{(because the conjugate of a product is the product of conjugates)}$$

$$= \sum_{k=0}^{n} a_k (\bar{z})^k \qquad \text{(since } a_0, a_1, \ldots, a_n \text{ are real numbers)}$$

$$= f(\bar{z})$$

Hence, $\bar{z}$ is a complex zero of f.

Example 3

$1 + 2i$ is a complex zero of the function f given by $f(z) = z^4 - 2z^3 + 6z^2 - 2z + 5$, for all complex numbers z. Find the other three complex zeros.

Solution

Since $1 + 2i$ is a zero, it follows that $\overline{1 + 2i} = 1 - 2i$ is also a zero. Thus, for the appropriate numbers a, b, and c, we have for any complex number z,

$$
\begin{aligned}
z^4 - 2z^3 &+ 6z^2 - 2z + 5 \\
&= [z - (1 + 2i)][z - (1 - 2i)](az^2 + bz + c) \\
&= [(z - 1) - 2i][(z - 1) + 2i](az^2 + bz + c) \\
&= (z^2 - 2z + 5)(az^2 + bz + c)
\end{aligned}
$$

Dividing, we obtain

$$az^2 + bz + c = z^2 + 1$$

and thus $1 + 2i$, $1 -- 2i$, i and $-i$ (i.e., $\bar{\imath}$) are the required zeros.

Example 4

Find whether there exists a complex polynomial function with real coefficients of degree 3 whose zeros include $3 - 2i$ and $5 + 7i$.

Solution

The answer is *no*. To see this, note that any polynomial function with real coefficients whose zeros include $3 - 2i$ and $5 + 7i$ must also have $\overline{3 - 2i} = 3 + 2i$ and $\overline{5 + 7i} = 5 - 7i$ among its zeros. Hence, it cannot be of degree 3.

□

Theorem 3 can be strengthened as follows (although we will not prove it): *If c is zero of f of multiplicity m, then so is $\bar{c}$.*

Observe that, for any complex number $c = a + bi$, the quadratic function q defined by $q(z) = (z - c)(z - \bar{c})$ has real coefficients, since

$$q(z) = z^2 - (c + \bar{c})z + c\bar{c} = z^2 - 2az + (a^2 + b^2)$$

We use this fact in proving the main result on the factorization of real polynomial functions.

Theorem 4

Any real polynomial function is a product of (real) linear and (real) quadratic factors.

Proof Note that f can be regarded as a complex polynomial function with real coefficients and that it suffices to prove the theorem in this context. By Theorem 2, f is a product of linear factors; that is, $f(z) = a(z - c_1) \cdots (z - c_n)$.

The factors associated with real zeros are already linear with real coefficients. The product of the factors associated with any conjugate pair of nonreal zeros is quadratic with real coefficients, as we have just seen. This accounts for all the factors since the nonreal zeros occur in conjugate pairs by the strengthened form of Theorem 3. We have then the asserted factorization.

As promised earlier, we now prove the following:

Theorem 5

A polynomial function

$$x \rightarrow a_0 + a_1 x + \cdots + a_n x^n, \qquad a_n \neq 0,$$

determines $a_0, a_1, \ldots, a_n$.

Proof We will proceed indirectly. Suppose there are numbers $b_0, \ldots, b_m$, with $b_m \neq 0$, not identical to $a_0, \ldots, a_n$ for which

$$a_0 + a_1 x + \cdots + a_n x^n = b_0 + b_1 x + \cdots + b_m x^m$$

for all numbers x.

Subtracting, for certain numbers $c_0, \ldots, c_l$, we obtain $c_0 + \cdots + c_l x^l = 0$ for all numbers x, and hence

$$x^l \left(c_l + \frac{c_{l-1}}{x} + \cdots + \frac{c_0}{x^l} \right) = 0 \quad \text{for all numbers } x \neq 0$$

or

$$x^l [c_l + r(x)] = 0 \quad \text{for all numbers } x \neq 0$$

where

$$r(x) = \frac{c_{l-1}}{x} + \cdots + \frac{c_0}{x^l} \quad \text{for all numbers } x \neq 0$$

To see that this is impossible, observe that for x sufficiently large, $r(x)$ is arbitrarily close to 0, and hence $c_l + r(x) \neq 0$. Since $x^l \neq 0$, the product is $\neq 0$, a contradiction.

Exercises 5-6

1. Let $f(z) = 3z^3 - 7z^2 + 27z - 63$, for all complex numbers z. Show that $3i$ is a zero of f, and find the other zeros of f.

2. Let $f(z) = 3z^3 - 10z^2 + 7z + 10$, for all complex numbers z. Show that $2 + i$ is a zero of f, and find the other zeros of f.

3. Let $f(z) = z^4 - 4z^3 + 11z^2 - 14z + 10$. Given that $1 + i$ is a zero of f, find the other zeros of f.

4. For all complex numbers z, let $f(z) = z^4 - 1$ and $g(z) = z^6 - 1$. (a) Show that f and g have precisely two real zeros, 1 and -1. (*Hint:* Show that $f(z) = (z^2 - 1)h(z)$, where h has no real zeros.) (b) Find all the zeros of f and g.

5. Find a complex polynomial function f of degree 4 which has i and $1 - i$ as two of its zeros.

6. Let f be a complex polynomial of degree 5 with real number coefficients. Suppose that $-3 + 4i$ and $2 - 7i$ are complex zeros of f. What can be said about the other zeros of f?

7. Let f be the complex polynomial function given by $f(z) = z^3 - 2iz^2 + z - 2i$. Show that $-i$ and i are complex zeros of f. Find another complex zero, say r. Is $\bar{r}$ also a zero? Does this contradict the result about zeros and conjugates of zeros?

8. Let a, b, c, and d be real numbers, with $a \neq 0$, and let $f(x) = ax^3 + bx^2 + cx + d$ for any real number x. Suppose -3 is a zero of f of multiplicity 2. How many times does the graph of f intersect the horizontal axis? Justify your answer.

9. Let f be a polynomial function (whose coefficients are real numbers and whose domain is the real line) of degree 11. If 1, 2, and 3 are zeros of multiplicity 2, 4, and 4, respectively, how many times does the graph of f intersect the horizontal axis?

◊5-7 BOUNDS AND RATIONAL ZEROS*

The following theorem is often used to estimate the "spread" of the zeros of a polynomial function.

Theorem 1

If $f(x) = (x - c)g(x) + f(c)$ and if $f(c)$ and the coefficients of g are ≥ 0, but not all $= 0$, and if $c > 0$, then no zeros of f lie to the right of c (that is, c is an **upper bound** for the zeros of f).

Proof If $x > c > 0$, then $x - c > 0$, $g(x) \geq 0$ (since the coefficients of g are ≥ 0), and $f(c) \geq 0$ (by assumption). Hence, since not all the coefficients of g together with $f(c)$ can $= 0$,

$$f(x) = (x - c)g(x) + f(c) > 0 \qquad \text{for } x > c$$

Thus, f has no zeros to the right of c.

Remark The conclusion of Theorem 1 remains correct if $f(c)$ and the coefficients of g are ≤ 0. The details of the proof are left as an exercise.

Example 1

We know that $x \to x - 1$ is a factor of $f: x \to x^5 - x^4 + x^3 - x^2 + x - 1$, since (using synthetic division)

$$
\begin{array}{r|rrrrr|r}
1 & 1 & -1 & 1 & -1 & 1 & -1 \\
  &   & 1 & 0 & 1 & 0 & 1 & 0 \\
\hline
  & 1 & 0 & 1 & 0 & 1 & & 0
\end{array}
$$

so that

$$x^5 - x^4 + x^3 - x^2 + x - 1 = (x - 1)(x^4 + x^2 + 1)$$

Here $b_4 = b_2 = b_0 = 1 > 0$, $b_3 = b_1 = f(c) = 0$, $c = 1 > 0$. Hence, f has no zeros larger than 1.

*This section can be omitted without loss of continuity.

Example 2

Find an upper bound for the zeros of $f: x \to x^4 + \frac{5}{2}x^3 - \frac{3}{2}x^2 - 4x + 2$.

Solution

Some trial and error might suggest 2 as a possibility:

$$
\underline{2\rfloor} \quad
\begin{array}{cccc|c}
1 & \frac{5}{2} & -\frac{3}{2} & -4 & 2 \\
1 & \frac{9}{2} & \frac{15}{2} & 11 & 24
\end{array}
$$

Thus, 2 is an upper bound. How about 1?

$$
\underline{1\rfloor} \quad
\begin{array}{cccc|c}
1 & \frac{5}{2} & -\frac{3}{2} & -4 & 2 \\
1 & \frac{7}{2} & 2 & -2 & 0
\end{array}
$$

We see that 1 is a zero of f, but the test fails to show whether or not it is the largest one.

Remark   Recall that the graph of $x \to f(-x)$ is obtained by reflecting the graph of f in the vertical axis, and hence that the smallest zero of f is the negative of the largest zero of $x \to f(-x)$. Thus, a number c is a **lower bound** for the zeros of f (i.e., there are no zeros of f to the left of c) if and only if $-c$ is an **upper bound** for the zeros of $x \to f(-x)$. See Fig. 5-16.

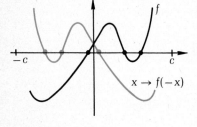

Figure 5-16

Example 3

Since 1 is an upper bound for the zeros of $f: x \to x^5 - x^4 + x^3 - x^2 + x - 1$, -1 is a lower bound for the zeros of $x \to f(-x) = -x^5 - x^4 - x^3 - x^2 - x - 1$ and hence, of $x \to x^5 + x^4 + x^3 + x^2 + x + 1$.

Example 4

Show that 5 and -3 are, respectively, upper and lower bounds for the zeros of $f: x \to 2x^4 - 3x^3 - 17x^2 - 55$.

Solution

$$
\underline{5\rfloor} \quad
\begin{array}{ccccc|c}
2 & -3 & -17 & 0 & & -55 \\
2 & 7 & 18 & 90 & & 395 = f(5) > 0
\end{array}
$$

Hence, 5 is an upper bound for the zeros of f.

$$
\underline{3\rfloor} \quad
\begin{array}{ccccc|c}
2 & 3 & -17 & 0 & & -55 \\
2 & 9 & 10 & 30 & & 35
\end{array}
$$

Hence, 3 is an upper bound for the zeros of $x \to f(-x)$, and so -3 is a lower bound for the zeros of f.

□

The next theorem is helpful in locating the *rational* zeros of certain polynomial functions. It is based on the following fact, usually studied in introductions to number theory:

If p and q are integers with no common divisors other than 1 and -1, and if b is an integer for which p divides $b \cdot q$, then p divides b.

Example 5

The only common divisors of 8 and -3 are 1 and -1; since 8 divides $312 = (-3)(-104)$, it also divides -104.

□

Theorem 2

If

$$f: x \rightarrow a_n x^n + a_{n-1} x^{n-1} + \cdots + a_1 x + a_0$$

is a polynomial function, where $a_n, a_{n-1}, \ldots, a_1, a_0$ are integers, and if $r = p/q$ is a zero of f with p and q integers whose only common divisors are 1 and -1, then p divides a_0 and q divides a_n.

Proof   We have

$$f\left(\frac{p}{q}\right) = a_n \frac{p^n}{q^n} + a_{n-1} \frac{p^{n-1}}{q^{n-1}} + \cdots + a_1 \frac{p}{q} + a_0 = 0$$

and so

$$p(a_n p^{n-1} + a_{n-1} p^{n-2} q + \cdots + a_1 q^{n-1}) = -a_0 q^n$$

Thus, p divides $a_0 q^n = (a_0 q^{n-1})q$, and hence, by the foregoing remark, it also divides $a_0 q^{n-1} = (a_0 q^{n-2})q$. Thus, it divides $a_0 q^{n-2} = (a_0 q^{n-3})q$, and so on. Continuing in this manner, we find that p divides $a_0 q$, and hence a_0. The proof that q divides a_n is similar, and is left as an exercise. It uses the fact that

$$a_n p^n = -q(a_{n-1} p^{n-1} + \cdots + a_1 p q^{n-2} + a_0 q^{n-1})$$

Example 6

Find the rational zeros of $f: x \rightarrow 2x^4 + 5x^3 - 3x^2 - 8x + 4$.

Solution

We try $r = p/q$ with the possibilities $p = \pm 1, \pm 2, \pm 4$ (the divisors of 4) and $q = \pm 1, \pm 2$ (the divisors of 2); that is, the possibilities for r are $r = \pm 1,\ \pm\frac{1}{2},\ \pm 2$, and ± 4. We use synthetic division to check these possibilities:

$$
\begin{array}{r|rrrr|r}
\underline{1} & 2 & 5 & -3 & -8 & 4 \\
  & 2 & 7 & 4 & -4 & 0
\end{array}
$$

Thus, $f(1) = 0$, and so 1 is a zero of f, while the other zeros of f must also be zeros of h: $x \to 2x^3 + 7x^2 + 4x - 4$:

$$
\begin{array}{r|rrr|r}
\underline{-1} & 2 & 7 & 4 & -4 \\
  & 2 & 5 & -1 & -3 \neq 0
\end{array}
$$

Therefore, -1 is neither a zero of h nor of f. However,

$$
\begin{array}{r|rrr|r}
\underline{\frac{1}{2}} & 2 & 7 & 4 & -4 \\
  & 2 & 8 & 8 & 0
\end{array}
$$

and so $\frac{1}{2}$ is a zero of h, and hence of f.

We now have the following factorization: $f(x) = (x - 1)(x - \frac{1}{2})(2x^2 + 8x + 8)$, and the last factor of f is a quadratic function whose zeros are the same as those of $x \to x^2 + 4x + 4 = (x + 2)^2$. Hence, f has only rational zeros, and they are 1, $\frac{1}{2}$, and -2.

Example 7

Find all the positive rational zeros of f: $x \to 2x^6 - x^5 + 3x^3 - 2x^2 - x + 18$.

Solution

The divisors of 18 are $\pm 1,\ \pm 2,\ \pm 3,\ \pm 6,\ \pm 9$, and ± 18, while the divisors of 2 are ± 1 and ± 2. Hence, the only possible positive rational zeros are 1, $\frac{1}{2}$, 2, 3, $\frac{3}{2}$, 6, 9, $\frac{9}{2}$, and 18. Trying 1, we have

$$
\begin{array}{r|rrrrrr|r}
\underline{1} & 2 & -1 & 0 & 3 & -2 & -1 & 18 \\
  & 2 & 1 & 1 & 4 & 2 & 1 & 19 > 0
\end{array}
$$

and using the theorem on bounds (Theorem 1), we see that not only is 1 not a zero of f, but that all the zeros of f are smaller than 1. Furthermore,

$$
\begin{array}{r|rrrrrr|r}
\underline{\frac{1}{2}} & 2 & -1 & 0 & 3 & -2 & -1 & 18 \\
  & 2 & 0 & 0 & 3 & -\frac{1}{2} & -\frac{5}{4} & 18 - \frac{5}{8} \neq 0
\end{array}
$$

and so f has *no* positive rational roots.

Exercises 5-7

1. (a) Show that 2 and -6 are, respectively, upper and lower bounds for the zeros of the function $f: x \rightarrow x^3 + 4x^2 - 4x + 12$. (b) Show that the function f does not have any rational zeros.

2. Find all the rational zeros (if any) of the function $f: x \rightarrow 2x^5 - x^4 - 12x^3 - 13x^2 - 14x - 12$.

In Exercises 3–11, find any rational zeros that the given function f may have, also find the other zeros when it is possible to do so by solving a quadratic equation.

3. $f(x) = 4x^3 + 14x^2 - 6x - 21$

4. $f(x) = 2x^3 + 11x^2 - 7x + 6$

5. $f(x) = 4x^4 + 16x^3 + x^2 + 6x + 8$

6. $f(x) = 2x^3 + 7x^2 + 2x - 6$

7. $f(x) = x^4 + x^2 + 2x - 4$

8. $f(x) = 4x^4 - 13x^3 - 7x^2 + 41x - 14$

9. $f(x) = 3x^3 - 5x^2 - 14x - 4$

10. $f(x) = 2x^4 + 2x^3 + 9x^2 - x - 5$

11. $f(x) = 2x^5 + 3x^3 + 7$

In Exercises 12–15, find an upper bound and a lower bound for the zeros of f where

12. $f(x) = 18x^3 - 12x^2 - 11x + 25$

13. $f(x) = x^5 - 3x^3 + 24$

14. $f(x) = x^4 - 2x^3 - 11x^2 - 2x - 1$

15. $f(x) = x^6 - 2x + 1$

5-8 GRAPHS OF POLYNOMIAL FUNCTIONS

As we saw in the previous section, a polynomial function of degree n has at most n zeros, and hence its graph cannot cross or touch the horizontal axis at more than n points. A useful generalization of this, usually proved in calculus courses, is the fact that the graph of such a function has *at most $n - 1$ turning points* (i.e., "peaks" or "valleys"). Another result essential to graphing polynomial functions (and whose proof also depends on calculus) is the fact that these graphs are *smooth, continuous curves;* that is, they have no "holes" or "jumps" or "corners." (These facts have already been observed in the case of linear and quadratic functions, whose graphs are straight lines and parabolas, respectively.)

Let $f: x \rightarrow a_n x^n + a_{n-1} x^{n-1} + \cdots + a_1 x + a_0$ be a polynomial function of degree n; then for $x \neq 0$ we have

$$f(x) = a_n x^n \left(1 + \frac{a_{n-1}}{a_n} \cdot \frac{1}{x} + \cdots + \frac{a_1}{a_n} \cdot \frac{1}{x^{n-1}} + \frac{a_0}{a_n} \frac{1}{x^n} \right)$$

Now, for $|x|$ very large, the numbers

$$\frac{a_{n-1}}{a_n} \cdot \frac{1}{x}, \ldots, \frac{a_1}{a_n} \cdot \frac{1}{x^{n-1}}, \frac{a_0}{a_n} \cdot \frac{1}{x^n}$$

are all very small in absolute value, and, therefore, so is their sum. Hence, $f(x) \sim a_n x^n$ for $|x|$ very large.* The implications of this are illustrated in Figure 5-17. (Note that a polynomial function of odd degree must have at least one zero. Explain.)

As the next example shows, factoring is also very helpful in graphing polynomial functions.

*That is, $\dfrac{f(x)}{a_n x^n}$ is close to 1 for $|x|$ large.

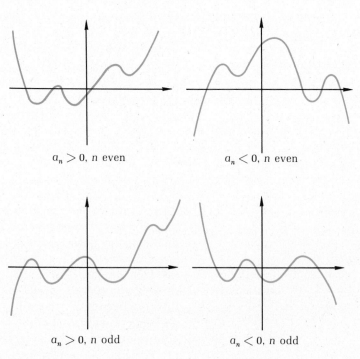

$a_n > 0$, n even $a_n < 0$, n even

$a_n > 0$, n odd $a_n < 0$, n odd

Figure 5-17

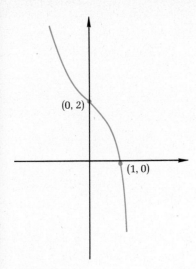

Figure 5-18

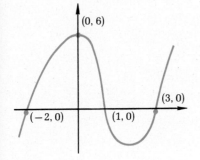

Figure 5-19

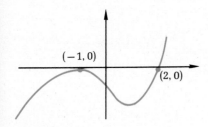

Figure 5-20

Example 1
Graph $f: x \to -x^3 + x^2 - 2x + 2$.

Solution
For x positive and large, $f(x)$ is negative and large in absolute value, and for x negative and large in absolute value, $f(x)$ is positive and large, since, for large $|x|$, we have

$$f(x) = -x^3 + x^2 - 2x + 2$$

$$= -x^3 \left(1 - \frac{1}{x} + \frac{2}{x^2} - \frac{2}{x^3}\right) \sim -x^3$$

Furthermore, $f(x) = -x^2(x-1) - 2(x-1) = -(x-1)(x^2+2)$ for all numbers x. Since $x^2 + 2$ is always positive, $f(x) > 0$ to the left of 1, $f(x) < 0$ to the right of 1, and $f(1) = 0$. See Figure 5-18.

Example 2
Graph $f: x \to (x+2)(x-1)(x-3)$.

Solution
Here we reason as follows: To the left of -2 (i.e., for $x < -2$), $x + 2$, $x - 1$, and $x - 3$ are all negative, and hence $f(x) < 0$. Between -2 and 1 (i.e., for $-2 < x < 1$), $x + 2 > 0$, but $x - 1 < 0$ and $x - 3 < 0$; hence, $f(x) > 0$. Between 1 and 3, $x + 2 > 0$, $x - 1 > 0$, and $x - 3 < 0$, so $f(x) < 0$; and to the right of 3 all the factors are positive, and so $f(x) > 0$. See Figure 5-19.

It is important to realize that without calculus we have no precise way of locating the peaks and valleys, so that the foregoing analysis yields at best a rough approximation. However, plotting additional points on the graph increases the accuracy of this approximation.

Example 3
Let $f(x) = (x+1)^2(x-2)$.

Solution
The zeros of f are -1 and 2. Since $(x+1)^2$ is positive or zero for all numbers x, $f(x) \leq 0$ when $x < 2$, and $f(x) > 0$ for $x > 2$. See Figure 5-20.

Exercises 5-8

In Exercises 1–19, sketch the graph of the given functions.

1. $f(x) = x^3 - 1$ **2.** $f(x) = -x^3 + 1$

3. $f(x) = x^3 + 2x^2 - 8x$ **4.** $f(x) = -x^3 - 2x^2 + 8x$

5. $f(x) = x^3 - 9x$ **6.** $f(x) = x^3 + 4x^2 + 4x$

7. $f(x) = x^3 - 2x^2 + x$ **8.** $f(x) = x^3 + 2x^2 - x + 2$

9. $f(x) = x^3 - 6x^2 + 11x - 6$ (*Hint:* Show that 1 is a zero of f.)

10. $f(x) = x^3 - 5x^2 - 2x + 24$ (*Hint:* Show that 3 is a zero of f.)

11. $f(x) = 4x^3 - 8x^2 - 11x - 3$ (*Hint:* Show that 3 is a zero of f.)

12. $f(x) = 9x^3 - \frac{3}{2}x^2 - 2x + \frac{1}{2}$ (*Hint:* Show that $-\frac{1}{2}$ is a zero of f.)

13. $f(x) = x^3 - 3x^2 + 3x - 1$

14. $f(x) = x^3 - 3x^2 + 3x + 5$ [*Hint:* Use Exercise 13 after observing that for every real number x, $x^3 - 3x^2 + 3x + 5 = (x^3 - 3x^2 + 3x - 1) + 6$]

15. $f(x) = x^4$ **16.** $f(x) = 2x^4 + 2$

17. $f(x) = -x^4 + 1$ **18.** $f(x) = (x + 2)^4 + 3$

19. $f(x) = -(x - 2)^4 + 3$

In Exercises 20–27, find all numbers for which $f(x) \geq 0$.

20. $f(x) = x - 8$ **21.** $f(x) = x^2 + 2$

22. $f(x) = x^2 + 3x + 2$ **23.** $f(x) = x^3 + 3x^2 - 4x$

24. $f(x) = x^3 - 2x^2 + 4x - 8$ **25.** $f(x) = x^3 - 7x + 6$

26. $f(x) = x^4 - 6x^3 + 5x^2$

27. $f(x) = x^4 + 2x^3 + x^2 - 2x - 2$

In Exercises 28–35, find all numbers for which $f(x) < 0$.

28. $f(x) = x + 2$ **29.** $f(x) = x^2$

30. $f(x) = x^2 + 4x + 3$ **31.** $f(x) = x^3 + 2x^2 - 8x$

32. $f(x) = x^3 - 3x^2 + 4x - 12$ **33.** $f(x) = x^3 + 5x - 6$

34. $f(x) = x^4 - 4x^3 - 12x^2$

35. $f(x) = x^4 + 2x^3 - 7x^2 - 8x + 12$

Miscellaneous Problems

1. Find a polynomial function of degree 2 for which the sum of the zeros is twice their product. (See Problems 39 and 40 of Section 4.)

2. What can be concluded about the numbers a_0, a_1, a_2, a_3 and a_4 if

$$a_0 + a_1x + a_2x^2 + a_3x^3 + a_4x^4 = 0$$

for

$$x = 1, 2, 3, 4, 5.$$

3. Let $f(x) = x^3 - 3x^2 + t$ and $g(x) = 9x^2 + 6x + 1$. Find t for which the polynomial functions f and g have a common factor.

4. Let f be a polynomial function of degree 6. Can some straight line intersect the graph of f seven times? Explain.

5. Suppose f is a polynomial function of degree ≥ 1. Explain why there must be some interval on the line that contains *all* the zeros of f.

6. Let $f(x) = ax^3 + bx^2 + cx + d$. Show that 0 is a zero of f if and only if $d = 0$. In that case, how does its multiplicity depend upon a, b, c?

◇7. Suppose $c^3 = 1$, with c a complex nonreal number. Show that $1 + c + c^2 = 0$.

8. Let $f(x) = x^2 + bx + c$, where b and c are any real numbers. Find a quadratic function $g(x) = x^2 + sx + t$ whose zeros are the squares of the zeros of f. (See Problems 39 and 40 of Section 4.)

◇9. Let $f(x) = \frac{1}{3}x^3 + \frac{1}{2}x^2 + \frac{1}{6}x$. Show by induction that $f(x)$ is a positive integer for every positive integer x.

10. Let $f(x) = x^3 - x^2$, x being real. Show that $|f(x)| > 4$ for $|x| > 2$. (Hint: Factor f.)

◇11. Let $f(x) = x^3 + bx^2 + cx + d$, where b, c, and d are real numbers. Suppose w is a zero of f, where $w = u + iv$ with $v \neq 0$. Show that if $b = 0$ and $d > 0$ then $u > 0$. (Hint: Express b and d in terms of the zeros of f.)

12. Let C be the circle given by $(x - 1)^2 + (y + 2)^2 = 2$. Find an equation for the line tangent to C at the point $(2, -1)$.

6 Rational Functions

6-1 RATIONAL FUNCTIONS

Consider the polynomial functions $x \to x^2 - x$ and $x \to x - 1$, and their "quotient" $x \to \dfrac{x^2 - x}{x - 1}$. By the *domain convention* established in Chapter 3, the domain of $x \to \dfrac{x^2 - x}{x - 1}$ is the set of all numbers x for which $x - 1 \neq 0$, that is, for which $x \neq 1$.

In general, let f and g be any functions. The function f/g: $x \to f(x)/g(x)$ is called the **quotient** of f and g, and its domain is the set of all x in R for which $g(x) \neq 0$. A quotient of two polynomial functions is called a **rational function.**

Example 1

If $x \neq 1$, then $\dfrac{x^2 - x}{x - 1} = \dfrac{x(x - 1)}{x - 1} = x$. Hence, the graph of the rational function $x \to \dfrac{x^2 - x}{x - 1}$ is the same as the graph of the polynomial function $x \to x$, except for a "hole" above 1, which is not in the domain. See Figure 6-1.
□

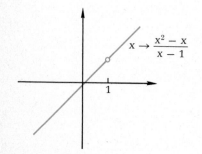

Figure 6-1

As Example 1 indicates, the zeros of g can lead to some peculiarities in the graph of the rational function f/g. We shall now investigate such graphs, making repeated use of the observation that if $f(c) \neq 0$ and $g(c) = 0$ then the graph of f/g "blows up" near c. That is, if x is sufficiently close to c, then $|f(x)/g(x)|$ is very large, because $f(x)$ is close to the nonzero number $f(c)$, while $g(x)$ is close to $g(c) = 0$. In fact, $|f(x)/g(x)|$ becomes "arbitrarily large" as x "approaches" c.

This statement includes the cases where

1. $f(x)/g(x)$ becomes arbitrarily large, in which case we write

$$\lim_{x \to c^+} \frac{f(x)}{g(x)} = \infty \quad \text{or} \quad \lim_{x \to c^-} \frac{f(x)}{g(x)} = \infty$$

depending on whether this occurs as x approaches c from the *right* or from the *left;* and

2. $f(x)/g(x)$ becomes arbitrarily large in absolute value, but *negative,* in which case we write

$$\lim_{x \to c^+} \frac{f(x)}{g(x)} = -\infty \quad \text{or} \quad \lim_{x \to c^-} \frac{f(x)}{g(x)} = -\infty$$

depending on whether this occurs as x approaches c from the *right* or from the *left.* See Figure 6-2; here, $h = f/g$.

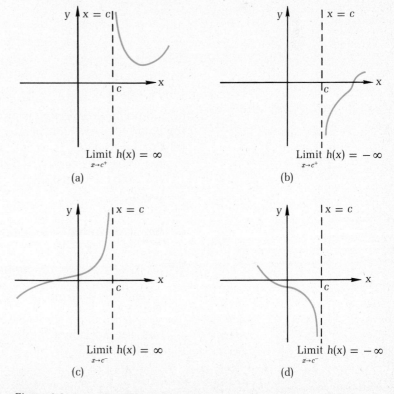

(a) Limit $h(x) = \infty$, $x \to c^+$

(b) Limit $h(x) = -\infty$, $x \to c^+$

(c) Limit $h(x) = \infty$, $x \to c^-$

(d) Limit $h(x) = -\infty$, $x \to c^-$

Figure 6-2

Figure 6-3 indicates some cases (and notation) that may occur as |x| increases beyond bound (i.e., increases indefinitely); again, $h = f/g$. *Note:* The lines given by $x = c$ in Figure 6-2 and by $y = c$ in Figure 6-3 are called the **asymptotes** of the given graphs.

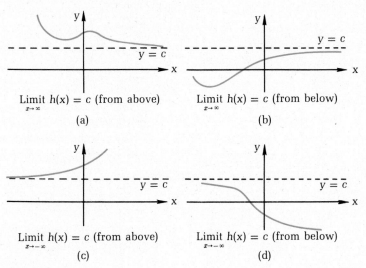

Limit $h(x) = c$ (from above)
$x \to \infty$

(a)

Limit $h(x) = c$ (from below)
$x \to \infty$

(b)

Limit $h(x) = c$ (from above)
$x \to -\infty$

(c)

Limit $h(x) = c$ (from below)
$x \to -\infty$

(d)

Figure 6-3

Example 2

To graph the rational function $f: x \to 1/x$, we consider the following facts:

(a) No point of the graph is on the line given by $x = 0$. (Why?)
(b) x and $1/x$ are both positive or both negative.
(c) For |x| "small," |1/x| is "large"; and for |x| "large," |1/x| is "small" (e.g., $f(.0001) = 10,000$; $f(10,000) = .0001$; $f(-.0001) = -10,000$; $f(-10,000) = -.0001$; etc.); in other words,

$$\lim_{x \to 0^+} \frac{1}{x} = \infty, \lim_{x \to 0^-} \frac{1}{x} = -\infty, \lim_{x \to \infty} \frac{1}{x} = 0 \quad \text{from above}$$

and

$$\lim_{x \to -\infty} \frac{1}{x} = 0 \quad \text{from below}$$

Thus, the horizontal and vertical axes are asymptotes. See Figure 6-4.

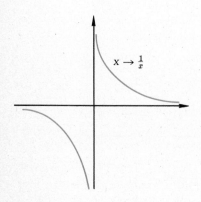

$x \to \frac{1}{x}$

Figure 6-4

Example 3

Let $f(x) = \dfrac{x}{x+1}$. Sketch the graph of f.

Solution

The domain of f is (by our domain convention) the set of all real numbers except -1. We shall first investigate $f(x)$ for x near -1. Note that for x near -1 and to the right of -1, $f(x)$ is negative with $|f(x)|$ large; for example,

$$f(-1 + .01) = -99, \; f(-1 + .001)$$
$$= -999, \; f(-1 + .0001) = -9999$$

Thus,

$$\lim_{x \to -1^+} f(x) = -\infty$$

On the other hand, for x near -1 and to the left of -1, $f(x)$ is positive and large; for example,

$$f(-1 - .01) = 101, \quad f(-1 - .001) = 1001,$$
$$f(-1 - .0001) = 10{,}001$$

Thus,

$$\lim_{x \to -1^-} f(x) = \infty$$

To investigate the numbers $f(x)$ for large $|x|$, observe that

$$\frac{x}{x+1} = \frac{1}{1 + (1/x)},$$

and hence

$$\lim_{x \to \pm\infty} \frac{x}{x+1} = \lim_{x \to \pm\infty} \frac{1}{1 + (1/x)} = 1$$

$\left(\text{since, for } \left|\dfrac{1}{x}\right| \text{ very small, } \dfrac{1}{1 + (1/x)} \text{ is very close to } 1\right)$.
Note also that for $x > -1$ (i.e., $x + 1 > 0$), we have $f(x) < 1\left(\text{since } x + 1 > x \text{ for any real number } x, \text{ and}\right.$

thus, since $x + 1 > 0$, we have $\left. 1 > \dfrac{x}{x+1}\right)$. Similarly, for $x < -1$ (i.e., $x + 1 < 0$), we have $f(x) > 1$. (Why?) Consequently, to the right of -1, the graph of f is below the horizontal line given by $y = 1$, and to the left of -1, the graph of f is above that line. See Figure 6-5.

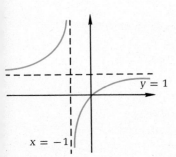

Figure 6-5

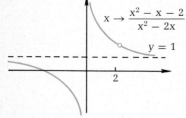

Figure 6-6

Example 4

Graph $x \to \dfrac{x^2 - x - 2}{x^2 - 2x}$.

Solution

For $x^2 - 2x \neq 0$ (i.e., $x \neq 0$ and $x \neq 2$), we have

$$\frac{x^2 - x - 2}{x^2 - 2x} = \frac{(x - 2)(x + 1)}{(x - 2)x} = \frac{x + 1}{x} = 1 + \frac{1}{x}$$

Hence, the graph of $x \to \dfrac{x^2 - x - 2}{x^2 - 2x}$ is the same as that of $x \to 1 + \dfrac{1}{x}$ except for a "hole" above 2. Furthermore, the graph of $x \to 1 + \dfrac{1}{x}$ is the vertical shift by 1 unit of the graph of $x \to \dfrac{1}{x}$. Hence (see Figure 6-4), the graph is as given in Figure 6-6.

□

The following general observations are helpful in finding the limit of a rational function as $x \to \pm\infty$. Let f and g be polynomial functions

1. If the degree of f is the same as the degree of g, then $\lim\limits_{x \to \pm\infty} f(x)/g(x)$ is equal to the quotient of the leading coefficients.

2. If the degree of f is smaller than the degree of g, then $\lim\limits_{x \to \pm\infty} f(x)/g(x) = 0$.

3. If the degree of f is larger than the degree of g, then $\lim\limits_{x \to \pm\infty} f(x)/g(x) = \infty$ or $-\infty$, depending on whether $f(x)/g(x)$ is ultimately positive or negative.

The next three examples illustrate these observations.

Example 5

$$\lim_{x \to \pm\infty} \frac{4x^3 - 3x^2 + 1}{3x^3 + x} = \lim_{x \to \pm\infty} \frac{4 - \dfrac{3}{x} + \dfrac{1}{x^3}}{3 + \dfrac{1}{x^2}} = \frac{4}{3}$$

since $3/x$, $1/x^3$, and $1/x^2$ all approach 0 as $x \to \pm\infty$.

Example 6

$$\lim_{x \to \pm\infty} \frac{2x^2 - x + 17}{x^3 + 3x^2} = \lim_{x \to \pm\infty} \frac{\dfrac{2}{x} - \dfrac{1}{x^2} + \dfrac{17}{x^3}}{1 + \dfrac{3}{x}} = \frac{0}{1} = 0$$

since $2/x$, $1/x^2$, $17/x^3$, and $3/x$ all approach 0 as $x \to \pm\infty$.

Example 7

$$\lim_{x \to \infty} \frac{x^3 + 3x^2}{2x^2 - x + 17} = \lim_{x \to \infty} \frac{x + 3}{2 - \dfrac{1}{x} + \dfrac{17}{x^2}} = \infty$$

and

$$\lim_{x \to -\infty} \frac{x^3 + 3x^2}{2x^2 - x + 17} = \lim_{x \to -\infty} \frac{x}{2} = -\infty$$

□

The graph of a function can cross its horizontal asymptote. The following example illustrates this and shows how to find the crossing point.

Example 8

Let $f(x) = \dfrac{4x^2}{4x^2 + 3x - 1}$. Graph f.

Solution

We see that

$$\lim_{x \to \infty} \frac{4x^2}{4x^2 + 3x - 1} = \lim_{x \to \infty} \frac{1}{1 + \dfrac{3}{4x^2} - \dfrac{1}{4x^2}} = 1$$

Thus the line $y = 1$ is a horizontal asymptote. Is there a number x for which $f(x) = 1$? Yes, since $\dfrac{4x^2}{4x^2 + 3x - 1} = 1$ if $4x^2 = 4x^2 + 3x - 1$, that is, $3x - 1 = 0$ or $x = \frac{1}{3}$. So $f(\frac{1}{3}) = 1$ and the point $(\frac{1}{3}, 1)$ is on the graph of f. The entire graph is given in Figure 6-7.

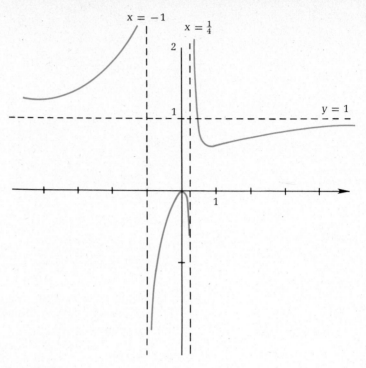

Figure 6-7

Exercises 6-1

In Exercises 1–27, find $\lim\limits_{x \to \infty} f(x)$ and $\lim\limits_{x \to -\infty} f(x)$; then find the asymptotes, if any, and graph f.

1. $f(x) = \dfrac{1}{x + 2}$

2. $f(x) = \dfrac{1}{x - 3}$

3. $f(x) = \dfrac{4}{x^2 - 2x + 1}$

4. $f(x) = \dfrac{-2}{x^2 + 8x + 16}$

5. $f(x) = \dfrac{5}{x^2 + 1}$

6. $f(x) = \dfrac{5}{x^2 - 4x + 5}$ (*Hint:* Complete the square and shift.)

7. $f(x) = \dfrac{6x^2 + 1}{(3x - 2)(x + 1)}$

8. $f(x) = \dfrac{x - 1}{2x^2 + 3x + 1}$

9. $f(x) = \dfrac{3x - 5}{x^2 - 2x - 8}$

10. $f(x) = \dfrac{2x^2 - 5x - 3}{x^2 - x - 20}$

11. $f(x) = \dfrac{2x^2 + 6x + 4}{-x^2 + x + 12}$

12. $f(x) = \dfrac{x^2 - 4}{2x^2 - 5x - 3}$

13. $f(x) = \dfrac{x^2}{x^2 - x - 6}$

14. $f(x) = \dfrac{3x^2 - 2x}{x^2 - 9}$

15. $f(x) = \dfrac{x^3 - 27}{x^2 - 4}$

16. $f(x) = \dfrac{-x^3 + 4x}{x^2 + 6x - 9}$

17. $f(x) = \dfrac{x - 1}{x^3 - 4x}$

18. $f(x) = \dfrac{x^2 + x - 2}{x^3 - 9x}$

19. $f(x) = \dfrac{x + 2}{x^2 + 4x + 4}$

20. $f(x) = \dfrac{x^2 + 2x}{x^2 - 3x}$

21. $f(x) = \dfrac{x^2 - 4}{x^2 - 3x + 2}$

22. $f(x) = \dfrac{x^2 + x}{x^3 - x^2 - 2x}$

23. $f(x) = \dfrac{x^3 + 3x^2 + 2x}{x^3 - x^2 - 2x}$

24. $f(x) = \dfrac{1}{x^4}$

25. $f(x) = \dfrac{1}{(x - 3)^4 - 1}$

26. $f(x) = \dfrac{x}{(x - 1)^4 - 4}$

27. $f(x) = \dfrac{(x + 1)(x - 1)(x - 3)}{(x + 1)(x - 1)(x - 3)}$

28. Let a be a positive real number. Find all real numbers x for which

$$\frac{x}{x + a} - \frac{2a^2}{x^2 - a^2} = 0$$

29. Let a be a positive real number. Find all real numbers x for which

$$\frac{x}{4x^2 - a^2} + \frac{x + a}{2x^2 + xa} = 0$$

30. Let a be a nonzero real number. Find all real numbers x for which

$$\frac{\dfrac{a^2}{x} - \dfrac{x^2}{a}}{\dfrac{a}{x} + \dfrac{x}{a}} = 0$$

31. Let $f(x) = \dfrac{x^3 + 2x^2 - 3x}{x^2 - 6x + 5}$. Find the zeros of f.

32. Let

$$f(x) = \frac{2x - 3}{x^2 - 4} \qquad x \neq 2, \ x \neq -2$$

and

$$g(x) = \frac{x + 1}{x + 2} \qquad x \neq -2$$

(a) Show that the function $f + g$ is also a rational function.
(b) Show that the function f/g is also a rational function.
(c) Show that the function $f + (g/f)$ is a rational function.

33. Let

$$f(x) = \frac{2x + 5}{x^2 + 6x + 9}, \quad g(x) = \frac{x}{x^2 - 9}$$

Find the zeros of the rational function $f + g$.

34. Let

$$f(x) = \frac{7}{2 + x}, \quad g(x) = \frac{3}{4 - 2x}, \quad \text{and} \quad h(x) = \frac{6}{4 - x^2}$$

Find all points where the graph of $g + h$ intersects the graph of f.

35. Let

$$f(x) = \frac{4}{3x + 4}, \quad g(x) = \frac{3}{3x - 4}, \quad \text{and} \quad h(x) = \frac{5}{9x^2 - 16}$$

Find all points where the graph of g intersects the graph of $f - h$.

36. Let

$$f(x) = \frac{3x^2 - 5x - 2}{x^3 + 4x} \quad \text{and} \quad g(x) = \frac{x^4 + 4x^2}{9x^2 - 1}$$

Find all numbers x for which $f(x) \cdot g(x) = 0$.

37. Let

$$f(x) = \frac{36 - 4x^2}{x^3 - 1} \quad \text{and} \quad g(x) = \frac{x^4 + x^3 + x^2}{6 + 8x + 2x^2}$$

Find all numbers x for which $f(x) \cdot g(x) = 0$

38. Let

$$f(x) = \frac{x - 4}{x + 4} - \frac{x + 4}{x - 4} \quad \text{and} \quad g(x) = \frac{x^3 - 16x}{16}$$

Find all numbers x for which $f(x) \cdot g(x) = 0$.

39. Let

$$f(x) = x + \frac{x}{1 - \dfrac{1}{x + 2}} \quad \text{and} \quad g(x) = \frac{x^2 + x}{2x^2 + x - 3}$$

Find all numbers x for which $f(x) \cdot g(x) = 0$.

40. Let

$$f(x) = \frac{-x^2 + 6x - 9}{4x^2 - 4} \quad \text{and} \quad g(x) = \frac{2x - 2}{x - 3}$$

Find all numbers x for which $f(x) \cdot g(x) = 0$ (i.e., find the zeros of fg).

41. Let a be a real number, and let f and g be rational functions given by

$$f(x) = \frac{x + a}{x + 3a} \quad \text{and} \quad g(x) = \frac{x^2 - 9a^2}{a^2 - x^2}$$

Find all numbers x such that $f(x) \cdot g(x) = 0$.

42. Let the rational functions f and g be defined by

$$f(x) = \frac{a}{a - x} - \frac{x}{a + x} \quad \text{and} \quad g(x) = \frac{a + x}{a^2 + x^2}$$

respectively (here, a is a real number). Find all numbers x such that $f(x) \cdot g(x) = 0$.

In Exercises 43–53, find the real numbers x for which $f(x) > 0$.
Note that if $f(x) = g(x)/h(x)$ then $f(x) > 0$ if and only if $g(x) > 0$ and $h(x) > 0$, or, $g(x) < 0$ and $h(x) < 0$; that is, if and only if $g(x)h(x) > 0$.

43. $f(x) = \dfrac{3}{2x + 1}$

44. $f(x) = -\dfrac{5}{x - \frac{1}{2}}$

45. $f(x) = \dfrac{x}{x + 1}$

46. $f(x) = \dfrac{x - 3}{-2x}$

47. $f(x) = \dfrac{2x - 1}{1 - x}$

48. $f(x) = \dfrac{x - \frac{1}{2}}{2x + 1}$

49. $f(x) = \dfrac{x^2 - 4}{x + 1}$

50. $f(x) = \dfrac{x^2 + 2x - 3}{x}$

51. $f(x) = \dfrac{8 - 2x - x^2}{3x^2}$

52. $f(x) = \dfrac{x}{x^2 + 3x + 2}$

53. $f(x) = \dfrac{x + 1}{x + 2} - 3$

6-2 DIVISION

The Remainder Theorem, proved in Chapter 5, has the following extension to polynomial functions in general: *If f and g are polynomial functions with the degree of f greater than or equal to the degree of g, then*

$$f = g \cdot h + r$$

where h is a polynomial function whose degree is the difference between the degrees of f and g, and r (the **remainder**) is either a polynomial function whose degree is less than that of g, or r is the zero function. It follows that for $g(x) \neq 0$, we have:

$$\frac{f(x)}{g(x)} = h(x) + \frac{r(x)}{g(x)}$$

The process of obtaining h and r is called **division** (or the **division algorithm**). Its validity is usually proved in courses in *abstract algebra*. Here we shall limit ourselves to reminding the reader how this process actually works. Let

$$f(x) = a_n x^n + a_{n-1} x^{n-1} + a_{n-2} x^{n-2} + \cdots + a_0$$

and

$$g(x) = b_m x^m + b_{m-1} x^{m-1} + b_{m-2} x^{m-2} + \cdots + b_0$$

where $b_m \neq 0$, $a_n \neq 0$, and $n \geq 0$.

Step 1   Divide $b_m x^m$ into $a_n x^n$ to obtain $\dfrac{a_n}{b_m} x^{n-m}$; multiply this by $g(x)$, and subtract the resulting product from $f(x)$. The result is the first *remainder*, r_1, whose degree is $n - 1$ (or possibly less). If this degree is less than that of g, then we stop; if not, we continue.

Step 2   Divide $b_m x^m$ into the leading term of r_1. Multiply the result by $g(x)$, and subtract the resulting product from $r_1(x)$ to obtain $r_2(x)$. The *second remainder*, r_2, is of degree $n - 2$ or less; again, if the degree of r_2 is less than the degree of g, we stop. Otherwise, we continue by dividing $b_m x^m$ into the leading term of r_2, and so on.

This process continues until a remainder of degree less than that of g is obtained. The next few examples illustrate this.

Example 1

Divide $x \to 2x^3 + 3x - 1$ by $x \to x^2 + x - 2$.

Solution

$$
\begin{array}{r}
\overset{g(x)}{\searrow} \quad \overset{\overset{\displaystyle h(x)}{\diagup}}{2x - 2} \longleftarrow \\
x^2 + x - 2 \overline{\smash)2x^3 + 0x^2 + 3x - 1} \quad \leftarrow f(x) \\
2x^3 + 2x^2 - 4x \quad\quad \leftarrow (2x) \cdot g(x) \\
\hline
-2x^2 + 7x - 1 \quad \leftarrow r_1(x) \\
-2x^2 - 2x + 4 \quad \leftarrow (-2) \cdot g(x) \\
\hline
9x - 5 \quad \leftarrow r_2(x) = r(x)
\end{array}
$$

We stop at this point because the degree of r is less than that of g. The result is:

$$2x^3 + 3x - 1 = (x^2 + x - 2)(2x - 2) + 9x - 5$$

for all numbers x and hence,

$$\frac{2x^3 + 3x - 1}{x^2 + x - 2} = 2x - 2 + \frac{9x - 5}{x^2 + x - 2}$$

for $x^2 + x - 2 \neq 0$.

Example 2

Divide $x \to 3x^6 + x^4 - x + 3$ by $x \to x^3 + 2x^2 + x$.

Solution

$$
\begin{array}{l}
\overset{g(x)}{\searrow} \quad\quad 3x^3 - 6x^2 + 10x \; - 14 \longleftarrow \overset{h(x)}{} \\
x^3 + 2x^2 + x \overline{\smash)3x^6 + 0x^5 + \;\; x^4 + \; 0x^3 + \;\; 0x^2 - \;\;\;\; x + 3} \quad \leftarrow f(x) \\
 3x^6 + 6x^5 + \;\; 3x^4 \quad\quad\quad\quad\quad\quad\quad \leftarrow (3x^3) \cdot g(x) \\
\hline
 -6x^5 - \;\; 2x^4 + \; 0x^3 + \cdots \quad\quad\quad \leftarrow r_1(x) \\
 -6x^5 - 12x^4 - \;\; 6x^3 \quad\quad\quad\quad \leftarrow (-6x^2) \cdot g(x) \\
\hline
 10x^4 + \;\; 6x^3 + \; 0x^2 - \cdots \quad \leftarrow r_2(x) \\
 10x^4 + 20x^3 + 10x^2 \quad\quad\quad \leftarrow (10x) \cdot g(x) \\
\hline
 -14x^3 - 10x^2 - \;\;\; x \cdots \quad \leftarrow r_3(x) \\
 -14x^3 - 28x^2 - 14x \quad\quad \leftarrow (-14) \cdot g(x) \\
\hline
 18x^2 + 13x + 3 \quad \leftarrow r(x)
\end{array}
$$

Thus,

$$\frac{3x^6 + x^4 - x + 3}{x^3 + 2x^2 + x}$$

$$= 3x^3 - 6x^2 + 10x - 14 + \frac{18x^2 + 13x + 3}{x^3 + 2x^2 + x}$$

for $x^3 + 2x^2 + x \neq 0$

$\square$

The results of Examples 1 and 2 may be verified by direct multiplication, and the reader is urged to do so.

Exercises 6-2

In Exercises 1–24, find functions h and r for which $f = h \cdot g + r$.

1. $f(x) = 2x^3 - 4x^2 + x - 2, \quad g(x) = x^3 - x^2 - x - 2$

2. $f(x) = x^4 + x^3 + x^2 + x + 1, \quad g(x) = x^3 - 1$

3. $f(x) = x^5 + x^4 + 2x^3 - x^2 - x - 2, \quad g(x) = x^2 + x + 1$

4. $f(x) = x^4 + 2x^3 + 5x^2 + 4x + 4, \quad g(x) = x^3 - 2x^2 + x + 4$

5. $f(x) = x^3 + \frac{1}{2}x^2 + \frac{1}{3}x + \frac{1}{6}, \quad g(x) = x - \frac{1}{2}$

6. $f(x) = x^3 + 2, \quad g(x) = x^2 + 15$

7. $f(x) = x^5 + x^3 + x, \quad g(x) = x^3 + 2x + 6$

8. $f(x) = x^3 + x + 1, \quad g(x) = x + 15$

9. $f(x) = x^3 + 3x^2 + 6x + 2, \quad g(x) = x - 4$

10. $f(x) = 3x^3 + 12x^2 + 8x - 2, \quad g(x) = x + 3$

11. $f(x) = 2x^5 + 5x^3 + 3x^2 - 12x + 5, \quad g(x) = x^2 - 2$

12. $f(x) = x^4 - 2x^3 + 6x^2 + 5x + 12, \quad g(x) = x^2 + x + 2$

13. $f(x) = 2x^5 - x^4 + x^3 + 5x^2 - 2x + 3, \quad g(x) = x^2 - x + 1$

14. $f(x) = 2x^5 + x^4 + x^3 - 13x + 7, \quad g(x) = 2x^2 + x - 1$

15. $f(x) = x^3 + 2x^2 + 3x + 2, \quad g(x) = x^2 + 4$

16. $f(x) = x^3 + (1 - \sqrt{2})x - \sqrt{2}, \quad g(x) = x^2 - 2$

17. $f(x) = x^3 + \sqrt{7}x + 3\sqrt{7}, \quad g(x) = x - \sqrt{7}$

18. $f(x) = x^3 + (2i + 1)x^2 + ix + i + 1,$
$$g(x) = x^2 + (i - 1)x - 2i - 2$$

19. $f(x) = x^4 + 9, \quad g(x) = x + i,$

20. $f(x) = x^3 - 3xa + 2a^3, \quad g(x) = x + 2a$

21. $f(x) = 2x^3 - 3x^2a - 7xa^2 + 3a^3, \quad g(x) = 2x + 3a$

22. $f(x) = 3x^4 - x^3a + 3x^2a - xa^2 - 6xa^3 + 3a^4,$
$$g(x) = 3x - a$$

23. $f(x) = x^4 - x^2a^2 - xa^3 + 3x^3a + 8a^4, \quad g(x) = x + 3a$

24. $f(x) = 3x^2 + 12xa + 12a^2 + 2x + 4a - 21,$
$$g(x) = x + 2a + 3$$

Miscellaneous Problems

1. Let $f(x) = \dfrac{x^2 + 2x - 15}{x^3 - x^2}$ and let $g(x) = \dfrac{x}{100}$. For how many positive real numbers x is $f(x) = g(x)$? (*Hint:* Consider the graphs.)

2. Let $f(x) = \dfrac{3x - 1}{4x - 5}$. Find a function g for which $f(g(x)) = x$ for all x in the range of f, that is, for $x \neq \frac{3}{4}$.

3. Find a polynomial function g such that

(a) $\displaystyle\lim_{x \to 1} \dfrac{(x - 1)(x + 3)}{g(x)} = \infty$

(b) $\displaystyle\lim_{x \to 1} \dfrac{(x - 1)(x + 3)}{g(x)} = 4$

(c) $x \to \dfrac{(x - 1)(x + 3)}{g(x)}$ has no zeros.

4. Let $h(x) = \dfrac{5x - 3}{x^2 - 2x - 3}$. Find real numbers a and b for which $h(x) = \dfrac{a}{x + 1} + \dfrac{b}{x - 3}$ for $x \neq -1$ and $x \neq 3$.

5. Let $h(x) = \dfrac{1}{x^3(x^2 + 1)}$. Find polynomial functions f and g of degrees 1 and 2, respectively, such that $h(x) = \dfrac{f(x)}{x^2 + 1} + \dfrac{g(x)}{x^3}$.

6. Let $f(x) = \dfrac{1}{x^3}$, $x \neq 0$. Find a function g for which $f(g(x)) = x$, $x \neq 0$. Is g a rational function?

7. Let a be any nonzero real number. Find quadratic functions f and g for which

$$\lim_{x \to \infty} \dfrac{f(x)}{g(x)} = a$$

Can you find quadratic functions f and g for which

$$\lim_{x \to \infty} \dfrac{f(x)}{g(x)} = \infty? \qquad \lim_{x \to \infty} \dfrac{f(x)}{g(x)} = 0?$$

Explain.

8. Let a, b, c and d be real numbers, with c and d not both zero, and let

$$f(x) = \dfrac{ax + b}{cx + d}$$

Show that f is a constant function if and only if $ad - bc = 0$.

9. Suppose f is a polynomial function of degree ≥ 1. Show that the x-axis is a horizontal asymptote for the function $x \to 1/f(x)$. What can be said about vertical asymptotes?

10. Let f be a rational function. Explain why either $\displaystyle\lim_{x \to \infty} |f(x)| = \infty$ or $\displaystyle\lim_{x \to \infty} |f(x)| = c$ for some real number c.

3

TRANSCENDENTAL FUNCTIONS

So far, our study has been limited to algebraic functions, that is, functions that may be obtained from the identity and constant functions by means of addition, multiplication, division, and the extraction of roots. Functions that may *not* be obtained this way are known as *transcendental functions*. They include the exponential, logarithmic, and trigonometric functions.

Exponential and logarithmic functions arose in connection with phenomena that involve natural growth or decay—for example, population growth, nuclear decay, chemical decomposition, cell division, decay of current in a circuit, and so on.

Trigonometric functions originated in the study of navigation, surveying, and other sciences that relied on the relationships between the angles and the sides of triangles. Today, however, the major application of these functions is in the study of wave phenomena such as sound, heat, light, and electricity, and in nuclear physics and biology.

The concept of the *inverse* of a function is particularly useful in the study of transcendental functions since, for example, the inverse of each exponential function is a logarithmic function, and the inverse of each logarithmic function is an exponential function. Hence, we begin this part of the book with a general introduction to inverses of functions and leave the study of inverses of specific functions to the appropriate sections. Part Three also contains applications of trigonometry to the study of complex numbers.

7 Exponential and Logarithmic Functions

7-1 INVERSE FUNCTIONS

We shall now consider pairs of operations, or functions, that undo each other. For example, doubling and halving are such operations; that is, if a number x is doubled, to obtain 2x, and this result is halved, to obtain $\frac{1}{2}(2x)$, we end up with the number x with which we began. The result, of course, will be the same if we halve first and *then* double.

Example 1

Consider the doubling function d, and the halving function h, given, respectively, by $d(x) = 2x$ and $h(x) = \frac{1}{2}x$. We have $d[h(x)] = 2(\frac{1}{2}x) = x$ and $h[d(x)] = \frac{1}{2}(2x) = x$. That is, d and h undo each other.

If two functions undo each other, then each is called the **inverse*** of the other. More precisely, we say that functions f and g are inverses of each other if

$$g[f(x)] = x \qquad \text{for every x in the domain of } f$$

and

$$f[g(x)] = x \qquad \text{for every x in the domain of } g$$

Thus, the two functions in Example 1 are inverses of each other.

---

*We shall see in the sequel that a function cannot have more than one inverse; it may, however, have none at all.

Example 2

Let $f(x) = x^3$ and $g(x) = \sqrt[3]{x}$. Then for any number x,

$$f[g(x)] = (\sqrt[3]{x})^3 = x \quad \text{and} \quad g[f(x)] = \sqrt[3]{x^3} = x$$

Hence, the cubing function and the cube root function are each other's inverses.

Example 3

Find the inverse of $f\colon x \to 3x - 2$.

Solution

Since f *multiplies* by 3 and then *subtracts* 2, its inverse first *adds* 2 and then *divides* by 3 (just as taking off a shoe and then a sock undoes putting on a sock and *then* a shoe). Hence, the inverse of f is $x \to \frac{1}{3}(x + 2)$.

Alternate Solution

Suppose g is the inverse of f. Then, for any number x, $f[g(x)] = 3g(x) - 2 = x$. Solving for $g(x)$, we obtain

$$g(x) = \frac{x + 2}{3}.$$

□

If f and g are inverses of each other, we write

$$g = f^{-1} \quad \text{and} \quad f = g^{-1}$$

x	$f(x) = x^3$	x	$f^{-1}(x) = \sqrt[3]{x}$
-2	$(-2)^3 = -8$	-8	$\sqrt[3]{-8} = -2$
-1	$(-1)^3 = -1$	-1	$\sqrt[3]{-1} = -1$
0	$0^3 = 0$	0	$\sqrt[3]{0} = 0$
1	$1^3 = 1$	1	$\sqrt[3]{1} = 1$
2	$2^3 = 8$	8	$\sqrt[3]{8} = 2$

x	$f(x) = 3x - 2$	x	$f^{-1}(x) = \frac{1}{3}(x + 2)$
-2	$3(-2) - 2 = -8$	-8	$\frac{1}{3}(-8 + 2) = -2$
-1	$3(-1) - 2 = -5$	-5	$\frac{1}{3}(-5 + 2) = -1$
0	$3(0) - 2 = -2$	-2	$\frac{1}{3}(-2 + 2) = 0$
1	$3(1) - 2 = 1$	1	$\frac{1}{3}(1 + 2) = 1$
2	$3(2) - 2 = 4$	4	$\frac{1}{3}(4 + 2) = 2$

Figure 7-1

The tables in Figure 7-1 illustrate the fact that if a point (a, b) is on the graph of a function, then (b, a) is on the graph of its inverse. Thus, for example, $(2, 8)$ is on the graph of $x \to x^3$ and $(8, 2)$ is on the graph of $x \to \sqrt[3]{x}$; $(0, -2)$ is on the graph of $x \to 3x - 2$ and $(-2, 0)$ is on the graph of $x \to \frac{1}{3}(x + 2)$. In general (a, b) is on the graph of f if and only if (b, a) is on the graph of f^{-1}. To see this observe that if (a, b) is on the graph of f, then $b = f(a)$, and then (since $f^{-1}[f(a)] = a$) we have $f^{-1}(b) = a$, so (b, a) is on the graph of f^{-1}.

It follows that the graphs of f and f^{-1} are each other's reflections in the line given by $y = x$, since this line is the perpendicular bisector of the segment joining (a, b) and (b, a); see

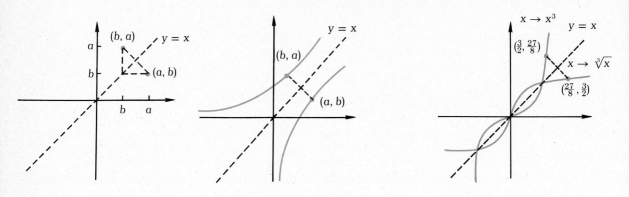

Figure 7-2.

Figure 7-3.

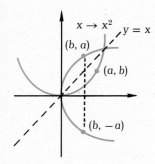

Figure 7-4.

Figure 7-2. In Figure 7-3, this is illustrated for the functions of Example 2.

The *uniqueness* of the inverse function follows now from the uniqueness of the *reflected graph;* in other words, since the reflected graph cannot be the graph of more than one function, *no function can have more than one inverse.*

On the other hand, not every function has an inverse. Consider, for example, the graph of $x \to x^2$, and the reflected graph (Figure 7-4). Here the *reflected graph* contains both (b, a) and $(b, -a)$ for every point (a, b) on the graph of $x \to x^2$, and hence the reflected graph is not the graph of a function. (Recall that the graph of any function g is characterized by the fact that if $a \neq c$, then the points (b, a) and (b, c) cannot both be on it.) Hence, $x \to x^2$ has no inverse function.

The squaring function is a prototype of functions that do not have inverses. The trouble with this function is that (b, a) and $(b, -a)$ are both on the reflected graph, which in turn follows from the fact that both (a, b) and $(-a, b)$ are on the graph of $x \to x^2$ [since $b = (-a)^2 = a^2$].

Now, a function f which satisfies the condition that *if $a \neq c$, then $f(a) \neq f(c)$ for all numbers a and c in its domain* is called **one-to-one,** or **1–1** (because precisely one number in the domain corresponds to each number in the range). (Equivalently, "f is 1–1" means that if $f(a) = f(c)$, then $a = c$.) Geometrically, f is 1–1 if and only if every horizontal line intersects the graph of f in at most one point. This brings us to the following result.

Theorem

A function f has an inverse if and only if it is 1–1.

Proof f has an inverse if and only if the reflection of its graph in the line given by $y = x$ is the graph of a function; that is, every *vertical* line intersects the graph of the reflection in at most one point, or, equivalently, every *horizontal* line intersects the graph of f in at most one point; that is, if and only if f is 1–1.

By this criterion, then, the doubling, halving, cubing, and cube root functions are all 1–1, whereas the squaring function is not. The following examples show how to verify directly whether or not a function is 1–1.

Example 4

Show that $f: x \rightarrow 7x - 3$ is 1–1.

Solution

If a and b are numbers with $a \neq b$, then $7a \neq 7b$, and hence $7a - 3 \neq 7b - 3$; that is, $f(a) \neq f(b)$. Another way to do this is by showing that if $f(a) = f(b)$, then $a = b$. Thus, if $f(a) = f(b)$, then

$$7a - 3 = 7b - 3$$
$$7a = 7b$$

and

$$a = b$$

Hence, f is 1–1. [Yet another way of doing this is to compute the inverse $x \rightarrow \frac{1}{7}(x + 3)$].

Example 5

Show that $f: x \rightarrow x^3 + x - 3$ is 1–1.

Solution

If a and b are real numbers for which $f(a) = f(b)$, then $a^3 + a - 3 = b^3 + b - 3$, so that subtracting

$$a^3 - b^3 + a - b = (a - b)(a^2 + ab + b^2 + 1) = 0$$

Now,

$$a^2 + ab + b^2 + 1 = \left(a + \frac{b}{2}\right)^2 + \frac{3}{4}b^2 + 1 > 0$$

for all choices of a and b, and hence $a - b = 0$ or, equivalently, $a = b$. Hence, f is 1-1. (Thus, we see that f has an inverse, although it may be far from easy to compute it.)

Remark   Recall that if a function f is 1-1, then its graph may not be intersected more than once by any horizontal line. It follows from this remark that a polynomial function of *even* degree cannot be 1-1 (and hence cannot have an inverse). Why? It also follows that if the graph of a function steadily rises (or falls), then this function *is* 1-1 (and hence has an inverse).

Example 6

Let $f(x) = x^3 + 5$. Find f^{-1}.

Solution

It is easy to verify that f is 1-1 and hence f^{-1} exists. For any number x,

$$x = f[f^{-1}(x)] = [f^{-1}(x)]^3 + 5.$$

Hence, $f^{-1}(x) = \sqrt[3]{x - 5}$, for any number x.

Exercises 7-1

In Exercises 1–7, find and graph f^{-1}.

1. $f(x) = 5x$ 2. $f(x) = x^5$
3. $f(x) = 3x - 1$ 4. $f(x) = 4x + 2$
5. $f(x) = x^3 - 1$ 6. $f(x) = \dfrac{1}{x}$

7. $f(x) = \dfrac{1}{x^3}$

Answer Exercises 8–11 true or false and justify your answer.

8. The absolute value function $x \to |x|$ has an inverse.

9. Every first degree polynomial function has an inverse.

10. If the domain of f is the set of all real numbers and $f(f(x)) = 1$ for every x then f has an inverse.

11. If f is an increasing function then f^{-1} exists and is also an increasing function.

12. Let

$$f(x) = \begin{cases} x^2 - 4x + 4 & \text{for } x \geq 2 \\ x - 2 & \text{for } x \leq 2 \end{cases}$$

Graph f, and determine if f^{-1} exists. If it does exist, find f^{-1} and graph it.

13. Let

$$f(x) = \begin{cases} x^2 - x - 6 & \text{for } x \geq \frac{1}{2} \\ x - \frac{27}{4} & \text{for } x \leq \frac{1}{2} \end{cases}$$

Graph f, and determine if f^{-1} exists. If it does exist, graph it.

7-2 THE EXPONENTIAL FUNCTIONS

We have already defined a^x for any positive real number a and *rational* number x. For example, $4^{5/3} = \sqrt[3]{1024}$, $2^{3/5} = (\sqrt[5]{2})^3$, and so on. We shall now extend this definition to the case where x is *irrational*.

Let us begin with an example: How shall we define $3^{\sqrt{2}}$? Recall that the sequence 1.4, 1.41, 1.414, ... converges to $\sqrt{2}$. Since this sequence is *increasing* and *bounded*, so is the sequence $3^{1.4}$, $3^{1.41}$, $3^{1.414}$, ..., and it must therefore converge to some number that we shall call $3^{\sqrt{2}}$.

In general, suppose that a is a positive real number and that x is irrational. There is always a *bounded, increasing* sequence $\langle r_n \rangle$ of *rational* numbers converging to x; the corresponding sequence $\langle a^{r_n} \rangle$ is also *bounded*; it is *increasing* if $a > 1$ and *decreasing* if $a < 1$.* In either case, $\langle a^{r_n} \rangle$ must converge to some real number (see Section 4-6). Furthermore, if $\langle k_n \rangle$ is any other sequence converging to x, then the sequences $\langle a^{k_n} \rangle$ and $\langle a^{r_n} \rangle$ converge to the same number. We call this number a^x. Thus, by definition,

$$a^x = \lim_{n \to \infty} a^{r_n}$$

where $\langle r_n \rangle$ is any sequence of rationals converging to x, that is,

$$x = \lim_{n \to \infty} r_n$$

Graphically, this definition may be illustrated as follows: the graph of $x \to 3^x$, for x *rational*, steadily rises and has no "jumps" (i.e., if r_1 and r_2 are close to each other, then so are 3^{r_1} and 3^{r_2}). On the other hand, this graph has a "hole" above each irrational number x. By our definition, the second coordi-

---

*Recall that if $a > 0$ and $m < n$, then

$$a^m < a^n \quad \text{if } a > 1$$

and

$$a^m > a^n \quad \text{if } a < 1$$

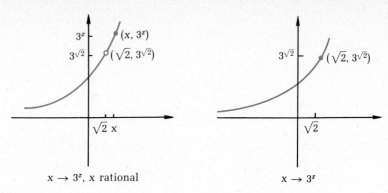

Figure 7-5

nate of this "hole" is the number 3^x. For example, $3^{\sqrt{2}}$ is the second coordinate of the "hole" above $\sqrt{2}$. See Figure 7-5.

The function $x \to 3^x$ is called the **exponential function, base 3** and is denoted by **$\exp_3$**. Thus, $\exp_3(\sqrt{2}) = 3^{\sqrt{2}}$, $\exp_3(4) = 3^4$, $\exp_3\left(-\frac{1}{2}\right) = 3^{-1/2}$, and, for any real number x, $\exp_3(x) = 3^x$.

Similarly, if a is any positive real number except 1, then the function $x \to a^x$ is called the **exponential function, base a,** and is denoted by **$\exp_a$**. Thus, $\exp_a(x) = a^x$ for any number x. The graph of $\exp_a$ steadily rises if $a > 1$ and falls if $a < 1$, because, for $x_1 < x_2$, $a^{x_1} < a^{x_2}$ if $a > 1$, and $a^{x_1} > a^{x_2}$ if $a < 1$. See Figure 7-6, which indicates the general behavior of exponential functions.

Example 1

Graph $\exp_{1/2}$.

Solution

$\exp_{1/2}$ is the exponential function base $\frac{1}{2}$: $x \to \left(\frac{1}{2}\right)^x$. The graph of this function steadily falls. When x is large and positive, $\left(\frac{1}{2}\right)^x$ is small and positive and approaches 0 as x increases indefinitely. Hence, the graph approaches the horizontal axis. See Figure 7-7.

The "laws" proved for rational exponents also hold for irrational exponents, and hence for all *real exponents*.* We restate them as follows. If a and b are positive real numbers and x and y any real numbers, then:

1. $a^{x+y} = a^x \cdot a^y$
2. $(a^x)^y = a^{xy}$
3. $(ab)^x = a^x \cdot b^x$
4. $\left(\dfrac{a}{b}\right)^x = \dfrac{a^x}{b^x}$

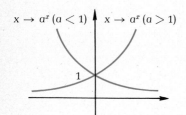

Figure 7-6

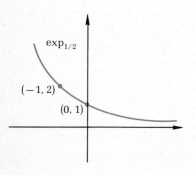

Figure 7-7

*This is proved in courses in Advanced Calculus.

Furthermore, since $a^x \neq a^y$ for $x \neq y$, we see that $\exp_a$ is 1–1; i.e.,

$$\text{if } a^x = a^y, \text{ then } x = y$$

Example 2

(a) $2^{\sqrt{2}} \cdot 2^{-\sqrt{3}} = 2^{\sqrt{2}-\sqrt{3}}$

(b) $(2^{\sqrt{2}})^{\sqrt{8}} = 2^{\sqrt{2}\cdot\sqrt{8}} = 2^4 = 16$

(c) $3^{\pi} \cdot 4^{\pi} = 12^{\pi}$

(d) $\left(\dfrac{3}{5}\right)^{\sqrt{5}} = \dfrac{3^{\sqrt{5}}}{5^{\sqrt{5}}}$

(e) Find all real numbers x for which $3^{x^2-4x} = 9$.

Solution

Since $9 = 3^2$, we have

$$x^2 - 4x = 2, \text{ hence } x^2 - 4x - 2 = 0$$

and so

$$x = \frac{4 \pm \sqrt{16 + 8}}{2} = 2 \pm \sqrt{6}$$

Exercises 7-2

In Exercises 1–11, graph f.

1. $f(x) = 2^x$ **2.** $f(x) = 2^{x+3}$

3. $f(x) = 2^{3x}$ **4.** $f(x) = 2^x + 1$

5. $f(x) = 2^{|x|}$ **6.** $f(x) = 2^{x^2}$

7. $f(x) = 2^{|x|^3}$ **8.** $f(x) = 2^{1/x}$

9. $f(x) = 2^{1/x^2}$ **10.** $f(x) = 2^{-1/x}$

11. $f(x) = 2^{-1/x^2}$

In Exercises 12–17, find all real numbers x for which:

12. $2^{3x-1} = 64$ **13.** $5^{4x+3} = 625$

14. $3^{x^2-x} = 729$ **15.** $5^{x^2+2x+3} = 25$

16. $2^{x^2+x} = 32$ **17.** $10^{x^2+3x+1} = \frac{1}{100}$

18. Let $f(x) = a \cdot 10^{kx}$, and suppose $f(0) = 5$ and $f(1) = 500$. Find the numbers a and k.

19. For how many real numbers x is $3^x = x^2$?

◊**20.** Given that $3^x = 3 - x$, locate x between consecutive integers.

◊**21.** Let $f(x) = 2^{x^2+x+1}$. Graph f and give both coordinates of the the lowest point on the curve.

7-3 THE LOGARITHMIC FUNCTIONS

Recall that the exponential functions are one-to-one. Hence, for any positive real number b other than 1, the exponential function, base b (i.e., the function $\exp_b$, or $x \rightarrow b^x$), has an inverse. This inverse is called the **logarithm function, base b,** and is denoted by $\log_b$. Thus,

$$\exp_b[\log_b(x)] = x \quad \text{and} \quad \log_b[\exp_b(x)] = x$$

or, equivalently,

$$b^{\log_b(x)} = x \quad \text{and} \quad \log_b(b^x) = x$$

for any positive real number x. That is,

$$y = \log_b(x) \quad \text{means} \quad b^y = x$$

(In words: $\log_b(x)$ is the *exponent* to which b, the *base*, must be raised to obtain x). Often $\log_b(x)$ is written as $\log_b x$.

Example 1

(a) $\log_3(3^2) = 2$ and $3^{\log_3(2)} = 2$
(b) $\log_2(32) = \log_2(2^5) = 5$
(c) $y = \log_5(125)$ *means* $5^y = 125$; hence, y = 3

Example 2

Let a and b be positive real numbers, and $a \neq 1$, $b \neq 1$. Show that

$$\log_a b = \frac{1}{\log_b a}$$

Solution

By the definition of log, we have

$$a^{\log_a b} = b$$

and hence

$$a = b^{1/\log_a b}$$

But (again using the definition of log) we also have

$$a = b^{\log_b a}$$

Since the exponential functions are one-to-one (i.e., if $b^x = b^y$, then x = y), it follows that

$$\log_b a = \frac{1}{\log_a b}$$

which is the desired result.

□

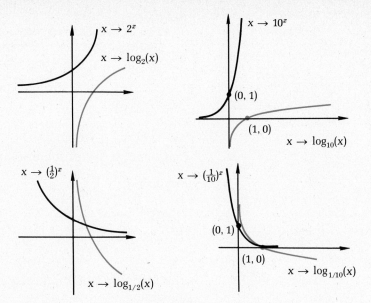

Figure 7-8

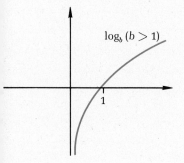

Figure 7-9

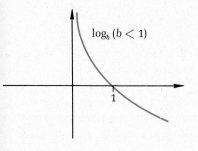

Figure 7-10

The graph of a log function is, of course, the reflection in the line $y = x$ of its inverse (exponential) function. See Figure 7-8.

If $b > 1$ then the graph of $\log_b$ is as given in Figure 7-9. The graph for $b < 1$ is given in Figure 7-10.

The basic properties of exponential functions are reflected in the basic properties of the log functions: If x, y, and b are positive numbers with $b \neq 1$, then

1. $\log_b (xy) = \log_b (x) + \log_b (y)$

and

2. $\log_b (x^t) = t \cdot \log_b (x)$, for any number t

For example

$$\log_4 (6) = \log_4 (2 \cdot 3)$$
$$= \log_4 (2) + \log_4 (3)$$

and

$$\log_7 (8) = \log_7 (2^3) = 3 \cdot \log_7 (2)$$

The proof of these properties depends on the fact that the exponential functions are 1–1 (and, of course, on the definition of $\log_b$).

Proof of 1

$$b^{\log_b(xy)} = xy \qquad \text{[by the definition of } \log_b(xy)]$$

$$= b^{\log_b(x)} \cdot b^{\log_b(y)} \qquad \text{[by the definition of } \log_b(x) \text{ and } \log_b(y)]$$

$$= b^{\log_b(x) + \log_b(y)} \qquad \text{[law of exponents]}$$

Since $\exp_b$ is 1–1, it follows that $\log_b(xy) = \log_b(x) + \log_b(y)$.

Proof of 2

$$b^{\log_b(x^t)} = x^t$$

$$= (b^{\log_b(x)})^t$$

$$= b^{t \cdot \log_b(x)}$$

The reader is urged to justify each step here and to complete the argument.

From Properties 1 and 2 it follows that

$$3.\ \log_b\left(\frac{x}{y}\right) = \log_b(x) - \log_b(y)$$

since

$$\log_b\left(\frac{x}{y}\right) = \log_b(x \cdot y^{-1}) = \log_b(x) + \log_b(y^{-1})$$

$$= \log_b(x) + (-1) \cdot \log_b(y)$$

Example 3

(a) $\log_{1/2}(46) = \log_{1/2}(2 \cdot 23) = \log_{1/2}(2) + \log_{1/2}(23)$

(b) $\log_7(125) = \log_7(5^3) = 3 \cdot \log_7(5)$

(c) $\log_2\left(\frac{16}{27}\right) = \log_2(2^4) - \log_2(3^3) = 4\log_2(2) - 3\log_2(3)$
$$= 4 - 3\log_2(3)$$

□

Since each log function has an inverse (namely, the corresponding exponential function), *the log functions are 1–1*. As the following examples illustrate, this fact is sometimes used, in conjunction with *tables of logarithms** (see page 460), for computation. Henceforth, for convenience, we shall omit parentheses when there is no chance of ambiguity.

---

*The tables of logarithms, base 10, give the logs of numbers between 1 and 10. Now, $\log_{10} 1 = 0$ and $\log_{10} 10 = 1$. Hence [since, if $x > y$, then $\log_{10} x > \log_{10} y$], all the entries in these tables are between 0 and 1.

Example 4

Find all numbers x for which $3^x = 7$.

Solution

x is such a number if and only if

$$\log_{10} 3^x = \log_{10} 7 \qquad \text{[by the 1-1 property}$$

or, equivalently,   *of $\log_{10}$]*

$$x \log_{10} 3 = \log_{10} 7 \qquad \text{[Property 2]}$$

Hence,

$$x = \frac{\log_{10} 7}{\log_{10} 3} \sim \frac{.8451}{.4771} \qquad \text{[Tables]}$$

$$= \tfrac{8451}{4771} \sim 1.77$$

Example 5

Using tables, find $\log_{10} 317$.

Solution

$$\log_{10} 317 = \log_{10} [(3.17)(100)] \qquad \text{[since (3.17)(100) = 317]}$$

$$= \log_{10} 3.17 + \log_{10} 100 \qquad \text{[Property 1]}$$

$$\sim .5011 + 2 \qquad \text{[Table]}$$

Hence,

$$\log_{10} 317 \sim 2.5011$$

Example 6

Using tables, find $\log_{10} .000632$.

Solution

$$\log_{10} .000632 = \log_{10} [(6.32)(10^{-4})] \qquad \text{[since (6.32)(10}^{-4}\text{) =}$$
$$\text{.00632]}$$

$$= \log_{10} 6.32 + \log_{10} 10^{-4} \qquad \text{[Property 1]}$$

$$= \log_{10} 6.32 - 4 \log_{10} 10 \qquad \text{[Property 2]}$$

$$\sim .8007 - 4 \qquad \text{[Tables]}$$

$$= -3.1993$$

□

Some treatments of logarithms emphasize their computational uses and require the reader to assimilate a host of new terms, such as *mantissa, characteristic,* and *antilog;* our purpose, however, is to stress the fact that nothing more is needed for computation than the definition, the basic Properties 1, 2, and 3 given above, and the tables. Our computations follow this principle.

Example 7

Compute $10^{.4355}$.

Solution

From the tables we see that $.4355 \sim \log_{10} 2.73$. Hence (by definition),

$$10^{.4355} \sim 10^{\log_{10} 2.73} = 2.73$$

Example 8

$$
\begin{aligned}
10^{-3.2958} &= 10^{-3.2958+4-4} && \textit{[Adding and subtracting 4 is} \\
&= 10^{.7042-4} && \textit{a device to overcome the} \\
&= 10^{.7042} \cdot 10^{-4} && \textit{difficulty of negative expo-} \\
&= 5.06 \cdot 10^{-4} && \textit{nents, since they do not ap-} \\
&= .000506 && \textit{pear in the tables]}
\end{aligned}
$$

Example 9

Compute $\sqrt{7.43}$.

Solution

$$
\begin{aligned}
\sqrt{7.43} &= (7.43)^{1/2} = 10^{\log_{10}(7.43)^{1/2}} && \textit{[definition of } \log_{10}] \\
&= 10^{(1/2)\log_{10} 7.43} && \textit{[Property 3]} \\
&\sim 10^{(1/2)(.8710)} && \textit{[Table]} \\
&= 10^{.4355} && \\
&\sim 2.73 && \textit{[by Example 7]}
\end{aligned}
$$

Example 10

Compute $(3.14)(57.4)$.

Solution

$$
\begin{aligned}
(3.14)(57.4) &= 10^{\log_{10}(3.14)(57.4)} && \textit{[definition of } \log_{10}] \\
&= 10^{\log_{10}(3.14)+\log_{10}(57.4)} && \textit{[Property 1]} \\
&\sim 10^{.4969+1.7589} && \textit{[Table]} \\
&= 10^{2.2558} && \textit{[arithmetic]} \\
&= 10^{2+.2558} && \textit{[arithmetic]} \\
&= 10^2 \times 10^{.2558} && \textit{[law of exponents]} \\
&\sim 10^2 \times 1.8 && \textit{[Table]} \\
&= 180
\end{aligned}
$$

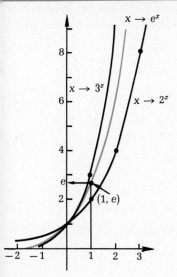

Figure 7-11

Remark There is an irrational number e that can be defined by

$$e = 1 + 1 + \frac{1}{2} + \frac{1}{2 \cdot 3} + \frac{1}{2 \cdot 3 \cdot 4} + \frac{1}{2 \cdot 3 \cdot 4 \cdot 5} + \cdots$$

or

$$e = \lim_{n \to \infty} \left(1 + \frac{1}{n}\right)^n$$

It turns out that e is approximately 2.718. Hence, since $2 < 2.718 < 3$, the graph of $x \to e^x$ lies between the graphs of $x \to 2^x$ and $x \to 3^x$. See Figure 7-11. Log_e is called the *natural log function* and occurs frequently in calculus.

□

The following fact, often referred to as *change of base*, is given in Exercise 48:

$$\log_a c = \frac{\log_b c}{\log_b a}$$

where a, b, c are positive numbers with $a \neq 1$, $b \neq 1$. For example,

$$\log_e x = \log_{10} x / \log_{10} e$$

Thus, using tables for $\log_{10}$, we can compute $\log_e x$ for any positive number x.

Example 11
Which is larger, 3^{57} or 241^{12}?

Solution
Let $a = 3^{57}$. Then

$$\log_{10} a = 57 \log_{10} 3 \sim 57 \times .4771 = 27.2947$$

Hence,

$$a = 10^{27.2947} = 10^{27} \times 10^{.2947} = 10^{27} \times 1.971$$

Similarly, let $b = 241^{12}$. Then

$$\log_{10} b = 12 \log_{10} 241 = 12 \times (2.382) = 28.584$$

Hence,

$$b = 10^{28} \times 10^{.584} = 10^{28} \times 3.837$$

Therefore, 241^{12} is larger than 3^{57}.

Example 12

Which is larger, e^{20} or 400,000,000?

Solution

Let $k = e^{20}$. Then

$$\log_{10} k = 20 \log_{10} e \sim 20 \log_{10} 2.718 \quad \text{(since } e \sim 2.718\text{)}$$
$$= 20 \times 0.4343 = 8.686$$

Hence,

$$k = 10^{8.686} = 10^8 \times 10^{.686} \cong 10^8 \times 4.6 = 460,000,000$$

Therefore e^{20} is larger than 400,000,000.

Example 13

Find $\log_e 5$.

Solution

$$\log_e 5 = \frac{\log_{10} 5}{\log_{10} e} = \frac{.6990}{.4343} = \frac{6990}{4343}$$

Example 14

$$\log_7 5 = \frac{\log_{10} 5}{\log_{10} 7} = \frac{.6990}{.8451} = \frac{6990}{8451}$$

Exercises 7-3

Note: If no base is specified, assume the base is 10. Graph the function f in Exercises 1–12.

1. $f(x) = \log x$
2. $f(x) = \log (x + 1)$
3. $f(x) = \log 3x$
4. $f(x) = 1 + \log x$
5. $f(x) = \log x^2, x > 0$
6. $f(x) = \log x^3$
7. $f(x) = \log |x|$
8. $f(x) = \log \dfrac{1}{x}$
9. $f(x) = |\log x|$
10. $f(x) = \log (3x - 2)$
11. $f(x) = \dfrac{\log x}{x}$
12. $f(x) = x \log x$

In Exercises 13–18, compute

13. $\log_2 16$
14. $\log_3 27$
15. $\log_5 \frac{1}{5}$
16. $\log_7 1$
17. $\log_{1/2} 4$
18. $\log_{1/3} 27$

Given that $\log 3 = .4771$ and $\log 5 = .6990$, do the computations in Exercises 19–25 without further use of tables.

19. $\log 15$

20. $\log \frac{1}{3}$

21. $\log \sqrt{5}$

22. $\log \frac{3}{5}$

23. $\log 25$

24. $\log 225$

25. $\log \frac{1}{2}$

In Exercises 26–29, find all real numbers x for which:

26. $\log_3 x + \log_3 (x - 1) = 6$ **27.** $\log (x + 1) - \log x = \frac{1}{2}$

28. $\log (x^2 - 1) - \log (x + 2) = 1$

29. $\log (x + 16) - \log (x + 1) = \log 6$

In Exercises 30–37, use logs to compute

30. $(.053)(53.75)$

31. $(181)(.0072)$

32. $\dfrac{.0014}{234.1}$

33. $\dfrac{147.5}{234.1}$

34. $\sqrt[3]{173}$

35. $\sqrt[5]{17.3}$

36. $\sqrt[4]{\dfrac{1.32}{.003}}$

37. $(4.32)^{17.2}$

38. Find all real numbers x for which $2^{x^2 - 2x + 6} = 32$.

39. Find the pairs (x, y) of real numbers for which $25^y = 8^x$ and $125^y = 4^{x-1}$.

40. Find the pairs (x, y) of real numbers for which $5^y = 7^x$ and $3^y = 7^{2-x}$.

41. At time t (in hours) the number of bacteria in a particular sample is ae^{kt} where a and k are real numbers. The number of bacteria at time $t = 0$ is 10^4, and the number after 3 hours is 10^8. (a) What is the size of the bacteria population after 5 hours? (b) When will the number of bacteria be 20,000?

42. A certain radioactive element decomposes so that after t hours the number of milligrams present is ae^{kt}, where a and k are real numbers. After 4 hours, half of the original amount remains. When does $\frac{1}{10}$ of the original amount remain?

◊**43.** At what interest rate compounded annually must we invest if we want to double our money in 10 years?

◊**44.** How many years will it be before $1000, left on deposit at 2% interest compounded annually, will have doubled?

◊**45.** Compute $\displaystyle\sum_{n=1}^{999} \log_{10}\left(1 + \frac{1}{n}\right)$.

◊**46.** Find all numbers x for which $2^{2x} - 2^{x+1} - 15 = 0$.

◇**47.** Find the smallest integer x for which $\left(\frac{100}{99}\right)^x$ is greater than 2500.

◇**48.** Show that, if a, b, and c are positive numbers with $a \neq 1$ and $b \neq 1$, then $\log_a c = \dfrac{\log_b c}{\log_b a}$.

◇**49.** Compute the product

$$(\log_2 3)(\log_3 4)(\log_4 5) \cdots (\log_{126} 127)(\log_{127} 128).$$

◇**50.** Show that $\displaystyle\sum_{n=2}^{100} \frac{1}{\log_n 5} = \frac{1}{\log_{100!} 5}$, where $100! = 100 \cdot 99 \cdot 98 \cdots 3 \cdot 2 \cdot 1$.

◇**51.** Show that if $a^2 + b^2 = c^2$, then

$$\log_{b+c} a + \log_{c-b} a = 2(\log_{b+c} a)(\log_{c-b} a).$$

◇**52.** For how many real numbers does $\log_{10} x = \dfrac{x}{100}$.

◇**53.** Find (with an error of less than $\frac{1}{10}$) all real numbers x for which $\log_{10} x = x/100$.

Miscellaneous Problems

1. Let $f(x) = \dfrac{1}{x^2 + 1}$, $x \geq 0$. Find f^{-1} and graph f and f^{-1}.

2. Find the smallest integer n for which $\left(\dfrac{99}{100}\right)^n < \dfrac{1}{100}$.

3. The diameter of an oxygen molecule is 2.97×10^{-8} centimeters. What is the number of oxygen molecules that can be laid side by side a distance of 1 inch? (1 inch = 2.54 centimeters)

4. For what natural numbers n does the function $x \to x^n$ have an inverse? Explain.

5. Let $f(x) = x^2 - 4x + 4$, $2 \leq x \leq 6$. Find f^{-1} and sketch its graph.

6. Find the real numbers x for which $2^x = 9$.

7. For how many real numbers is $(x + 1)^2 = 2^x$?

8. Let $f(x) = xe^{-x}$. Graph f.

9. Let $f(x) = xe^{-|x|}$. Graph f.

10. Let $f(x) = x^2 e^{-x}$. Graph f.

11. Compute $\displaystyle\lim_{n \to \infty} \frac{e^n + e^{2n}}{e^n - e^{2n}}$.

12. Find all real numbers x for which $x^{\log_3 x} = 9x$.

13. Find all numbers x for which

$$\left[\left(\frac{3^x - 3^{-x}}{2}\right)^2 + 1\right]^{1/2} = \frac{3^{-x} + 27}{2}$$

$\left(\textit{Hint:} \text{ Note that } \left[\left(\frac{3^x - 3^{-x}}{2}\right)^2 + 1\right]^{1/2} = \frac{1}{2}(3^x + 3^{-x}).\right)$

14. Which number is larger, 10^π or π^{10}? (*Hint:* Use logs.)

15. Find $\displaystyle\sum_{k=2}^{\infty} \log_{10}\left(\frac{k+1}{k}\right)$.

16. Use mathematical induction to show that for any positive integer,

$$\log x^n = n \cdot \log x \quad (x > 0).$$

Assume the other properties of log are already known.

17. Use mathematical induction to show that $2^{n-1} \leq 1 \cdot 2 \cdot 3 \cdots n$ for any positive integer n.

18. Use mathematical induction to show that $2^n > n$ for any positive integer n.

19. Suppose $I(t)$, the intensity of the sound of a note t seconds after being plucked on a violin string, is given by

$$I(t) = A \cdot 2^{-kt}$$

(Here, $A = I(0)$, the initial intensity of the sound.) Suppose that for a given violin, the intensity diminishes to $\frac{1}{2}$ the initial intensity after 4 seconds. In how many seconds will the intensity be $\frac{1}{40}$ of the initial intensity?

20. The wall of a building is being demolished. Each time the wall crusher pounds the wall, the wall loses $\frac{1}{9}$ of its strength. When the wall loses 50 percent of its initial strength, the wall will fall. How many times must the wall be hit before it falls?

8 Trigonometry

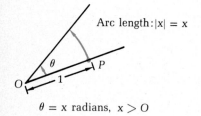

Arc length: $|x| = x$

$\theta = x$ radians, $x > 0$

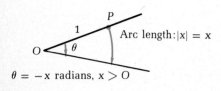

Arc length: $|x| = x$

$\theta = -x$ radians, $x > 0$

Figure 8-1

8-1 THE MEASUREMENT OF ANGLES

Consider a point O in the plane and a half-line originating from O. Let P be the point on this half-line whose distance from O is 1 unit. For *any* number x, rotate the half-line about O (counterclockwise if $x > 0$, clockwise if $x < 0$) until the point P traces out an arc whose length is $|x|$ units. Then, by definition, the half-line has been rotated by x **radians** and the angle so generated has radian measure x or, equivalently, is an **angle of x radians.** See Figure 8-1.

Recalling that the circumference of the unit circle has length 2π, we see in Figure 8-2 that an angle of 2π radians is a complete, counterclockwise revolution; that an angle of $5\pi/2$ radians is a counterclockwise one-and-a-quarter revolution; and that angles of $\pi/2$ and $-\pi/2$ radians are right angles (counterclockwise and clockwise, respectively).

In Figure 8-3, we consider the general case when P is r units from O. Since the arc length s traced out by P is proportional to the distance r of P from O, we have $x/1 = s/r$; that is,

$$x = \frac{s}{r} \quad \text{or, equivalently,} \quad s = x \cdot r$$

This can be seen another way. Since 2π radians gives a complete circle, an angle of x radians is the fraction $x/2\pi$ of a circle. Since the circumference of a circle of radius r is $2\pi r$, the arc length subtended by a central angle of x radians is given by $s = (x/2\pi)(2\pi r) = xr$, which is, of course, the same result as before.

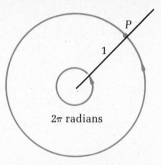

2π radians

Angle of 2π radians

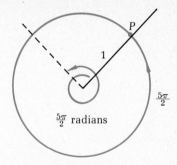

$\frac{5\pi}{2}$ radians

Angle of $\frac{5\pi}{2}$ radians

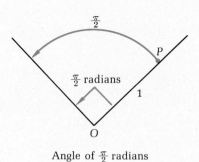

$\frac{\pi}{2}$ radians

Figure 8-2   Angle of $\frac{\pi}{2}$ radians

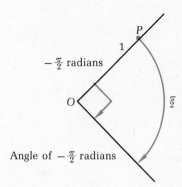

$-\frac{\pi}{2}$ radians

Angle of $-\frac{\pi}{2}$ radians

Similarly, for $0 \leq x \leq 2\pi$, the area A of the sector determined by this angle is

$$\frac{x}{2\pi} \cdot (\text{area of circle of radius } r)$$

and

$$\frac{x}{2\pi} \cdot (\text{area of circle of radius } r) = \frac{x}{2\pi} \cdot \pi r^2$$

that is,

$$A = \tfrac{1}{2}xr^2, \qquad 0 \leq x \leq 2\pi$$

Since $s = xr$, we also have

$$A = \tfrac{1}{2}sr, \qquad 0 \leq s \leq 2\pi r$$

[Note the similarity with: the area of a triangle $= \tfrac{1}{2}$ base $\times$ height.]

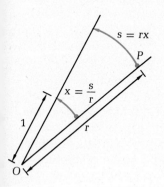

$s = rx$

$x = \dfrac{s}{r}$

Figure 8-3

Example 1

Consider a circle whose radius is 7 inches. Find the arc length subtended by a central angle of $3\pi/2$ radians, and then find the area of the corresponding sector.

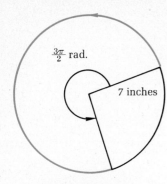

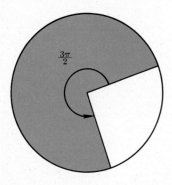

Figure 8-4

Solution

See Figure 8-4. For a central angle of $3\pi/2$ radians, we see that arc length s is

$$s = xr = \frac{3\pi}{2} \cdot 7 = \frac{21\pi}{2} \text{ (inches)}$$

and the area A of the corresponding sector is

$$A = \frac{1}{2} xr^2 = \frac{1}{2} \cdot \frac{3\pi}{2} \cdot 49 = \frac{147\pi}{4} \text{ (square inches)}$$

or

$$A = \frac{1}{2} sr = \frac{1}{2} \cdot \frac{21\pi}{2} \cdot 7 = \frac{147\pi}{4}$$

Example 2

The wheels of a bicycle are 1 foot in diameter. Suppose the bicycle travels at the constant rate of 10 miles per hour. Through how many radians will the wheels have turned at the end of 15 minutes?

Solution

Since the bicycle moves at 10 miles per hour, the wheels will have moved 10 miles in 1 hour; hence the wheels move $\frac{10}{4} = \frac{5}{2}$ miles in 15 minutes (i.e., $\frac{1}{4}$ of an hour); thus the arc length is $\frac{5}{2}$ miles or 13,200 feet. Since the radius r is $\frac{1}{2}$ foot, we see that the number of radians x through which the wheels have turned is

$$x = \frac{s}{r} = \frac{13,200}{\frac{1}{2}} = 26,400$$

The measurement of angles in **degrees** (°) simply amounts to a different choice of unit for measuring the corresponding rotation. Unlike the radian, however, the degree is not mathematically "natural," being merely a matter of historical convention. By definition, there are 360° in a circle. Therefore,

$$2\pi \text{ radians} = 360°$$

and hence

$$1 \text{ radian} = \left(\frac{360}{2\pi}\right)° = \left(\frac{180}{\pi}\right)° \qquad (\approx 57°)$$

and

$$1° = \frac{2\pi}{360} \text{ radians} = \frac{\pi}{180} \text{ radians} \qquad \left(\approx \frac{1}{57} \text{ radian}\right)$$

Example 3

$$45° = 45 \cdot \frac{\pi}{180} \text{ radians} = \frac{\pi}{4} \text{ radians}$$

$$120° = 120 \cdot \frac{\pi}{180} \text{ radians} = \frac{2\pi}{3} \text{ radians}$$

$$405° = 405 \cdot \frac{\pi}{180} \text{ radians} = \frac{9\pi}{4} \text{ radians}$$

$$\frac{\pi}{6} \text{ radians} = \frac{\pi}{6} \cdot \left(\frac{180}{\pi}\right)° = 30°$$

Exercises 8-1

In Exercises 1–8, assuming that $\pi \approx 22/7$, find the radian measure of the angle whose degree measure is:

1. 1.2	**2.** 18
3. 210	**4.** −900
5. 2π (be careful!)	**6.** −282
7. 100π (be careful!)	**8.** −1000

In the remaining problems, use $\pi \approx 3.14$.

In Exercises 9–17, find the degree measure of the angle whose radian measure is:

9. 1.2	**10.** 0.01
11. 2	**12.** −1.2
13. 100	**14.** 0.25
15. 360	**16.** −0.1
17. -14π	**18.** Express 3.5 degrees in radians.

19. Express 4.1 counterclockwise revolutions in radians.

20. Express 6 radians in counterclockwise revolutions.

21. Express 2.3 radians in degrees.

22. A car traveling at 60 miles per hour has tires 30 inches in diameter. Through how many radians does a tire turn in 1 second?

23. Through what angle does the earth rotate about its axis in 30 minutes?

24. Find the area of the sector *OPQ* (see figure) if angle *POQ* is: (a) 65° and $r = 5$; (b) $7\pi/12$ radians and $r = \sqrt{2}$; (c) 4 radians and $r = 3$.

25. Find the length of the arc determined by going counterclockwise 65° from *Q* to *P* in the figure in Exercise 24 if $r = 2$.

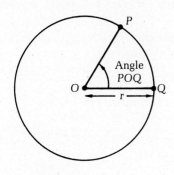

Exercise 24

26. An object is moving on a circle of radius 10 feet at the rate of 5 revolutions per minute. How many feet per minute is it moving on the circle?

27. Suppose we have a circle whose center is O and whose radius is 5 inches. Let A and B be points on the circle, and suppose the angle AOB subtends an arc on the circle whose length is 1 inch. Find the *degree* measure of the angle AOB.

28. What is the radian measure of the smaller of the angles between the hands of a clock at 10:30?

29. The outside border of an athletic field at a curved end is being laid out on a circle. If the surrounding track will curve 54° in 108 feet, what radius is required?

30. A wheel 3 feet in diameter rolls forward a distance of 456 ft. Through what angle does the wheel turn? How many revolutions does it make?

31. A piece of thread 100 inches long is to be wound around a circular spool whose diameter is 3 inches. Ignoring the thickness of the thread, how many turns are necessary to wind the thread?

32. Two wheels, one 4 inches in diameter and the other $\frac{1}{2}$ inch in diameter, are in contact. As the larger wheel turns, it turns the smaller wheel. Assuming there is no slippage, through how many radians has the smaller wheel turned when the large wheel has made 10.4 counterclockwise revolutions?

8-2 SINE AND COSINE

Two fundamental trigonometric functions are **sine** and **cosine,** usually abbreviated to **sin** and **cos.** They are defined as follows: Let O in Figure 8-5 be the origin, let P be the point $(1,0)$, and let x be any number. The rotation of x radians about O, when applied to P, determines a circular path ending at a point P_x, and having arc length $|x|$. By definition,

cos x is the first coordinate of P_x

and

sin x is the second coordinate of P_x

Note that since P_x is on the unit circle, we have

$$(\cos x)^2 + (\sin x)^2 = 1$$

for any number x, by the Theorem of Pythagoras. This fact will be used extensively.

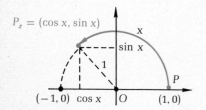

Figure 8-5

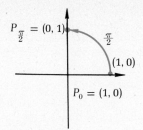

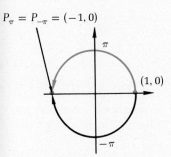

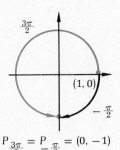

Figure 8-6

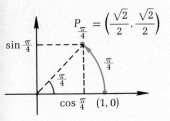

Figure 8-7

Remark "$(\cos x)^2$" and "$(\sin x)^2$" are usually written "$\cos^2 x$" and "$\sin^2 x$," respectively. Hence, we have, for any number x,

$$\sin^2 x + \cos^2 x = 1$$

Example 1

Compute

(a) $\cos 0$ (b) $\sin 0$ (c) $\cos \dfrac{\pi}{2}$ (d) $\sin \dfrac{\pi}{2}$

(e) $\cos \pi$ (f) $\sin \pi$ (g) $\cos \dfrac{3\pi}{2}$ (h) $\sin \dfrac{3\pi}{2}$

Solution

Recall that for each x, $\cos x$ and $\sin x$ are, respectively, the first and the second coordinates of P_x. Refer to Figure 8-6. Thus,

(a) $\cos 0 = 1$, since $\cos 0$ is the first coordinate of $P_0 = (1, 0)$.

(b) $\sin 0 = 0$, since $\sin 0$ is the second coordinate of $P_0 = (1, 0)$.

Similarly,

(c) $\cos \dfrac{\pi}{2} = 0$ and (d) $\sin \dfrac{\pi}{2} = 1$, since $P_{\pi/2} = (0, 1)$.

(e) $\cos \pi = -1$ and (f) $\sin \pi = 0$, since $P_\pi = (-1, 0)$.

(g) $\cos \dfrac{3\pi}{2} = 0$ and (h) $\sin \dfrac{3\pi}{2} = -1$, since $P_{3\pi/2} = (0, -1)$.

Note also that, since $P_\pi = P_{-\pi}$, we have $\cos (-\pi) = -1$ and $\sin (-\pi) = 0$.

Example 2

Observe that $\cos \dfrac{\pi}{4} = \sin \dfrac{\pi}{4}$ (see Figure 8-7). Hence,

$$1 = \cos^2 \frac{\pi}{4} + \sin^2 \frac{\pi}{4} = 2 \cos^2 \frac{\pi}{4}$$

and so

$$\cos \frac{\pi}{4} = \frac{\sqrt{2}}{2} = \sin \frac{\pi}{4}$$

Example 3

Compute $\sin \dfrac{\pi}{6}$ and $\cos \dfrac{\pi}{6}$.

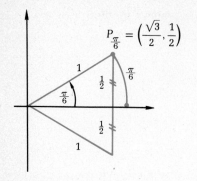

Figure 8-8

Solution

Consider the equilateral triangle in Figure 8-8. Since the horizontal axis bisects the angle at the vertex into two angles of $\pi/6$ radians each, it also bisects the opposite side into two segments of length $\frac{1}{2}$. Hence, the triangle above the axis has sides of lengths $\frac{1}{2}$ and $\sqrt{1 - \left(\frac{1}{2}\right)^2} = \frac{\sqrt{3}}{2}$, respectively. Therefore,

$$\cos \frac{\pi}{6} = \frac{\sqrt{3}}{2} \quad \text{and} \quad \sin \frac{\pi}{6} = \frac{1}{2}$$

□

It is frequently convenient to look at trigonometric functions as functions whose domains consist of *angles* rather than numbers. Hence, *we define* **the cosine of an angle of x radians** *to be the cosine of the number x. A similar definition may be made for the sine.*

Example 4

$$\sin 30° = \sin \left(\frac{\pi}{6} \text{radians}\right) = \sin \frac{\pi}{6} = \frac{1}{2}$$

Most of the computations listed in the following table have already been done. The angles considered here are the so-called standard angles.

Degrees	$-180°$	$-90°$	0	30°	45°	60°	90°	180°	270°	360°
Radians	$-\pi$	$-\dfrac{\pi}{2}$	0	$\dfrac{\pi}{6}$	$\dfrac{\pi}{4}$	$\dfrac{\pi}{3}$	$\dfrac{\pi}{2}$	π	$\dfrac{3}{2}\pi$	2π
sin	0	-1	0	$\dfrac{1}{2}$	$\dfrac{\sqrt{2}}{2}$	$\dfrac{\sqrt{3}}{2}$	1	0	-1	0
cos	-1	0	1	$\dfrac{\sqrt{3}}{2}$	$\dfrac{\sqrt{2}}{2}$	$\dfrac{1}{2}$	0	-1	0	1

In Example 4, we saw that $\sin 30° = \sin \dfrac{\pi}{6} = \dfrac{1}{2}$, but 30° is not the only angle whose sine is $\dfrac{1}{2}$.* The next example

---

*We say that "θ is an angle whose sine is q" if $q = \sin \theta$. This convenient terminology is also used in relation to the cosine and the other trigonometric functions; see Section 8-5.

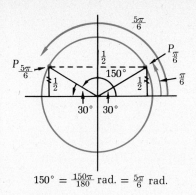

$150° = \frac{150\pi}{180}$ rad. $= \frac{5\pi}{6}$ rad.

Figure 8-9

shows this and, at the same time, illustrates a method for handling such problems.

Example 5

Find two numbers x for which $\sin x = \frac{1}{2}$.

Solution

As we have already observed, $\frac{1}{2}$ is the second coordinate of $P_{\pi/6}$ (see Figure 8-8) and, thus, $\sin\frac{\pi}{6} = \frac{1}{2}$. Notice (Figure 8-9) that another point on the unit circle whose second coordinate is $\frac{1}{2}$ is $P_{\pi-(\pi/6)} = P_{5\pi/6}$. Thus, $5\pi/6$ is also a solution.

Example 6

Find a real number x for which $\cos 7x = \frac{\sqrt{2}}{2}$.

Solution

From the table above we see that $\cos 7x = \sqrt{2}/2$ if $7x = \pi/4$, or, equivalently, $x = \pi/28$. Toward the end of this section, we shall explain how to find other such numbers x.
□

Note that since the unit circle is 2π units long, two angles whose radian measures differ by an integral multiple of 2π determine exactly the same point on that circle. In other words, if x is a number, then

$$P_x = P_{x+2n\pi}$$

for any integer n. See Figure 8-10. Now, since $\cos x$ and $\sin x$ are the coordinates of P_x, it follows that

$$\cos x = \cos(x + 2n\pi) \quad \text{and} \quad \sin x = \sin(x + 2n\pi)$$

Thus, to compute the sine or cosine of a number x with $x < 0$ or $x \geq 2\pi$, one may add to or subtract from x an appropriate multiple of 2π in order to obtain a number between 0 and 2π, and then compute the sine or cosine of *that* number.

Example 7

Compute

(a) $\sin\frac{9\pi}{4}$

(b) $\cos\frac{13\pi}{3}$

(c) $\sin(-690°)$

(d) $\cos(1260°)$

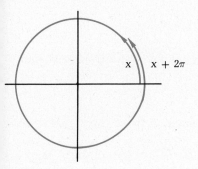

Figure 8-10

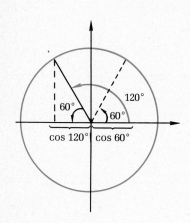

Figure 8-11a

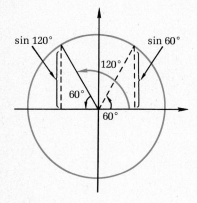

Figure 8-11b

Solution

(a) $\sin \dfrac{9\pi}{4} = \sin\left(2\pi + \dfrac{\pi}{4}\right) = \sin \dfrac{\pi}{4} = \dfrac{\sqrt{2}}{2}$

(b) $\cos \dfrac{13\pi}{3} = \cos\left(4\pi + \dfrac{\pi}{3}\right) = \cos \dfrac{\pi}{3} = \dfrac{1}{2}$

(c) $\sin -690° = \sin\left(-690° + 720°\right) = \sin 30° = \dfrac{1}{2}$

(d) $\cos 1260° = \cos\left(3 \cdot 360° + 180°\right) = \cos 180° = -1$

□

To facilitate further discussion, we shall say that *an angle of x radians is in a given quadrant if the corresponding point* P_x *is in that quadrant.** For example, the angles of $\pi/4$, $3\pi/4$, $-3\pi/4$, and $-\pi/4$ radians are in the first, second, third, and fourth quadrants, respectively. We shall next see that to compute the sine or cosine of an angle in any quadrant, *it is sufficient to compute the sine or cosine of an appropriate angle in the first quadrant.* This is very important in practice, because trigonometric tables list only the sines and cosines of angles in the first quadrant.

Example 8

Compute: (a) $\cos 120°$, (b) $\sin 120°$, (c) $\cos 330°$, (d) $\sin 330°$, and (e) $\cos(-135)$.

Solution

(a) Refer to Figure 8-11(a). It is clear from the figure that $\cos 120°$ is the same as $-\cos 60°$, and hence $\cos 120° = -1/2$. To answer (b), we see from Figure 8-11(b) that $\sin 120°$ is exactly the same as $\sin 60°$, and thus $\sin 120° = \sqrt{3}/2$.
(c) We shall now find $\cos 330°$. Notice from Figure 8-12(a) that $\cos 330° = \cos 30° = \sqrt{3}/2$, whereas in (d), to find $\sin 330°$, we see from Figure 8-12(b) that $\sin 330° = -\sin 30° = -1/2$.
(e) To find $\cos(-135°)$, we see from Figure 8-13 that $\cos(-135°) = -\cos 45° = -\sqrt{2}/2$.

□

It is possible to generalize the results of Example 8 to obtain: for any real number x,

1. $\cos(\pi - x) = -\cos x$ and $\sin(\pi - x) = \sin x$
2. $\cos(x \pm \pi) = -\cos x$ and $\sin(x \pm \pi) = -\sin x$
3. $\cos(-x) = \cos x$ and $\sin(-x) = -\sin x$

An alternative way to compute the sine and cosine of angles larger than 90° is to use formulas 1, 2, and 3. We shall illustrate this in the next example.

*See page 45.

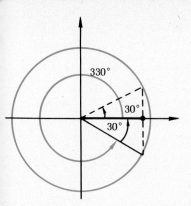

Figure 8-12a

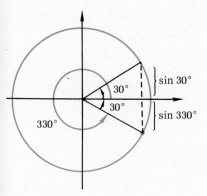

Figure 8-12b

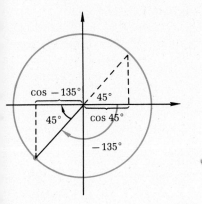

Figure 8-13

Example 9

(a) $\cos(-30°) = \cos 30° = \dfrac{\sqrt{3}}{2}$

(b) $\cos \dfrac{5\pi}{4} = \cos\left(\pi + \dfrac{\pi}{4}\right) = -\cos\dfrac{\pi}{4} = -\dfrac{\sqrt{2}}{2}$

(c) $\sin(-240°) = \sin(-240° + 360°) = \sin(120°)$

$$= \sin(180° - 60°) = \sin 60° = \dfrac{\sqrt{3}}{2}$$

or

$$\sin(-240°) = -\sin 240° = -\sin(180° + 60°)$$

$$= -(-\sin 60°) = \sin 60° = \dfrac{\sqrt{3}}{2}$$

Example 10

Use the tables to check (rounded off to two decimal places)

(a) $\sin 123° \approx .84$ (b) $\cos 261° \approx -.16$

(c) $\cos(-13°) \approx .97$ (d) $\sin\dfrac{25}{7}\pi \approx -.97$

(e) $\cos 23.1 \approx \cos 4.26 = -\cos 1.12 \approx -.44$

Exercises 8-2

1. Let $f(x) = \cos 2x$. Compute $f(0)$, $f\left(\dfrac{\pi}{4}\right)$, $f\left(\dfrac{\pi}{6}\right)$, $f\left(\dfrac{\pi}{8}\right)$, $f\left(\dfrac{\pi}{3}\right)$, $f\left(\pi + \dfrac{\pi}{8}\right)$, $f\left(\dfrac{-\pi}{2}\right)$, $f\left(2\pi + \dfrac{\pi}{3}\right)$, $f(10\pi)$, $f(-10\pi)$.

2. Compute $\cos 315°$.

3. Let $f(x) = \sin x - \cos x$. Find all zeros of f.

4. Designate which of the following numbers are positive: $\cos 2$, $\sin 5$, $\cos 20$, $\sin 17$.

5. Find $\cos\left(-\dfrac{\pi}{6}\right)$ and $\sin\left(-\dfrac{\pi}{6}\right)$.

6. Find $\cos\dfrac{5\pi}{2}$ and $\sin\dfrac{5\pi}{2}$.

7. Find $\sin\dfrac{2\pi}{3}$ and $\cos\dfrac{2\pi}{3}$.   **8.** Find $\cos\dfrac{5\pi}{6}$ and $\sin\dfrac{5\pi}{6}$.

9. Find $\sin\dfrac{25\pi}{4}$.   **10.** Find $\cos\dfrac{43\pi}{4}$.

11. Find $\sin 930°$.   **12.** Find $\cos(-750°)$.

In Exercises 13–16, answer true or false (and justify). For every number x:

13. $\sin\left(\dfrac{3\pi}{2} + x\right) = -\cos x$ **14.** $\sin\left(\dfrac{3\pi}{2} - x\right) = -\cos x$

15. $\cos\left(\dfrac{3\pi}{2} + x\right) = \sin x$ **16.** $\cos\left(\dfrac{3\pi}{2} - x\right) = -\sin x$

In Exercises 17–24, use the tables on pages 462–65 and take π to be 3.14; find:

17. $\sin 1$ **18.** $\cos 1$

19. $\sin 5$ **20.** $\cos 5$

21. $\sin 30$ **22.** $\cos 30$

23. $\sin\left(\dfrac{\pi}{2} + 5\right)$ **24.** $\cos\left(2\pi + 1\right)$

In Exercises 25–29, use tables, if necessary, to find a number t in $[0, \pi/2]$ for which:

25. $\sin t = \frac{1}{2}$ **26.** $\cos t = .15$

27. $\sin t = .05$ **28.** $\sin\left(2t + 1\right) = \dfrac{\sqrt{3}}{2}$

29. $\cos\left(3t - 2\right) = .8021$

30. (a) If $0 < x < \dfrac{\pi}{2}$ and $\sin x = \dfrac{1}{4}$, find $\cos x$.

    (b) If $\dfrac{\pi}{2} < x < \pi$ and $\cos x = -\dfrac{7}{10}$, find $\sin x$.

In Exercises 31–36, find all numbers x for which:

31. $\sin x = \dfrac{\sqrt{3}}{2}$ **32.** $\cos x = \frac{1}{2}$

33. $\sin x = -\frac{1}{2}$ **34.** $\cos x = -\dfrac{\sqrt{3}}{2}$

35. $\sin 3x = \dfrac{\sqrt{2}}{2}$ **36.** $\cos\left(\dfrac{x + \pi}{2}\right) = \dfrac{1}{2}$

Answer Exercises 37–40 true or false.

37. $\sin 60° < 2 \sin 30°$

38. $\sin 30° + \sin 60° > \sin 90°$

39. $2 \sin 45° \cos 45° < \sin 90°$

40. $\cos 210° < \cos 150° \cos 60° - \sin 150° \sin 60°$

In Exercises 41–44, find all real numbers x for which:

41. $\cos x = \sin x$ **42.** $\cos x = -\sin x$

43. $\cos x = \cos\left(-x\right)$ **44.** $\sin x = \sin\left(-x\right)$

45. Find all real numbers x for which $\sin\left(\pi - x\right) = \cos x$.

46. True or false: $\cos^2\left(a^2 - b^2\right) + \sin^2\left(b^2 - a^2\right) = 1$ for a, b any real numbers?

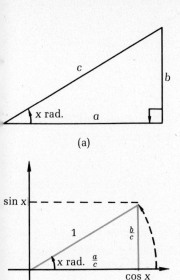

(a)

(b)

Figure 8-14

8-3 APPLICATIONS TO RIGHT TRIANGLES

A useful interpretation of the sine or cosine of an angle whose radian measure is between 0 and $\pi/2$ (i.e., of an *acute angle*) is arrived at through a consideration of *similar right triangles,* as follows: A right triangle, one of whose angles is x radians, as in Figure 8-14(a), is similar to a right triangle with hypotenuse 1 unit and located as in Figure 8-14(b). It follows that

$$\cos x = \frac{a}{c} = \frac{\text{length of the adjacent side}}{\text{length of the hypotenuse}}$$

and

$$\sin x = \frac{b}{c} = \frac{\text{length of the opposite side}}{\text{length of the hypotenuse}}$$

Before going on, it is useful to recall some basic facts about 45°–45°–90° and 30°–60°–90° triangles; such triangles are shown in Figure 8-15. It is a standard exercise in plane geometry to show that the three lengths indicated in each triangle are in the correct proportions for the given triangles. From Figure 8-15 and our previous observation that sin x and cos x can be given in terms of the ratios of the lengths of the sides of a right triangle, we have

$$\cos 45° = \frac{1}{\sqrt{2}} \qquad \sin 45° = \frac{1}{\sqrt{2}}$$

$$\cos 30° = \frac{\sqrt{3}}{2} \qquad \sin 30° = \frac{1}{2}$$

$$\cos 60° = \frac{1}{2} \qquad \sin 60° = \frac{\sqrt{3}}{2}$$

These results were noted previously in Examples 2 and 3 of Section 8-2.

Example 1

One of the sides of a right triangle is 5 inches long, and the angle opposite that side is 30°. Find the lengths of the other sides.

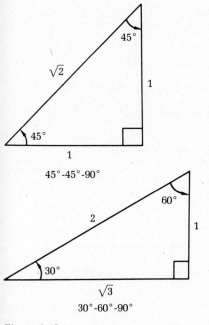

45°-45°-90°

30°-60°-90°

Figure 8-15

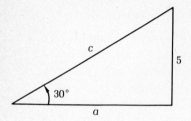

Figure 8-16

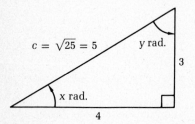

Figure 8-17

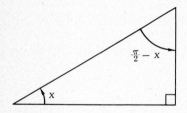

Figure 8-18

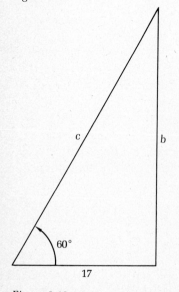

Figure 8-19

Solution

From Figure 8-16, we see that $5/c = \sin 30° = \frac{1}{2}$. Hence, $c = 10$ and $a = \sqrt{100 - 25} = \sqrt{75} = 5\sqrt{3}$ (by the Theorem of Pythagoras).

Example 2

Find the sines and cosines of the acute angles of the right triangle whose perpendicular sides are 3 and 4 feet long.

Solution

From Figure 8-17, we see that $\sin x = \frac{3}{5}$, $\cos x = \frac{4}{5}$, $\sin y = \frac{4}{5}$, and $\cos y = \frac{3}{5}$.

□

Since the sum of the angles of any triangle is an angle of π radians, the sum of the two acute angles of a *right triangle* is an angle of $\pi/2$ radians. Hence, if the radian measure of one of the acute angles is x, then the radian measure of the other is $(\pi/2) - x$ (see Figure 8-18). Furthermore, the sine of one is the cosine of the other (since the side opposite one is the side adjacent to the other). Thus we arrive at another result to be added to the list following Example 8:

$$4. \quad \cos\left(\frac{\pi}{2} - x\right) = \sin x \quad \text{and} \quad \sin\left(\frac{\pi}{2} - x\right) = \cos x$$

(Again, the reader is asked to verify that this relationship holds for all real numbers x.)

Example 3

If the shadow cast by a tree is 17 feet long when the elevation of the sun above the earth is 60°, how tall is the tree?

Solution

See Figure 8-19. We have $\dfrac{17}{c} = \cos 60° = \dfrac{1}{2}$, and therefore $c = 34$. Hence,

$$\frac{b}{34} = \sin 60° = \frac{\sqrt{3}}{2} \quad \text{and} \quad b = 17\sqrt{3}$$

or

$$b = \sqrt{34^2 - 17^2} = \sqrt{2^2 \cdot 17^2 - 17^2} = \sqrt{17^2(2^2 - 1)} = 17\sqrt{3}$$

Thus, the tree is $17\sqrt{3}$ feet tall.

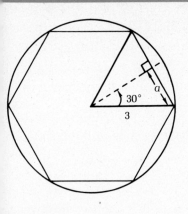

Figure 8-20

Example 4

Find the perimeter of a 6-sided regular polygon inscribed in a circle of radius 3.

Solution

Refer to Figure 8-20. Suppose each edge of the polygon has length $2a$. Since there are 6 edges, we see that each edge of the triangle subtends an angle of 60° (that is, $\frac{1}{6}$ of 360°). From the right triangle in Figure 8-20, we see that

$$\frac{1}{2} = \sin 30° = \frac{a}{3}$$

and hence $a = \frac{3}{2}$. The perimeter of the polygon is, therefore,

$$6 \cdot 2a = 6 \cdot 2 \cdot \frac{3}{2} = 18$$

Exercises 8-3

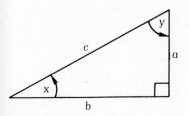

Exercises 1–6

In Exercises 1–6, refer to the accompanying figure and find:

1. a given that $y = 30°$ and $c = 5$

2. y given that $b = \sqrt{3}$ and $c = 2$

3. x given that $b = 4$ and $c = 8$

4. a given that $y = 60°$ and $b = 1$

5. a given that $y = 10°$ and $b = 3$

6. y given that $b = 3$ and $a = 4$

7. Find the area of a 5-sided regular polygon inscribed in a circle of radius 4.

8. A regular polygon with 10 sides is inscribed in a circle whose radius is 10 inches. What is the area of the polygon? Round off your answer to one decimal place.

9. Approximate π by finding the length of a 12-sided regular polygon inscribed in a circle of radius 1.

10. A point P is located 10 feet from the base of a vertical post. The angle at P between the ground and the top of the post is found to be 72°. (a) How far is the top of the pole from the point P? (b) How tall is the post?

8-4 PERIODIC FUNCTIONS. GRAPHS OF SINE AND COSINE

Recall that for any real number x, we have sin $(x + 2\pi)$ = sin x, and cos $(x + 2\pi)$ = cos x. This illustrates a property known as **periodicity.**

In general, suppose f is a function and p a positive real number with the property that whenever x is in the domain of f, so is $x + p$, and

$$f(x + p) = f(x)$$

Then f is called **periodic;** and if p is the smallest positive number for which this relationship holds, then p is called the **period** of f.

Example 1

Sin and cos are periodic functions of period 2π.

We now turn our attention to graphs of periodic functions. Let f be a function with period p. Since for *every* x in the domain of f, $x + p$ is also in the domain of f and $f(x) = f(x + p)$, it follows that for every x in the domain of f the numbers $x + 2p$, $x + 3p$, and so on, are also in the domain of f and $f(x) = f(x + p) = f(x + 2p) = f(x + 3p) = \cdots$. That is, for every natural number n we have

$$f(x) = f(x + np)$$

Thus, for every x in the domain of f, the points $(x + np, f(x + np))$, $n = 1, 2, 3, \ldots$, on the graph of f are at precisely the same height above (or below) the horizontal axis as the point $[x, f(x)]$, and each is p units away from its predecessor. See Figure 8-21. Note, therefore, that the graph of f is itself periodic, in the sense that the entire graph is composed of a series of replicas of any portion of it corresponding to an interval of length p on the horizontal axis.

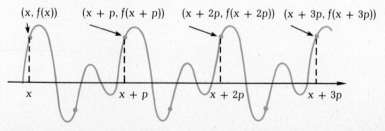

Figure 8-21

Example 2

Graph $x \to \cos x$.

Solution

Since 2π is the period of this function, it is sufficient to graph it for x between $-\pi$ and π. Referring once more to the unit circle or to the table in Section 8-2, we see that as increasing numbers x are chosen between 0 and $\pi/2$, cos x decreases from 1 to 0. For example,

$$\cos 0 = 1, \quad \cos\frac{\pi}{6} = \frac{\sqrt{3}}{2}, \quad \cos\frac{\pi}{4} = \frac{\sqrt{2}}{2}, \quad \cos\frac{\pi}{3} = \frac{1}{2},$$

$$\cos\frac{\pi}{2} = 0$$

See the upper part of the colored curve in Figure 8-22.

From this piece, we can obtain the graph for x between $\pi/2$ and π by using the fact that $\cos x = -\cos(\pi - x)$ (see Figure 8-22). Lastly, since $\cos(-x) = \cos x$, we obtain the graph for x between $-\pi$ and 0. By periodicity, the graph can now be sketched over any interval. See Figure 8-23.

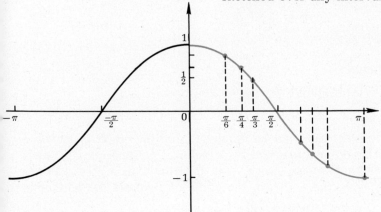

Figure 8-22

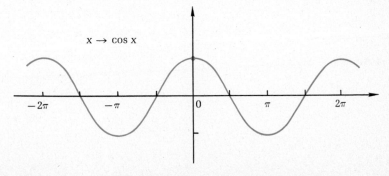

Figure 8-23

Example 3

Graph $x \to \sin x$.

Solution

Since for all real numbers x we have

$$\sin x = \cos\left(\frac{\pi}{2} - x\right) = \cos\left(x - \frac{\pi}{2}\right)$$

the graph of **sin** is a shift by $\pi/2$ (to the right) of the graph of **cos**. See Figure 8-24.

In the following examples, we sketch the graphs of some shifts and stretches of sin and cos.

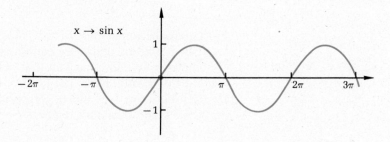

Figure 8-24

Example 4

Let $f(x) = \sin 4x$. Sketch the graph of f.

Solution

The graph of f can be obtained from the graph of sin by a horizontal compression by a factor of $\frac{1}{4}$. Consequently, since the graph of sin crosses the horizontal axis at the points $n\pi$, with n any integer, the graph of f crosses the horizontal axis at the points $\frac{n\pi}{4}$. See Figure 8-25.

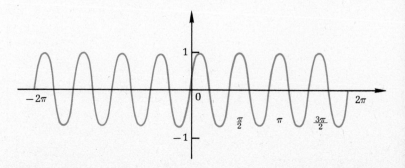

Figure 8-25

Example 5

Let $f(x) = 3 \sin 2x$. Sketch the graph of f.

Solution

The graph of f is obtained from the graph of sin by a vertical stretch by a factor of 3 and a horizontal compression by a factor of $\frac{1}{2}$. Another way to graph f is to note that $\sin t = 0$, if and only if $t = 0$, $\pm\pi$, $\pm 2\pi$, ... And so we see that $3 \sin 2x = 0$, if and only if $2x = 0$, $\pm\pi$, $\pm 2\pi$, ..., that is, if and only if $x = 0$, $\pm\pi/2$, $\pm\pi$, ... Since $\sin t$ is largest (and equals 1) when $t = \pi/2$, $\pm 2\pi + \pi/2$, $\pm 4\pi + \pi/2$, ..., we see that $3 \sin 2x$ is largest (and equals 3) when $2x = \pi/2$, $\pm 2\pi + \pi/2$, $\pm 4\pi + \pi/2$, ..., that is, when $x = \pi/4$, $\pm\pi + \pi/4$, $\pm 2\pi + \pi/4$, ... See Figure 8-26.

□

In general, if a, b, and c are real numbers, and if $f(x) = c \sin (ax + b)$, then the graph of f can be obtained from the graph of sin by stretches, shifts, and possible reflections. Hence, as the next two examples illustrate, the graph of f can be obtained by locating its zeros (i.e., the crossing points) and its high and low points.

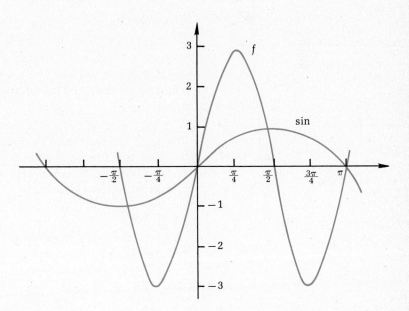

Figure 8-26

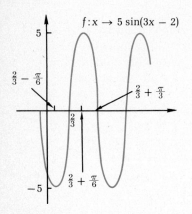

$f : x \rightarrow 5 \sin(3x - 2)$

$\frac{2}{3} - \frac{\pi}{6}$

$\frac{2}{3} + \frac{\pi}{3}$

$\frac{2}{3}$

$\frac{2}{3} + \frac{\pi}{6}$

Figure 8-27

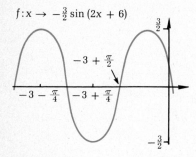

$f : x \rightarrow -\frac{3}{2} \sin(2x + 6)$

$-3 + \frac{\pi}{2}$

$-3 - \frac{\pi}{4}$ $-3 + \frac{\pi}{4}$

$\frac{3}{2}$

$-\frac{3}{2}$

Figure 8-28

Example 6

Let $f(x) = 5 \sin(3x - 2)$. Sketch the graph of f.

Solution

To find the zeros of f, note that $\sin t = 0$ if $t = n\pi$, where n is any integer; thus, $5 \sin(3x - 2) = 0$ if $3x - 2 = n\pi$, or, equivalently, $x = \frac{2}{3} + n\pi/3$, with n any integer. Furthermore, since $\sin t$ is positive for t between 0 and π, attaining its maximum of 1 at the midpoint $\pi/2$, it follows that $f(x)$ is positive for $3x - 2$ between 0 and π, that is, for x between $\frac{2}{3}$ and $\frac{2}{3} + \pi/3$, attaining its maximum of 5 midway between $\frac{2}{3}$ and $\frac{2}{3} + \pi/3$, that is, at $\frac{2}{3} + \pi/6$. See Figure 8-27. Another way to obtain the graph is to note that the graph of

$$x \rightarrow 5 \sin(3x - 2) = 5 \sin 3(x - \tfrac{2}{3})$$

is the graph of sin stretched horizontally by a factor of $\frac{1}{3}$ and then shifted by $\frac{2}{3}$ units and stretched vertically by a factor of 5.

Example 7

Let $f(x) = -\frac{3}{2} \sin(2x + 6)$. Sketch the graph of f.

Solution

Note that $\sin(2x + 6) = 0$ for $2x + 6 = n\pi$ or, equivalently, $x = -3 + n\pi/2$, n any integer. Also, $-\frac{3}{2} \sin(2x + 6)$ is negative for x between -3 and $-3 + \pi/2$ (that is, for $2x + 6$ between 0 and π), attaining its minimum of $-\frac{3}{2}$ at $-3 + \pi/4$. See Figure 8-28.

□

We now illustrate how to graph the function $x \rightarrow \sin[g(x)]$, where g is a function. Again, the graph will be similar to the graph of sin and can be sketched by finding crossing points and high and low points.

Exercises 8-4

In Exercises 1–10, graph f and give its period.

1. $f(x) = \sin 2x$ **2.** $f(x) = \cos 6x$

3. $f(x) = \sin(x - 2)$ **4.** $f(x) = \sin \dfrac{x}{2}$

5. $f(x) = 2 + \sin x$ **6.** $f(x) = \sin \dfrac{x}{6}$

7. $f(x) = \cos \dfrac{x}{8}$ **8.** $f(x) = -2 \sin(3x + 1)$

9. $f(x) = -\sin(2x - 4)$ **10.** $f(x) = 3 \sin\left(\dfrac{x}{2} + 1\right)$

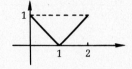

11. Suppose the graph of f in $[0, 2]$ is

and f is periodic with period 2. Graph f in $[-6, 6]$.

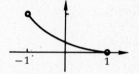

12. Suppose the graph of f in $(-1, 1)$ is

and f is periodic with period 2. Graph f in $(-3, 3)$.

13. Let f be a periodic function, with period 2. Suppose $f(1) = -3$ and $f(2) = 1$. Compute $2f(-11) - 5f(26)$.

14. If f is a function with the property that for all numbers x, $f(x + 2) = f(x)$, $f(-x) = f(x)$, and $f(\frac{1}{2}) = 3$, find:

(a) $f(\frac{9}{2})$,   (b) $f(\frac{7}{2})$,   (c) $f(-\frac{9}{2}) + f(-\frac{7}{2})$.

In Exercises 15–24, graph the given functions.

15. $f(x) = -2 \cos(3x + 1)$ **16.** $f(x) = \sin |x|$

17. $f(x) = x^2 \sin x$ **18.** $f(x) = |\sin x|$

19. $f(x) = 1 + 3 \cos\left(x - \dfrac{\pi}{2}\right)$ **20.** $f(x) = 1 - 2 \sin(3x + 1)$

21. $f(x) = \sin(x - \pi)$ ◊**22.** $f(x) = \sin^2 x$

◊**23.** $f(x) = \sqrt{\sin x}$ ◊**24.** $f(x) = \dfrac{1}{\sin x}$

◊**25.** Let $f(x) = \log_{10} x - \sin x$, with $x > 0$. How many zeros does f have? (*Hint:* Consider separately the graphs of $\log_{10}$ and sin.)

◊**26.** Let $f(x) = \sin x - \sin 2x$. Find the number of zeros of f in the interval $[-\pi, \pi]$.

◊**27.** Let $f(x) = (x/100) - \sin x$. How many zeros does f have?

8-5 THE ADDITION FORMULAS

Many computations involving sin and cos are based on the **addition formulas** given in the following theorem.

Theorem

If x and y are real numbers, then

$$\text{(a) } \cos (x \pm y) = \cos x \cos y \mp \sin x \sin y$$

and

$$\text{(b) } \sin (x \pm y) = \sin x \cos y \pm \cos x \sin y$$

Proof

(a) Referring to Figure 8-29, note that the distance between the points $(\cos y, \sin y)$ and $(\cos x, \sin x)$ is the same as between $(\cos (y - x), \sin (y - x))$ and $(\cos 0, \sin 0) = (1, 0)$, since the corresponding arcs have the same length. Hence.

$$(\cos y - \cos x)^2 + (\sin y - \sin x)^2$$
$$= [\cos (y - x) - 1]^2 + \sin^2 (y - x)$$

and hence,

$$\cos^2 y - 2 \cos y \cos x + \cos^2 x + \sin^2 y$$
$$-2 \sin y \sin x + \sin^2 x$$
$$= \cos^2 (y - x) - 2 \cos (y - x) + 1 + \sin^2 (y - x)$$

Using the facts that $\cos^2 t + \sin^2 t = 1$ for all real numbers t and that $\cos (y - x) = \cos (x - y)$, and simplifying, we obtain

$$\cos (x - y) = \cos x \cos y + \sin x \sin y$$

If we now replace y by $-y$, we obtain

$$\cos [x - (-y)] = \cos x \cos (-y) + \sin x \sin (-y)$$

and recalling that $\cos (-y) = \cos y$ and $\sin (-y) = -\sin y$, we see that

$$\cos (x + y) = \cos x \cos y - \sin x \sin y$$

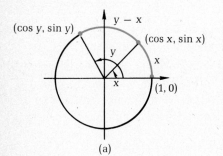

(a)

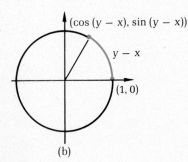

(b)

Figure 8-29

(b) Here we use the fact (see Section 8-2) that for all real numbers t, $\sin t = -\cos(t + \pi/2)$. Hence, if x and y are numbers, then

$$\sin(x + y) = -\cos\left(x + y + \frac{\pi}{2}\right)$$

$$= -\cos\left[x + \left(y + \frac{\pi}{2}\right)\right]$$

$$= -\left[\cos x \cos\left(y + \frac{\pi}{2}\right) - \sin x \sin\left(y + \frac{\pi}{2}\right)\right]$$

Since $\cos(y + \pi/2) = -\sin y$ and $\sin(y + \pi/2) = \cos y$, we have

$$\sin(x + y) = \sin x \cos y + \cos x \sin y$$

Again, replacing y with $-y$ and noting that $\cos(-y) = \cos y$ and $\sin(-y) = -\sin y$, we obtain

$$\sin(x - y) = \sin x \cos y - \cos x \sin y$$

Example 1

Show that for any number x,

$$\sin(x + \pi/6) = \tfrac{1}{2}(\sqrt{3}\sin x + \cos x)$$

Solution

Using the addition formula for sine, we see that for any number x,

$$\sin\left(x + \frac{\pi}{6}\right) = \sin x \cos\frac{\pi}{6} + \cos x \sin\frac{\pi}{6}$$

$$= \frac{\sqrt{3}}{2}\sin x + \frac{1}{2}\cos x$$

$$= \tfrac{1}{2}(\sqrt{3}\sin x + \cos x)$$

Example 2

$$\cos\frac{\pi}{12} = \cos\left(\frac{\pi}{3} - \frac{\pi}{4}\right)$$

$$= \left(\cos\frac{\pi}{3}\right)\left(\cos\frac{\pi}{4}\right) + \left(\sin\frac{\pi}{3}\right)\left(\sin\frac{\pi}{4}\right)$$

$$= \frac{1}{2}\cdot\frac{1}{\sqrt{2}} + \frac{\sqrt{3}}{2}\cdot\frac{1}{\sqrt{2}}$$

$$= \frac{1 + \sqrt{3}}{2\sqrt{2}}$$

Example 3

Find $\sin 195°$ without using tables.

Solution

$$\sin 195° = \sin (135° + 60°)$$
$$= \sin 135° \cos 60° + \sin 60° \cos 135°$$
$$= \frac{1}{\sqrt{2}} \cdot \frac{1}{2} + \frac{\sqrt{3}}{2}\left(-\frac{1}{\sqrt{2}}\right) = \frac{1 - \sqrt{3}}{2\sqrt{2}}$$

Example 4

Show that $\sin (x + 60°) - \cos (x + 30°) = \sin x$ for any angle x (in degrees).

Solution

$$\sin (x + 60°) - \cos (x + 30°) = \sin x \cos 60°$$
$$+ \cos x \sin 60° - \cos x \cos 30° + \sin x \sin 30°$$
$$= \frac{1}{2} \sin x + \frac{\sqrt{3}}{2} \cos x - \frac{\sqrt{3}}{2} \cos x + \frac{1}{2} \sin x$$
$$= \sin x$$

Example 5

Find the degree measure of x if $0° \leq x \leq 360°$ and $\sin (x + 60°) = 2 \sin x$.

Solution

$$\sin (x + 60°) = \sin x \cos 60° + \sin 60° \cos x$$
$$= \frac{1}{2} \sin x + \frac{\sqrt{3}}{2} \cos x$$

Thus, x is a solution if and only if

$$\frac{1}{2} \sin x + \frac{\sqrt{3}}{2} \cos x = 2 \sin x$$
$$\frac{\sqrt{3}}{2} \cos x = \frac{3}{2} \sin x$$
$$\cos x = \sqrt{3} \sin x$$

Thus, after squaring, we obtain

$$\cos^2 x = 3 \sin^2 x^*$$

$$1 - \sin^2 x = 3 \sin^2 x$$

$$4 \sin^2 x - 1 = 0$$

and finally

$$(2 \sin x + 1)(2 \sin x - 1) = 0$$

Thus, $\sin x = \frac{1}{2}$ or $\sin x = -\frac{1}{2}$; hence, $x = 30°$, $150°$, $210°$, or $330°$ are possible solutions. Checking, we find that the solutions are $30°$ and $210°$.

□

From the addition formulas, we can derive the following two important corollaries, called the **double-angle formulas** for sin and cos:

For any number x,

$$\cos 2x = \cos^2 x - \sin^2 x$$

and

$$\sin 2x = 2 \sin x \cos x$$

To see this, let $x = y$ in the addition formulas. Note, further, that since $\cos^2 x = 1 - \sin^2 x$, and $\sin^2 x = 1 - \cos^2 x$, we also have

$$\textbf{cos } 2x = 1 - 2 \sin^2 x = 2 \cos^2 x - 1$$

It follows from this that

$$\cos^2 x = \frac{1 + \cos 2x}{2} \quad \text{and} \quad \sin^2 x = \frac{1 - \cos 2x}{2}$$

Equivalently, replacing x by x/2, we obtain

$$\cos^2 \frac{x}{2} = \frac{1}{2}(1 + \cos x) \quad \text{and} \quad \sin^2 \frac{x}{2} = \frac{1}{2}(1 - \cos x)$$

These are the so-called **half-angle formulas.**

Example 6

Find $\sin 15°$.

---

*We must now be careful, since squaring may give rise to numbers that are not solutions of the original equation.

Solution

From the half-angle formula,

$$\sin^2 15° = \sin^2 \frac{30}{2} = \frac{1}{2}(1 - \cos 30°) = \frac{1}{2}\left(1 - \frac{\sqrt{3}}{2}\right)$$

Thus,

$$\sin^2 15° = \frac{(2 - \sqrt{3})}{4}$$

and thus,

$$\sin 15° = \frac{1}{2}\sqrt{2 - \sqrt{3}}$$

Exercises 8-5

In Exercises 1–6, use the addition formula to compute:

1. $\cos(45° + 30°)$
2. $\sin(45° - 30°)$
3. $\cos\left(\frac{2}{3}\pi + \frac{\pi}{4}\right)$
4. $\sin\left(\frac{2}{3}\pi - \frac{\pi}{4}\right)$
5. $\sin 285°$ *(Hint: 285 = 240 + 45.)*
6. $\sin 195°$
7. Find all numbers x such that $\cos x \cos 2x - \sin x \sin 2x = 0$.
8. Find $\cos 75°$ and $\sin 15°$.

Answer Exercises 9–11 true or false, and justify your answer. For any angle x:

9. $\cos(x - 30°) - \cos(x + 30°) = \sin x$
10. $\cos(x + 45°) + \cos(x - 45°) = \sqrt{2}\cos x$
11. $\sin(x + 30°) + \sin(x - 30°) = \sqrt{3}\sin x$
12. Show that for any number x,

$$\sin\left(x + \frac{\pi}{4}\right)\cos\left(x + \frac{\pi}{4}\right) = \frac{1}{2}[\cos^2 x - \sin^2 x]$$

13. Find all x, y in $[0, 2\pi]$ for which

$$\sin(x + y)\cos y - \cos(x + y)\sin y = \tfrac{1}{2}$$

14. Find all numbers x in $[0, 2\pi]$ for which

$$\sin\left(x + \frac{\pi}{6}\right) = \frac{1}{2}\cos x + \frac{\sqrt{3}}{4}$$

15. Show that for any number x,

$$\sin^2 5x = \frac{1 - \cos 10x}{2}$$

16. Show that for any numbers x and y,

$$\sin(x + y)\sin(x - y) = \sin^2 x - \sin^2 y$$

17. Show that $\cos(x + y)\cos(x - y) = \cos^2 x - \sin^2 y$ for any real numbers x and y.

18. Show that

$$\frac{\sin^2 2x}{1 + \cos 2x} = 2\sin^2 x, \text{ for } 1 + \cos 2x \neq 0$$

19. Find $\cos x$ if $\sin x + \cos x = \frac{2}{3}$. (*Hint:* Note that $\sin x = \frac{2}{3} - \cos x$, and square.)

20. Find $\sin x$ if $2\cos x + \sin x = -1$.

21. Find all numbers x in $[0, 2\pi]$ for which

$$\cos x - \sqrt{3}\sin x = 1$$

8-6 SOLVING EQUATIONS

One interesting application of the double-angle and half-angle formulas is in solving certain types of equations involving terms like sin 2x, cos 2x, sin 3x, cos 3x, etc. The following examples illustrate such applications.

Example 1

Show that for any real number x, $\sin 3x = 3\sin x - 4\sin^3 x$.

Solution

Let x be a number. Then,

$$
\begin{aligned}
\sin 3x &= \sin(2x + x) \\
&= \sin 2x \cos x + \cos 2x \sin x \quad [\textit{addition formulas}] \\
&= (2\sin x \cos x)\cos x + (1 - 2\sin^2 x)\sin x \\
&= 2\sin x \cos^2 x + \sin x - 2\sin^3 x \\
&= 2\sin x(1 - \sin^2 x) + \sin x - 2\sin^3 x \\
&= 3\sin x - 4\sin^3 x
\end{aligned}
$$

Example 2

Find all numbers x in $(0, 2\pi)$ for which $\cos 2x - 5 \sin x + 2 = 0$.

Solution

First, rewrite the problem so that only the sine function is present. To do this, recall (using the double-angle formula) that for every number x, $\cos 2x = 1 - 2 \sin^2 x$. Thus, we must find all x in $(0, 2\pi)$ for which

$$(1 - 2 \sin^2 x) - 5 \sin x + 2 = 0$$

or, equivalently,

$$2 \sin^2 x + 5 \sin x - 3 = 0$$

or

$$(2 \sin x - 1)(\sin x + 3) = 0$$

Since there are no numbers x for which $\sin x = -3$, we see that x is a solution if and only if

$$\sin x = \tfrac{1}{2}$$

that is, $x = \pi/6$ or $5\pi/6$.

Example 3

Find all numbers x for which $\sin 3x + 2 \cos 2x - 2 = 0$.

Solution

Recall that for any real number x,

$$\sin 3x = 3 \sin x - 4 \sin^3 x$$

(see Example 1) and

$$2 \cos 2x = 2(1 - 2 \sin^2 x) = 2 - 4 \sin^2 x$$

Thus, x is a solution if and only if

$$(3 \sin x - 4 \sin^3 x) + (2 - 4 \sin^2 x) - 2 = 0$$

or, equivalently,

$$-4 \sin^3 x - 4 \sin^2 x + 3 \sin x = 0$$
$$\sin x(4 \sin^2 x + 4 \sin x - 3) = 0$$
$$\sin x(2 \sin x + 3)(2 \sin x - 1) = 0$$

Now $\sin x = 0$ if and only if $x = n\pi$ for some integer n; there is no number x for which $\sin x = -\frac{3}{2}$, and $\sin x = \frac{1}{2}$ if and only if $x = (\pi/6) + 2m\pi$ or $x = (5\pi/6) + 2s\pi$, where m and s are integers. Thus, the solutions are

$$n\pi, \quad \frac{\pi}{6} + 2m\pi, \quad \text{and} \quad \frac{5\pi}{6} + 2s\pi$$

for any integers m, n, and s.

□

In the following two examples the zeros of a function are approximated by considering the graph of a related polynomial function.

◇**Example 4**

Find all real numbers x in $(0, 2\pi)$ for which

$$\sin 2x + 3 \sin x - 1 = 0$$

Solution

For any real number x,

$$\sin 2x + 3 \sin x - 1 = 2 \sin x \cos x + 3 \sin x - 1$$
$$= 2 \sin x(\pm \sqrt{1 - \sin^2 x}) + 3 \sin x - 1$$
$$= \pm 2w \sqrt{1 - w^2} + 3w - 1$$

where $w = \sin x$, and, hence, $-1 \leq w \leq 1$. Thus, we wish to find all numbers w in $[-1, 1]$ for which

$$\pm 2w \sqrt{1 - w^2} + 3w - 1 = 0$$

or, equivalently,

$$\pm 2w \sqrt{1 - w^2} = 1 - 3w$$

If w is such a number, then squaring (remember that this may introduce numbers that are not solutions)

$$4w^4 + 5w^2 - 6w + 1 = 0$$

To find these numbers, we sketch a graph of the function f given by

$$f(w) = 4w^4 + 5w^2 - 6w + 1, \qquad -1 \leq w \leq 1$$

and locate the zeros of f. For instance, to find $f(\frac{1}{10})$, *notice that**

$$.1 = \tfrac{1}{10} \begin{array}{|rrrrr} 4 & 0 & 5 & -6 & 1 \\ & .4 & .04 & .504 & -.5496 \\ \hline 4 & .4 & 5.04 & -5.496 & .4504 \end{array}$$

Hence, $f(.1) = .4504$. In this way, we construct a graph and a table, as in Figure 8-30, and find that the zeros of f are approximately 0.2 and 0.65. If x is a real number for which $\sin x = .2$, then $x \approx .21$ or $x \approx \pi - .21 \approx 2.93$. If $\sin x \approx .65$, then $x \approx .71$ or $x \approx \pi - .71 \approx 2.43$. Thus, .21, 2.93, .71 and 2.43 are the possible solutions for the equation $\sin 2x + 3 \sin x - 1 = 0$; checking, we find that .21 and 2.43 are the only solutions (approximate).

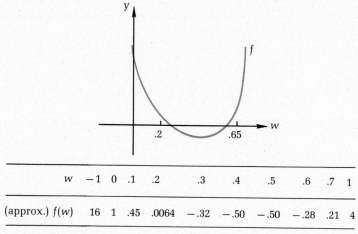

w	-1	0	.1	.2	.3	.4	.5	.6	.7	1
(approx.) $f(w)$	16	1	.45	.0064	$-.32$	$-.50$	$-.50$	$-.28$	.21	4

Figure 8-30

◊**Example 5**

Find all numbers x in $[0, 2\pi]$ for which $\sin 3x + 4 \sin^2 x - 3 \sin x + 1 = 0$.

Solution

Let x be such a number. Since $\sin 3x = 3 \sin x - 4 \sin^3 x$ (see Example 6), we have

$$0 = \sin 3x + 4 \sin^2 x - 3 \sin x + 1$$

$$= -4 \sin^3 x + 4 \sin^2 x + 1 = -4w^3 + 4w^2 + 1$$

where $w = \sin x$ (and therefore $-1 \le w \le 1$). Let

$$f(w) = 4w^3 - 4w^2 - 1, \qquad w \text{ in } [-1, 1]$$

*Synthetic division is discussed in Chapter 5.

The problem is to find the zeros of f. Note that for $w \leq 1$, $4w^3 - 4w^2 = 4w^2(w - 1) \leq 0$ (since $w^2 \geq 0$ and $w - 1 \leq 0$). Hence $4w^3 - 4w^2 - 1 \leq -1$ for $w \leq 1$, so f has no zeros in $[-1, 1]$; thus, the original equation has no solution.

Exercises 8-6

1. Let x be any number. Show that $\cos 3x = 4\cos^3 x - 3\cos x$.

2. Let x be a real number. Show that

$$\cos^4 x = \frac{3}{8} + \frac{\cos 2x}{2} + \frac{\cos 4x}{8}$$

$$\left(Hint\text{: } \cos^4 x = \frac{1 + 2\cos 2x + \cos^2 2x}{4}.\right)$$

3. Let x be a real number. Show that

$$\sin^4 x = \frac{\cos 4x - 4\cos 2x + 3}{8}$$

and

$$\sin^2 x \cos^2 x = \frac{1 - \cos 4x}{8}$$

(*Hint:* See Exercise 2.)

4. Find all numbers x in $[0, 2\pi]$ for which

$$\cos 2x - 3\sin x + 1 = 0$$

5. Find all real numbers in $[0, 2\pi]$ for which

$$4\sin^2 x + 4\cos 2x = 1$$

6. Find all numbers x in $[0, 2\pi]$ for which

$$\cos 2x - 4\cos x - 5 = 0$$

7. Find all real numbers x in $\left[0, \dfrac{\pi}{2}\right]$ for which $\cos 4x - 3\sin 2x + 4 = 0$. (*Hint:* First show that $\cos 4x = 1 - 2\sin^2 2x$.)

8. Show that for any real numbers x and y,

$$\cos(x + y) + \cos(x - y) = 2\cos x \cos y$$

Use this to show that if u and v are real numbers, then

$$\cos u + \cos v = 2[\cos \tfrac{1}{2}(u + v)][\cos \tfrac{1}{2}(u - v)]$$

(*Hint:* Let $u = x + y$, $v = x - y$.)

9. Find all real numbers x for which $\sin 2x = \cos 3x$. [*Hint:* For any real number y, $\cos y = \sin((\pi/2) - y)$, and $\sin x = \sin(\pi - x)$.]

10. Find the degree measure of x if x is between 0° and 360°, and $3\cos 2x + 4\sin 2x = 3$.

11. Find all real numbers x in $(0, 2\pi)$ for which

$$\cos 3x + \frac{1}{\cos^2 2x} = \frac{\sin^2 2x}{\cos^2 2x}$$

12. Show that for any real number x, with $\cos x + \sin x \neq 0$.

$$\frac{\cos^3 x + \sin^3 x}{\cos x + \sin x} = 1 - \frac{1}{2}\sin 2x$$

13. Show that for any real number x for which $\sin x \neq 0$,

$$(\cos x)(\cos 2x)(\cos 4x)(\cos 8x) \cdots (\cos 2^n x) = \frac{\sin 2^{n+1} x}{2^{n+1}\sin x}$$

(*Hint:* Multiply by $\sin x/\sin x$ and use the fact that $\sin x \cos x = \frac{1}{2}\sin 2x$.)

8-7 OTHER TRIGONOMETRIC FUNCTIONS. APPLICATIONS

The sine and cosine functions give rise to four other trigonometric functions, known as *tan* (tangent), *cot* (cotangent), *sec* (secant), and *csc* (cosecant). The function tan is defined by

$$\tan x = \frac{\sin x}{\cos x}$$

for all numbers x for which $\cos x \neq 0$ (i.e., $x \neq \pi/2 + n\pi$ for all integers n). It follows that, for any number x, either x and $x + \pi$ are both in the domain of tan, or neither is. In the former case, we have

$$\tan(x + \pi) = \frac{\sin(x + \pi)}{\cos(x + \pi)} = \frac{-\sin x}{-\cos x} = \tan x$$

so that tan is periodic with period $p \leq \pi$. The period of tan is actually π.[*]

---

[*]This can be seen from the fact that successive zeros of tan differ by π (the zeros of tan being the zeros of sin, namely, $n\pi$ where n is any integer).

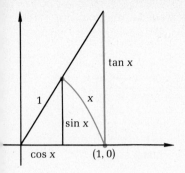

Figure 8-31

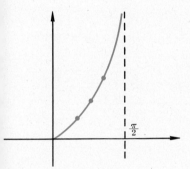

Figure 8-32

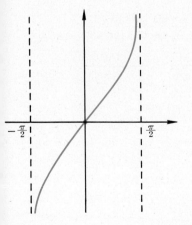

Figure 8-33

Note that if x is in the domain of tan, then so is −x [since, if cos x ≠ 0, then also cos (−x) ≠ 0] and

$$\tan(-x) = \frac{\sin(-x)}{\cos(-x)} = \frac{-\sin x}{\cos x} = -\tan x$$

that is,

tan (−x) = −tan x

This fact will be useful in graphing tan. Another fact useful in graphing can be inferred from Figure 8-31: tan 0 = 0, and as increasing numbers x are chosen between 0 and π/2, tan x is positive and increases indefinitely, so that the line given by x = π/2 is a vertical asymptote. See Figure 8-32. We now use the fact that tan (−x) = −tan x to obtain the graph of tan between −π/2 and π/2 (Figure 8-33), and then obtain the rest by periodicity (Figure 8-34).

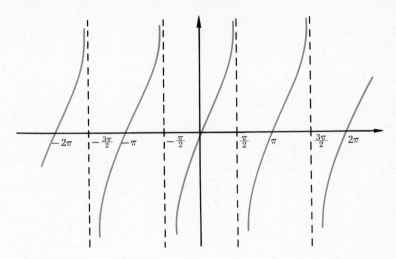

Figure 8-34

The *addition formulas* for sin and cos, and their corollaries, can be now applied to derive the following corresponding formulas for the tangent function:

1. $\tan(x + y) = \dfrac{\tan x + \tan y}{1 - \tan x \tan y}$

2. $\tan 2x = \dfrac{2\tan x}{1 - \tan^2 x}$

3. $\tan^2 \dfrac{x}{2} = \dfrac{1 - \cos x}{1 + \cos x}$

where (1), (2), and (3) hold for those real numbers x and y for which the denominators are not zero. We verify (1) by observing that

$$\tan (x + y) = \frac{\sin (x + y)}{\cos (x + y)} = \frac{\sin x \cos y + \cos x \sin y}{\cos x \cos y - \sin x \sin y}$$

$$= \frac{\dfrac{\sin x}{\cos x} + \dfrac{\sin y}{\cos y}}{1 - \dfrac{\sin x \sin y}{\cos x \cos y}} \qquad \begin{array}{l} [\textit{dividing through by} \\ \cos x \cos y] \end{array}$$

$$= \frac{\tan x + \tan y}{1 - \tan x \tan y}$$

Formula (2) is verified by taking x = y in (1), and (3) is verified by noting that

$$\tan^2 \frac{x}{2} = \frac{\sin^2 \dfrac{x}{2}}{\cos^2 \dfrac{x}{2}} = \frac{\dfrac{1}{2}(1 - \cos x)}{\dfrac{1}{2}(1 + \cos x)} = \frac{1 - \cos x}{1 + \cos x}$$

Example 1

Compute tan 105°.

Solution

$$\tan 105° = \tan (60° + 45°) = \frac{\tan 60° + \tan 45°}{1 - \tan 60° \tan 45°}$$

$$= \frac{\sqrt{3} + 1}{1 - \sqrt{3} \cdot 1} = \frac{1 + \sqrt{3}}{1 - \sqrt{3}} = \frac{(1 + \sqrt{3})^2}{(1 - \sqrt{3})(1 + \sqrt{3})}$$

$$= \frac{1 + 2\sqrt{3} + 3}{1 - 3} = -2 - \sqrt{3}$$

Example 2

Find tan 15°.

Solution

$$\tan^2 15° = \tan^2 \frac{30°}{2} = \frac{1 - \cos 30°}{1 + \cos 30°}$$

$$= \frac{1 - \dfrac{\sqrt{3}}{2}}{1 + \dfrac{\sqrt{3}}{2}} = \frac{2 - \sqrt{3}}{2 + \sqrt{3}} = 7 - 4\sqrt{3}$$

Hence, tan 15° = $\sqrt{7 - 4\sqrt{3}}$ (since tan x > 0 if 0 < x < π/2).

□

There are three other trigonometric functions, defined as follows:

$$\cot x = \frac{1}{\tan x} \qquad \left(= \frac{\cos x}{\sin x}\right)$$

$$\sec x = \frac{1}{\cos x}$$

and

$$\csc x = \frac{1}{\sin x}$$

Note that the zeros of sin (i.e., all integral multiples of π) are excluded from the domains of cot and csc, and the zeros of cos (i.e., the odd integral multiples of $\pi/2$) are excluded from the domain of sec.

Since for all x in the domain of cot we have

$$\cot x = \frac{\cos x}{\sin x} = \frac{\sin\left(x + \dfrac{\pi}{2}\right)}{-\cos\left(x + \dfrac{\pi}{2}\right)} = -\tan\left(x + \frac{\pi}{2}\right)$$

we see that the graph of cot is the graph of tan shifted horizontally (to the left) by $\pi/2$ units and then reflected in the horizontal axis. See Figure 8-35.

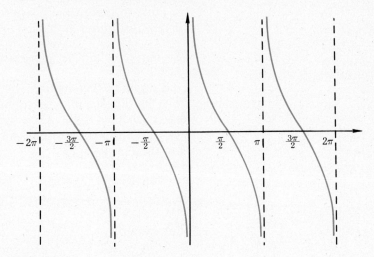

Figure 8-35

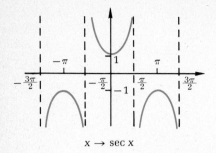

$x \rightarrow \sec x$

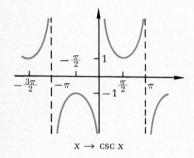

$x \rightarrow \csc x$

Figure 8-36

The periodicity and symmetry properties of cos and sin obviously hold for sec and csc, respectively. The graphs of sec and csc are shown in Figure 8-36. The reader should sketch these graphs for himself, using his knowledge of the behavior of sin and cos.

The following should be added to the list of fundamental formulas already derived:

4. $1 + \tan^2 x = \sec^2 x$

and

5. $1 + \cot^2 x = \csc^2 x$

where (4) and (5) hold for all real numbers in the domains of tan and cot, respectively. Formula (4) is verified by the following simple computation:

$$1 + \tan^2 x = 1 + \frac{\sin^2 x}{\cos^2 x} = \frac{\cos^2 x + \sin^2 x}{\cos^2 x} = \frac{1}{\cos^2 x}$$

$$= \sec^2 x$$

The verification of (5) is equally simple and is left to the reader.

As in the case of the sin and cos, the tables usually give tan x only for x between 0 and $\pi/2$. Now, since the period of tan is π, we have, for any number y, tan y = tan x for some x such that $0 \leq x \leq \pi$, and thus the addition or subtraction of an appropriate multiple of π allows us to reduce the computation of tan y to that of tan x where $0 \leq x < \pi$ (i.e., with the angle in question either in the first or second quadrant). For example,

$$\tan \tfrac{17}{3}\pi = \tan (5\pi + \tfrac{2}{3}\pi) = \tan \tfrac{2}{3}\pi \quad \text{and} \quad 0 \leq \tfrac{2}{3}\pi < \pi$$

If upon such reduction we arrive at a first-quadrant angle, then its *tangent* may be found in the tables. If, on the other hand, we have $\pi/2 < x < \pi$ (as in the foregoing example), then $(\pi - x)$ is in the first quadrant (i.e., $0 \leq \pi - x < \pi/2$), and we use the fact that

$$\tan x = -\tan (-x) \qquad [\textit{see page 227}]$$

$$= -\tan°(\pi - x) \qquad [\textit{by periodicity}]$$

For example,

$$\tan \frac{2}{3}\pi = -\tan \left(\pi - \frac{2}{3}\pi\right) = -\tan \frac{\pi}{3} = -\sqrt{3}$$

Example 3

Find all numbers t in $[0, 2\pi)$ for which
(a) $\tan t = 5.8$ (b) $\tan t = -5.7$ (approximately)

Solution

(a) From the table on p. 463, $\tan 1.4 = 5.8$, so 1.4 is the unique solution in $(0, \pi)$. Hence, also, $\tan (1.4 + \pi) = 5.8$. It follows that 1.4 and $1.4 + \pi$ are all the solutions between 0 and 2π.
(b) We have $\tan (\pi - 1.4) = -\tan 1.4 = -5.8$. Hence, $\pi - 1.4$ and $\pi - 1.4 + \pi = 2\pi - 1.4$ are the required numbers. A simple interpretation of these facts follows from the graph of tan given in Figure 8-37.

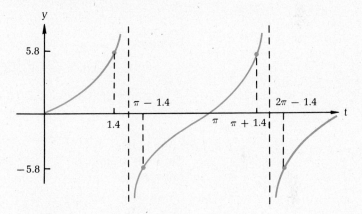

Figure 8-37

Example 4

Find all real numbers t in $[0, 2\pi)$ for which $\csc 2t + \cot 2t = 1$.

Solution

For $\sin 2t \neq 0$, we have

$$\csc 2t + \cot 2t = \frac{1}{\sin 2t} + \frac{\cos 2t}{\sin 2t}$$

$$= \frac{1 + \cos 2t}{\sin 2t}$$

$$= \frac{2 \cos^2 t}{2 \sin t \cos t} \qquad \left[since\ \frac{1 + \cos 2t}{2} = \cos^2 t\ \ and\ \sin 2t = 2 \sin t \cos t \right]$$

$$= \frac{\cos t}{\sin t}$$

$$= \cot t$$

Hence, $\csc 2t + \cot 2t = 1$ if and only if $\cot t = 1$, and this holds if and only if $t = \pi/4$ or $5\pi/4$.

Example 5

Find all numbers x in $[0, 2\pi)$ for which $\sqrt{3} \tan^2 x - \tan x = 0$.

Solution

$$\sqrt{3} \tan^2 x - \tan x = \tan x(-1 + \sqrt{3} \tan x) = 0$$

if and only if

$$\tan x = 0 \quad \text{or} \quad \tan x = \frac{1}{\sqrt{3}}$$

Hence, the solutions are 0, π, $\pi/6$, and $7\pi/6$.

Example 6

Find all numbers x in $[0, 2\pi)$ for which $\sec^2 x + 2 \tan x = 0$.

Solution

$$\sec^2 x + 2 \tan x = 1 + \tan^2 x + 2 \tan x = (1 + \tan x)^2 = 0$$

if and only if $\tan x = -1$. Hence, the solutions are $3\pi/4$ and $7\pi/4$.

Example 7

Find all numbers x in $[0, 2\pi)$ for which $\tan^3 x + \tan^2 x - \tan x = 0$.

Solution

$$\tan^3 x + \tan^2 x - \tan x = \tan x(\tan^2 x + \tan x - 1) = 0$$

if and only if

$$\tan x = 0$$

or

$$\tan x = \frac{-1 \pm \sqrt{5}}{2} \approx \frac{-1 \pm 2.2}{2} \qquad \begin{array}{l}\textit{[by the Quadratic} \\ \textit{Formula]}\end{array}$$

From this, using tables, we have

$$0, \quad \pi, \quad .54, \quad \pi + .54, \quad \pi - 1.01, \quad \text{and} \quad 2\pi - 1.01$$

(The last four solutions are, of course, approximations.)

◊**Example 8**

Find all numbers t in $(-\pi/2, \pi/2)$ for which $\tan t = t + 1$.

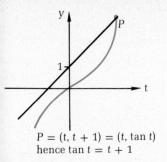

$P = (t, t + 1) = (t, \tan t)$
hence $\tan t = t + 1$

Figure 8-38

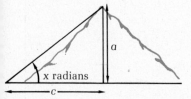

Figure 8-39

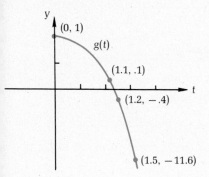

Figure 8-40

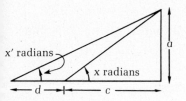

Figure 8-41

Solution

Let $f(t) = t + 1$. The graphs of tan and f are given in Figure 8-38, from which we see that there is exactly one solution, namely the first coordinate of the point of intersection. To find the solution (approximately), let $g(t) = t + 1 - \tan t$ and find where the graph of g crosses the horizontal axis. Thus, since $g(0) = 1$ and $g(1.5) \approx 1.5 + 1 - 14.1 = -11.6$, the solution is between 0 and 1.5. Continuing in this way, we obtain the graph in Figure 8-39, which shows that the solution is between 1.1 and 1.2.

Example 9

Find the height a of the mountain shown in Figure 8-40.

Solution

We first measure the angle of elevation. In Figure 8-40 the radian measure of that angle is x. (We cannot, however, directly measure c; therefore it is unknown.)* We then move out a known distance d and measure another angle of elevation, say x' radians. See Figure 8-41.

We now have

$$\tan x = \frac{a}{c} \quad \text{and} \quad \tan x' = \frac{a}{c + d}$$

We then get a condition on a alone, as follows:

$$\frac{d}{a} = \frac{c + d}{a} - \frac{c}{a} = \frac{1}{\tan x'} - \frac{1}{\tan x}$$

Therefore,

$$a = \frac{d}{\dfrac{1}{\tan x'} - \dfrac{1}{\tan x}} = \frac{d}{\cot x' - \cot x}$$

Thus, we are able to find a, the height of the mountain, even though we do not know c. In this problem, if $x' = \pi/6$ radians, $x = \pi/3$ radians, and $d = 2000$ feet, then (measured in feet)

$$a = \frac{2000}{\dfrac{1}{\tan(\pi/6)} - \dfrac{1}{\tan(\pi/3)}} = \frac{2000}{\dfrac{1}{\dfrac{1}{\sqrt{3}}} - \dfrac{1}{\sqrt{3}}} = \frac{2000}{\sqrt{3} - \dfrac{1}{\sqrt{3}}},$$

$$= \frac{2000\sqrt{3}}{2} = 1000\sqrt{3}$$

Therefore, the height of the mountain is approximately $1000(1.7) = 1700$ feet.

*Note that *if* we had some way to measure c, then we could find a simply by observing that $a = c \tan x$.

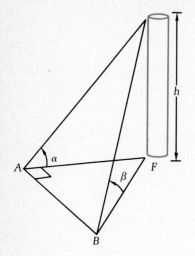

Figure 8-42

Example 10

Find the height h of the inaccessible tower shown in Figure 8-42. The line $\overline{AB}$ is a "base line" and perpendicular to the line $\overline{AF}$. The angles α and β are known, as well as the length of $\overline{AB}$.

Solution

We see that $\dfrac{h}{\overline{AF}} = \tan \alpha$, $\dfrac{h}{\overline{BF}} = \tan \beta$, and $\overline{AB^2} + \overline{AF^2} = \overline{BF^2}$.

Hence, $h = \overline{AF} \tan \alpha = \overline{BF} \tan \beta$. Squaring and substituting, we obtain

$$\overline{AF^2} \tan^2 \alpha = (\overline{AB^2} + \overline{AF^2}) \tan^2 \beta$$

Since $\overline{AB}$, α, and β are known, we can solve for $\overline{AF}$. We can then find h using the fact that $h = \overline{AF} \tan \alpha$.

Exercises 8-7

In Exercises 1–12, compute tan x, sec x, cot x, and csc x if x is:

1. $\dfrac{\pi}{6}$ **2.** $\dfrac{\pi}{4}$

3. $\dfrac{5\pi}{6}$ **4.** $\dfrac{4\pi}{3}$

5. $\dfrac{27\pi}{4}$ **6.** $\dfrac{-5\pi}{6}$

7. $-\dfrac{7\pi}{3}$ **8.** 4π

9. $\dfrac{\pi}{2}$ **10.** $\dfrac{7\pi}{2}$

11. $-\dfrac{\pi}{2}$ **12.** $-\dfrac{17\pi}{2}$

In Exercises 13–15, use the addition formula for tangent to find:

13. $\tan 75°$

14. $\tan \dfrac{\pi}{12}$ $\left(\text{Hint: } \dfrac{\pi}{12} = \dfrac{\pi}{3} - \dfrac{\pi}{4}.\right)$

15. $\tan \dfrac{7\pi}{12}$

In Exercises 16–21, use tables to find all real numbers t in $(0, 2\pi)$ for which

16. $\tan t = 1.3$ **17.** $\tan t = -1.3$

18. $\sec t = 3.4$ **19.** $\sec t = -3.4$

20. $\cot t = 11$ **21.** $\cot t = -11$

22. (a) If $\tan x = 2$ and $x + y = 135°$, find $\tan y$.

(b) If $\tan x = \dfrac{1}{3}$ and $x - y = \dfrac{3\pi}{4}$, find $\tan y$.

23. Find all numbers x in $(-\pi/2, \pi/2)$ for which
$$2\tan^2 x - \tan x - 3 = 0$$

24. Find all numbers x in $(0, \pi/2)$ for which
$$\tan^2 x - (1 + \sqrt{3})\tan x + \sqrt{3} = 0$$

In Exercises 25–32, show that the given equations hold for all values of x for which they are defined.

25. $\tan\left(\dfrac{x}{2} + \dfrac{\pi}{4}\right) - \tan\left(\dfrac{x}{2} - \dfrac{\pi}{4}\right) = 2\sec x$

26. $\tan\left(\dfrac{x}{2} + \dfrac{\pi}{4}\right) + \tan\left(\dfrac{x}{2} - \dfrac{\pi}{4}\right) = 2\tan x$

27. $\tan\dfrac{x}{2} + \cot\dfrac{x}{2} = 2\csc x$

28. $\csc x - \cot x = \tan\dfrac{x}{2}$

29. $\sec^2\left(\dfrac{x}{2} - \dfrac{\pi}{4}\right) = \dfrac{2}{1 + \sin x}$

30. $\tan\dfrac{x}{2} = \dfrac{\sin x}{1 + \cos x}$

31. $\dfrac{1 - \tan x}{1 + \tan x} = \tan\left(\dfrac{\pi}{4} - x\right)$

32. $\dfrac{2\tan x}{\tan x + \sin x} = \sec^2\dfrac{x}{2}$

◊**33.** Find all numbers x between 0 and $\pi/2$ for which $\tan x = 1 - x^2$.

34. Show that for any real numbers x and y for which $1 - \tan x \cdot \tan y \neq 0$,

$$\sec(x + y) = \dfrac{\sec x \sec y}{1 - \tan x \tan y}$$

◊**35.** Find

$$\tan\dfrac{x}{2} \text{ if } \sin x = \dfrac{p^2 - q^2}{p^2 + q^2} \text{ and } \cos x = \dfrac{2pq}{p^2 + q^2}$$

where p and q are nonzero real numbers.

36. A circle is circumscribed about a regular decagon having sides 7 inches long. Find the radius of the circle. What is the radius if the circle is inscribed?

37. Find the angle of elevation of the sun when a vertical post casts a shadow on level ground that is (a) twice as long as the pole; (b) as long as the pole; (c) one-half as long as the pole. In each case the angle of elevation is measured from the tip of the shadow.

38. Find the perimeter of a regular pentagon inscribed in a circle of radius 20.

39. The height of the Empire State Building is 1250 feet. What is the angle of elevation as seen from a point on the ground one-half mile from its base?

40. Find the height of a vertical post whose shadow on level ground is 40 feet long when the angle of elevation of the sun is 15°.

41. The distance CB through a swamp is desired. A line AC of length 200 yards is laid off at right angles to CB. Angle BAC is 52.6°. Find the length of CB.

42. The angle of elevation from an observer to the top of a church 260 feet away is 25°, and the angle subtended by the spire above it is 10.5°, as shown in the accompanying figure. Find the height of the spire.

43. An observer at a point 0 on a straight coast running north and south sights a ship in a direction 29.6° west of north. The ship is reported at the same time to be directly west of a point M. The point M is 28.2 miles north of 0. Find the distance of the ship from the observer.

44. A tower 150 feet high is situated on the bank of a river. At the top of the tower, the angle of depression of an object directly across the river on the opposite bank is 32°16′, as shown in the figure. Find the width of the river.

45. A telephone pole casts a shadow 30.5 feet long when the angle of elevation of the sun is 16.5°. Find the height of the pole.

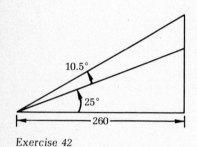

Exercise 42

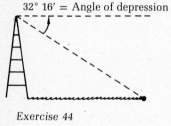

Exercise 44

Answer Exercises 46–53 true or false, and justify your answer.

46. $\sec A < 1$ if A is an acute angle

47. $\sin 15° = \cos 75°$

48. $\tan 60° > \tan 45°$

49. $\cos 47° = \csc 43°$

50. $2 \tan 30° = \tan 60°$

51. $\sin 30° + \sin 60° = \sin 90°$

52. $\sin A > 1$ if A is an acute angle

53. $\tan 50° = \dfrac{\cos 40°}{\cos 50°}$

8-8 IDENTITIES

We have already observed such facts as:

> *for every real number* x, $\cos 2x = \cos^2 x - \sin^2 x,$
> *for every real number* x, $\sin^2 x + \cos^2 x = 1,$

and

> *for every real number* x *for which the equation makes sense, that is, for which* $1 - \tan^2 x \neq 0$,
> $\tan 2x = (2 \tan x)/(1 - \tan^2 x).$

The preceding statements are in sharp contrast to a conditional statement like:

> *there are numbers* x *for which* $\sin x = 0$

since there are also many numbers x for which $\sin x \neq 0$ (in fact, if $x \neq n\pi$, where n is an integer, then $\sin x \neq 0$).

It has become customary to call conditions on numbers that hold for *every real number* x (or for every real number x for which the condition makes sense) *identities*. Thus,

> $\cos 2x = \cos^2 x - \sin^2 x$
>
> $\sin^2 x + \cos^2 x = 1$

and

> $\tan 2x = (2 \tan x)/(1 - \tan^2 x)$

are all examples of identities, whereas $\sin x = 0$ is *not*. We shall give some identities in the following examples.

Exercises 8-8

In Exercises 1–5, show that each equation is an identity.

1. $\sec x - \cos x = \tan x \sin x$

2. $\dfrac{\sin x - \cos x}{\tan x - 1} = \cos x$

3. $\tan x + \cot x = \sec x \csc x$

4. $\tan^4 x + \tan^2 x = \sec^4 x - \sec^2 x$

5. $\dfrac{1 - \tan^2 x}{\tan x} = \cot x - \tan x$

Verify that formulas 6–9 are identities.

6. $\sin 2s = \dfrac{2 \tan s}{1 + \tan^2 s}$

7. $\cos 2s = \dfrac{1 - \tan^2 s}{1 + \tan^2 s}$

8. $\dfrac{2}{\sin 2s} = \tan s + \cot s$

9. $\dfrac{1 + \cos 2s}{\sin 2s} = \cot s$

In Exercises 10–12, verify that each equation is an identity.

10. $\dfrac{1 - \tan x}{1 + \tan x} = \dfrac{\cot x - 1}{\cot x + 1}$

11. $\dfrac{\sin (x - y)}{\sin (x + y)} = \dfrac{\tan x - \tan y}{\tan x + \tan y}$

12. $\cos x - \tan y \sin x = \sec y \cos (x + y)$

In Exercises 13–15, show that the given equations hold for all values of x for which they are defined.

13. $\dfrac{1 - \cos 2x}{2 \sin x \cos x} = \tan x$

14. $\dfrac{1 - \cos 2x}{\sin 2x} = \tan x$

15. $1 + \tan x \tan 2x = \sec 2x$

Miscellaneous Problems

1. In order to measure the width of a river, a surveyor lays out a baseline PQ on one side of the river so that PQ and the line of sight from Q to a point R on the opposite bank form a right triangle. Find the width of the river if the baseline is 100 feet long and the angle QPR is 41°. See the accompanying figure.

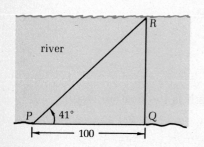

Problem 1

Now

$$\frac{\sin x}{1 - \cos x} - \frac{\cos x}{\sin x} = \frac{\sin^2 x - (1 - \cos x) \cos x}{(1 - \cos x) \sin x}$$

$$= \frac{\sin^2 x + \cos^2 x - \cos x}{(1 - \cos x) \sin x}$$

$$= \frac{1 - \cos x}{(1 - \cos x) \sin x}$$

$$= \frac{1}{\sin x}$$

since $1 - \cos x \neq 0$.

Example 3

Show that

$$\frac{2}{\sin 2 x} = \tan x + \cot x$$

is an identity.

Solution

We shall express all our functions in terms of sin and cos. Recall that for every number x, $\sin 2x = 2 \sin x \cos x$ (and thus, this equation is an identity). We see that for all x in the domains of both tan and cot, and for which $\sin 2x \neq 0$,

$$\frac{2}{\sin 2 x} = \tan x + \cot x$$

if and only if

$$\frac{1}{\sin x \cos x} = \frac{\sin x}{\cos x} + \frac{\cos x}{\sin x}$$

Now

$$\frac{\sin x}{\cos x} + \frac{\cos x}{\sin x} = \frac{\sin^2 x + \cos^2 x}{\sin x \cos x}$$

$$= \frac{1}{\sin x \cos x}$$

Hence, $\dfrac{2}{\sin 2x} = \tan x + \cot x$, which is the desired result.

Exercises 8-8

In Exercises 1–5, show that each equation is an identity.

1. $\sec x - \cos x = \tan x \sin x$

2. $\dfrac{\sin x - \cos x}{\tan x - 1} = \cos x$

3. $\tan x + \cot x = \sec x \csc x$

4. $\tan^4 x + \tan^2 x = \sec^4 x - \sec^2 x$

5. $\dfrac{1 - \tan^2 x}{\tan x} = \cot x - \tan x$

Verify that formulas 6–9 are identities.

6. $\sin 2s = \dfrac{2 \tan s}{1 + \tan^2 s}$

7. $\cos 2s = \dfrac{1 - \tan^2 s}{1 + \tan^2 s}$

8. $\dfrac{2}{\sin 2s} = \tan s + \cot s$

9. $\dfrac{1 + \cos 2s}{\sin 2s} = \cot s$

In Exercises 10–12, verify that each equation is an identity.

10. $\dfrac{1 - \tan x}{1 + \tan x} = \dfrac{\cot x - 1}{\cot x + 1}$

11. $\dfrac{\sin (x - y)}{\sin (x + y)} = \dfrac{\tan x - \tan y}{\tan x + \tan y}$

12. $\cos x - \tan y \sin x = \sec y \cos (x + y)$

In Exercises 13–15, show that the given equations hold for all values of x for which they are defined.

13. $\dfrac{1 - \cos 2x}{2 \sin x \cos x} = \tan x$

14. $\dfrac{1 - \cos 2x}{\sin 2x} = \tan x$

15. $1 + \tan x \tan 2x = \sec 2x$

Miscellaneous Problems

1. In order to measure the width of a river, a surveyor lays out a baseline PQ on one side of the river so that PQ and the line of sight from Q to a point R on the opposite bank form a right triangle. Find the width of the river if the baseline is 100 feet long and the angle QPR is 41°. See the accompanying figure.

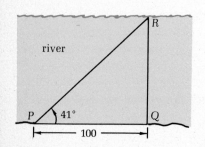

Problem 1

Answer Exercises 46–53 true or false, and justify your answer.

46. $\sec A < 1$ if A is an acute angle

47. $\sin 15° = \cos 75°$

48. $\tan 60° > \tan 45°$

49. $\cos 47° = \csc 43°$

50. $2 \tan 30° = \tan 60°$

51. $\sin 30° + \sin 60° = \sin 90°$

52. $\sin A > 1$ if A is an acute angle

53. $\tan 50° = \dfrac{\cos 40°}{\cos 50°}$

8-8 IDENTITIES

We have already observed such facts as:

for every real number x, $\cos 2x = \cos^2 x - \sin^2 x$,
for every real number x, $\sin^2 x + \cos^2 x = 1$,

and

for every real number x for which the equation makes sense, that is, for which $1 - \tan^2 x \neq 0$,
$\tan 2x = (2 \tan x)/(1 - \tan^2 x)$.

The preceding statements are in sharp contrast to a conditional statement like:

there are numbers x for which $\sin x = 0$

since there are also many numbers x for which $\sin x \neq 0$ (in fact, if $x \neq n\pi$, where n is an integer, then $\sin x \neq 0$).

It has become customary to call conditions on numbers that hold for *every real number x* (or for every real number x for which the condition makes sense) *identities.* Thus,

$$\cos 2x = \cos^2 x - \sin^2 x$$
$$\sin^2 x + \cos^2 x = 1$$

and

$$\tan 2x = (2 \tan x)/(1 - \tan^2 x)$$

are all examples of identities, whereas $\sin x = 0$ is *not.* We shall give some identities in the following examples.

Example 1

Show that for all real numbers x for which $1 + \sin x \neq 0$ and $\cos x \neq 0$, we have

$$\frac{\cos x}{1 + \sin x} = \frac{1 - \sin x}{\cos x}$$

that is, $\cos x/(1 + \sin x) = (1 - \sin x)/\cos x$ is an identity.

Solution

Note that for $1 + \sin x \neq 0$ and $\cos x \neq 0$

$$\frac{\cos x}{1 + \sin x} = \frac{1 - \sin x}{\cos x}$$

if and only if

$$(\cos x)(\cos x) = (1 - \sin x)(1 + \sin x),$$

that is,

$$\cos^2 x = 1 - \sin^2 x$$

or, finally,

$$\cos^2 x + \sin^2 x = 1$$

Since $\cos^2 x + \sin^2 x = 1$ is true for all x, it follows that $\cos x/(1 + \sin x) = (1 - \sin x)/\cos x$.

Example 2

Show that

$$\frac{\sin x}{1 - \cos x} - \cot x = \frac{1}{\sin x}$$

is an identity and thus holds for every real number x in the domain of cot for which $1 - \cos x \neq 0$, and $\sin x \neq 0$.

Solution

We shall rewrite our condition on x so that it involves only the functions sin and cos. To do this, we note the identity $\cot x = \cos x/\sin x$, and we see that

$$\frac{\sin x}{1 - \cos x} - \cot x = \frac{1}{\sin x}$$

is equivalent to

$$\frac{\sin x}{1 - \cos x} - \frac{\cos x}{\sin x} = \frac{1}{\sin x}$$

2. A straight road makes an angle of 6° with the horizontal. How much does the road rise in a distance of 1000 feet measured along the road?

3. Let $f(x) = x \sin 2x$. Graph f.

4. Let $f(x) = \dfrac{\sin x}{x}$, $x \geq \dfrac{\pi}{2}$. Graph f.

5. Let $f(x) = 2^{-x} \sin x$. Graph f.

◇**6.** For how many real numbers x is $\dfrac{x^2}{10} = \sin 2\pi x$?

(*Hint:* Consider the graphs.)

7. The arc and chord subtended by a small central angle are approximately the same length. Use this to compute the diameter of the sun given that the distance from earth to the sun is 93 million miles and the sun as seen from earth subtends an angle of .53 degree.

8. What is the area of a regular polygon of 12 sides which is inscribed in a circle of radius 5?

9. What is the perimeter of a regular polygon of 10 sides which is circumscribed about a circle of radius 1?

10. What angle does the diagonal of a cube make with each of the edges of the cube?

◇**11.** Use induction to show that $|\sin nx| \leq n|\sin x|$ for every real number x and every positive integer n.

◇**12.** Show that there is a polynomial function f with the property that $f(\cos x) = \cos 4x$.

◇**13.** Show that

(a) $\tan (\sin^{-1} x) = \dfrac{x\sqrt{1 - x^2}}{1 - x^2}$ (b) $\sin (2 \sin^{-1} x) = 2x\sqrt{1 - x^2}$

14. Three circles of radii 4, 5, and 6 are pairwise tangent to each other. A triangle is formed by connecting the centers of the circles. Find the angles of this triangle. (*Hint:* Find the lengths of the sides of the triangle.)

◇**15.** Show that if f is a nonconstant polynomial function, then f is not a periodic function.

In Problems 16–20, show that the equation holds for all x for which they are defined. That is, show that each equation is an identity.

16. $\sin x \tan x + \cos x = \sec x$

17. $\tan^2 x - \sin^2 x = \tan^2 x \sin^2 x$

18. $\dfrac{1}{\sec x - \tan x} = \sec x + \tan x$

19. $\cos^4 x - \sin^4 x = \cos 2x$

20. $\tan 2x = \dfrac{2}{\cot x - \tan x}$

9 Further Applications of Trigonometry

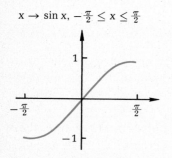

Figure 9-1

$x \to \sin x, -\frac{\pi}{2} \le x \le \frac{\pi}{2}$

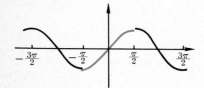

$x \to \arcsin x, -1 \le x \le 1$

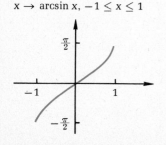

Figure 9-2

9-1 INVERSES OF TRIGONOMETRIC FUNCTIONS

Recall that a function has an inverse if and only if it is 1–1 (see Chapter 7). Since the trigonometric functions are periodic, they are not 1–1, and hence have no inverses. On the other hand, each trigonometric function is *sectionally 1–1;* that is, the real line is composed of intervals over each of which the function *is* one-to-one. For example, the sin function is 1–1 over the interval $[-\pi/2, \pi/2]$, and, in fact, over any of the intervals $[-(\pi/2) + k\pi, (\pi/2) + k\pi]$ for $k = 0, \pm 1, \pm 2, \pm 3, \ldots$. See Figure 9-1. Thus, when the sin function is restricted to one of these intervals, the resulting restricted function does have an inverse.

In the case of the sin function, we single out *the restriction whose domain is the interval* $[-\pi/2, \pi/2]$. The inverse of this function is denoted by **arcsin.** (See Figure 9-2 for the graph. *Recall that the graph of the inverse of a function f is the reflection of the graph of f in the line given by y = x.*) Note that the *domain of arcsin* is $[-1, 1]$, and its range is $[-\pi/2, \pi/2]$. Thus, if $-1 \le x \le 1$, then

$$y = \arcsin x \quad \text{if and only if} \quad \sin y = x$$

with $-\pi/2 \le y \le \pi/2$. In other words, for any number x between -1 and 1, arcsin x is between $-\pi/2$ and $\pi/2$, and **sin** $(\textbf{arcsin } x) = x$; furthermore, **arcsin** $(\textbf{sin } x) = x$ for every real number x between $-\pi/2$ and $\pi/2$.

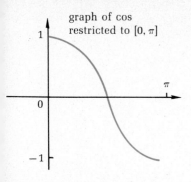

graph of cos
restricted to $[0, \pi]$

Figure 9-3

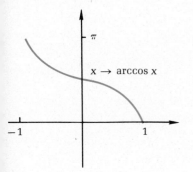

$x \to$ arccos x

Figure 9-4

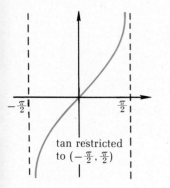

tan restricted
to $(-\frac{\pi}{2}, \frac{\pi}{2})$

Figure 9-5

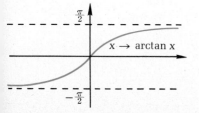

$x \to$ arctan x

Figure 9-6

Example 1

arcsin $(0) = 0$, arcsin $\frac{1}{2} = \pi/6$, and arcsin $(-\sqrt{3}/2) = -\pi/3$, because $\sin 0 = 0$, $\sin(\pi/6) = \frac{1}{2}$, $\sin(-\pi/3) = -\sqrt{3}/2$, and 0, $\pi/6$, $-\pi/3$ are all in $[-\pi/2, \pi/2]$.
□

When cos is restricted to the interval from 0 to π, it is one-to-one. The range of this restriction is the interval $[-1, 1]$. See Figure 9-3.

By definition, **arccos** is the inverse of cos; that is, for $-1 \le x \le 1$, arccos x is that number between 0 and π, inclusive, whose cos is x; hence, for $-1 \le x \le 1$ and $0 \le y \le \pi$,

$$y = \textbf{arccos } x \quad \textbf{if and only if} \quad \cos y = x$$

The graph of arccos is given in Figure 9-4.

Example 2

arccos $0 = \pi/2$, arccos $1 = 0$, arccos $\frac{1}{2} = \pi/3$, and arccos $(-\sqrt{3}/2) = 5\pi/6$ (since $\cos(\pi/2) = 0$, $\cos 0 = 1$, $\cos(\pi/3) = \frac{1}{2}$, and $\cos(5\pi/6) = -\sqrt{3}/2$, and $\pi/2$, 0, $\pi/3$, $5\pi/6$ are in $[0, \pi]$).
□

The function tan restricted to the interval $(-\pi/2, \pi/2)$ is one-to-one. The range of this restriction is the entire real line. See Figure 9-5.

By definition, **arctan** is the inverse of tan; that is, for any number x, arctan x is that number between $-\pi/2$ and $\pi/2$ whose tan is x; equivalently, for any real number x and $-\pi/2 < y < \pi/2$, we have

$$y = \textbf{arctan } x \quad \textbf{if and only if} \quad \tan y = x$$

The graph of arctan is given in Figure 9-6.

The functions **arcsec, arccsc,** and **arccot** are defined in a similar way:

arcsin is the inverse of the restriction of sin to the closed interval $[-\pi/2, \pi/2]$;

arccos is the inverse of the restriction of cos to the closed interval $[0, \pi]$;

arctan is the inverse of the restriction of tan to the open interval $(-\pi/2, \pi/2)$;

arccsc is the inverse of the restriction of csc to the closed interval $[-\pi/2, \pi/2]$; with 0 deleted.

arcsec is the inverse of the restriction of sec to the closed interval $[0, \pi]$; with $\pi/2$ deleted.

arccot is the inverse of the restriction of the cot to the open interval $(0, \pi)$.

Another notation commonly used for inverse trigonometric functions is as follows:

$$\arcsin = \sin^{-1}$$
$$\arccos = \cos^{-1}$$
$$\arctan = \tan^{-1}$$
$$\text{arcsec} = \sec^{-1}$$
$$\text{arccsc} = \csc^{-1}$$

Example 3

Compute $\sin^{-1}\left(\cos\frac{1}{8}\right)$.

Solution

$$\sin^{-1}\left(\cos\frac{1}{8}\right) = \sin^{-1}\left[\sin\left(\frac{\pi}{2} - \frac{1}{8}\right)\right] = \frac{\pi}{2} - \frac{1}{8}$$

Example 4

Compute $\cos[\arcsin(-1/\sqrt{2})$ [or, equivalently, $\cos[\sin^{-1}-1/\sqrt{2})]]$ and $\tan(\text{arcsec }2)$ [or, equivalently, $\tan(\sec^{-1}2)$].

Solution

$$\cos\left(\arcsin\left(-\frac{1}{\sqrt{2}}\right)\right) = \cos\left(-\frac{\pi}{4}\right) = \frac{1}{\sqrt{2}}$$

[since $\sin(-\pi/4) = -1/\sqrt{2}$], and

$$\tan(\text{arcsec }2) = \tan\frac{\pi}{3} = \sqrt{3} \quad \left(\text{since }\sec\frac{\pi}{3} = \frac{1}{\cos\dfrac{\pi}{3}} = 2\right)$$

Example 5

Show that for any x in $[-1, 1]$, $\sin(\cos^{-1}x) = \sqrt{1 - x^2}$.

Solution

Let $\alpha = \cos^{-1}x$. Then, by definition, $\cos\alpha = x$ (where $-1 \leq x \leq 1$ and $0 \leq \alpha \leq \pi$). In Figure 9-7, we construct a triangle for which $\cos\alpha = x$, with $x > 0$ and observe that

$$\sin\alpha = \frac{\sqrt{1 - x^2}}{1}. \quad \text{Thus,} \quad \sin\alpha = \sin(\cos^{-1}x) = \sqrt{1 - x^2}.$$

A similar argument holds for $x < 0$. (This time with a triangle in the second quadrant.)

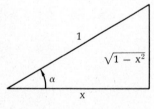

Figure 9-7

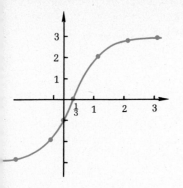

Figure 9-8

The graph of $f: x \rightarrow 2 \tan^{-1}(3x \rightarrow 1)$ compared to the graph of $x \rightarrow \tan^{-1} x$

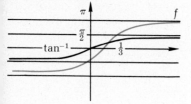

Figure 9-9

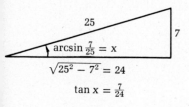

$$\sqrt{25^2 - 7^2} = 24$$

$$\tan x = \tfrac{7}{24}$$

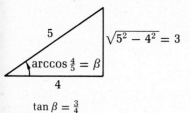

$$\tan \beta = \tfrac{3}{4}$$

Figure 9-10

Example 6

For any number x, let $f(x) = 2 \tan^{-1}(3x - 1)$. Graph f.

Solution

The graph of f has the same shape as the graph of $\tan^{-1}$. This may be seen by a direct computation (see the table and graph in Figure 9-8). Also, for any real number x, note that $f(x) = 2 \tan^{-1} 3(x - \tfrac{1}{3})$, and hence the graph of f may be obtained from the graph of $\tan^{-1}$ by a vertical stretch (by a factor of 2), a horizontal compression (by a factor of $\tfrac{1}{3}$), and then a horizontal shift $\tfrac{1}{3}$ unit to the right. See Figure 9-9.

Example 7

Find all real numbers x for which

$$\tan^{-1}(x + 2) = \sin^{-1} \tfrac{7}{25} + \cos^{-1} \tfrac{4}{5}$$

Solution

For any real number x,

$$\tan^{-1}(x + 2) = \sin^{-1} \tfrac{7}{25} + \cos^{-1} \tfrac{4}{5}$$

if and only if

$$\tan \tan^{-1}(x + 2) = \tan(\sin^{-1} \tfrac{7}{25} + \cos^{-1} \tfrac{4}{5})$$

that is,

$$x + 2 = \tan(\sin^{-1} \tfrac{7}{25} + \cos^{-1} \tfrac{4}{5})$$
$$= \frac{\tan(\sin^{-1} \tfrac{7}{25}) + \tan(\cos^{-1} \tfrac{4}{5})}{1 - \tan(\sin^{-1} \tfrac{7}{25}) \tan(\cos^{-1} \tfrac{4}{5})}$$
$$= \frac{\tfrac{7}{24} + \tfrac{3}{4}}{1 - \tfrac{7}{24} \cdot \tfrac{3}{4}} \quad \text{(see Figure 9-10)}$$
$$= \tfrac{4}{3}$$

or, equivalently,

$$x = -\tfrac{2}{3}$$

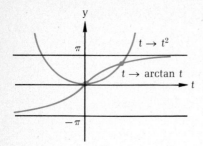

Figure 9-11

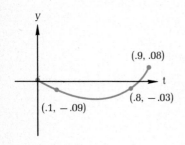

Figure 9-12

◊ **Example 8**

Find all real numbers t for which $\arctan t = t^2$.

Solution

From the graphs of arctan and the squaring function (see Figure 9-11) we suspect that there are two solutions. One solution is 0. (*Check:* $\arctan 0 = 0^2$.) To find the other solution (approximately), let $g(t) = t^2 - \arctan t$ and find where the graph of g crosses the horizontal axis. Since

$$g(\tfrac{1}{10}) = (\tfrac{1}{10})^2 - \arctan \tfrac{1}{10} \approx \tfrac{1}{100} - \tfrac{1}{10} < 0$$

and

$$g(1) = 1^2 - \arctan 1 = 1 - \frac{\pi}{4} > 0$$

the solution is between $\frac{1}{10}$ and 1. Further computations (see Figure 9-12) show that the second solution is between 0.8 and 0.9.

Exercises 9-1

In Exercises 1–9, compute $\sin^{-1} x$ and $\cos^{-1} x$ if x is

1. $\frac{1}{2}$ 2. $-\frac{1}{2}$

3. $\dfrac{1}{\sqrt{2}}$ 4. $-\dfrac{1}{\sqrt{2}}$

5. $\dfrac{\sqrt{3}}{2}$ 6. $-\dfrac{\sqrt{3}}{2}$

7. 1   8. -1

9. 0

In Exercises 10–16, compute $\tan^{-1} x$ and $\cot^{-1} x$ if x is

10. 0   11. $\sqrt{3}$

12. $-\sqrt{3}$ 13. $\dfrac{1}{\sqrt{3}}$

14. $-\dfrac{1}{\sqrt{3}}$ 15. 1

16. -1

In Exercises 17–33, find

17. $\cos\left(\arcsin \dfrac{1}{2}\right)$ 18. $\sin\left(\arccos \dfrac{1}{\sqrt{2}}\right)$

19. $\tan\left(\arccos \dfrac{\sqrt{3}}{2}\right)$ 20. $\sin(\cos^{-1} .73)$

21. $\cos(\sin^{-1}.47)$

22. $\sin[\cos^{-1}\frac{3}{5} + \sin^{-1}(-\frac{3}{5})]$

23. $\sin(\arccos .73)$

24. $\cos[\arcsin(-.47)]$

25. $\cos^{-1}(\sin\frac{1}{3})$

26. $\sin[2\cos^{-1}\frac{1}{3}]$

27. $\cot[\tan^{-1}\frac{1}{3}]$

28. $\cos(\sin^{-1}\frac{1}{10})$

29. $\sin(\cos^{-1}\frac{1}{5})$

30. $\tan(\sin^{-1}\frac{1}{3})$

31. $\cos[2\sin^{-1}\frac{1}{4}]$

32. $\cos[\sin^{-1}\frac{1}{3} + 2\sin^{-1}\frac{1}{5}]$ (*Hint:* Consider the addition formula for cos.)

33. $\sin[\sin^{-1}\frac{1}{5} + \cos^{-1}\frac{1}{4}]$ (*Hint:* Consider the addition formula for sin.)

34. Show that $\tan^{-1}\left(\cot\dfrac{1}{5}\right) = \dfrac{\pi}{2} - \dfrac{1}{5}$

Simplify Exercises 35–37 as shown below. In each case, x is a positive real number in the appropriate domain.

$$\sin(2\tan^{-1}x) = 2[\sin(\tan^{-1}x)][\cos(\tan^{-1}x)]$$

$$= 2 \cdot \frac{x}{\sqrt{1+x^2}} \cdot \frac{1}{\sqrt{1+x^2}} = \frac{2x}{1+x^2}$$

35. $\tan(2\tan^{-1}x)$ **36.** $\tan(\cos^{-1}x)$

37. $\sin(\sin^{-1}x + \cos^{-1}x)$

38. Compute (a) $\cos^{-1}\left[\dfrac{1}{2}\cot\left(\sin^{-1}\dfrac{1}{2}\right)\right]$ and

$$\text{(b) } \tan\left[\sin^{-1}\left(\cos\dfrac{\pi}{6}\right)\right]$$

39. Show that $\cos^{-1}\dfrac{4}{5} = 2\cos^{-1}\dfrac{3}{\sqrt{10}}$. (*Hint:* Use the fact that for x and y in $[0, \pi]$, $x = y$ if and only if $\cos x = \cos y$.)

In Exercises 40–44, show that

40. $\sin^{-1}\dfrac{7}{25} = 2\sin^{-1}\dfrac{\sqrt{2}}{10}$

41. $\cos^{-1}\dfrac{3}{\sqrt{10}} + \cos^{-1}\dfrac{2}{\sqrt{5}} = \dfrac{\pi}{4}$

42. $\sin^{-1}\frac{3}{5} = \sin^{-1}\frac{77}{85} - \sin^{-1}\frac{8}{17}$

43. $\cos^{-1}\frac{3}{5} = \sin^{-1}\frac{63}{65} - \sin^{-1}\frac{5}{13}$

44. $\cos^{-1}x = \dfrac{\pi}{2} - \sin^{-1}x$ if $-1 \le x \le 1$

In Exercises 45–52, find all numbers x for which

45. $\cos^{-1} 2x + \cos^{-1} 3x = \cos^{-1} \frac{2}{3}$ (*Hint:* Use the addition formula for cos.)

46. $\sin^{-1} x - \cos^{-1} x = \dfrac{5\pi}{6}$

47. $\sin^{-1} x = 2 \tan^{-1} x$

48. $\tan^{-1} 2x = 2 \cos^{-1} x$

49. $\tan^{-1} (x + 1) + \tan^{-1} (x - 1) = \dfrac{\pi}{4}$

50. $\tan^{-1} (x + 4) = \sin^{-1} \frac{7}{25} + \cos^{-1} \frac{4}{5}$

51. $\sin^{-1} (2x - 1) + \tan^{-1} \frac{5}{12} = \cos^{-1} \frac{3}{5}$

52. $\sin^{-1} (3x + 2) = 3 \sin^{-1} x$

53. Show that for any $x > 0$, $\tan^{-1} x + \tan^{-1} \dfrac{1}{x} = \dfrac{\pi}{2}$.

In Exercises 54–59, show that

54. $\tan^{-1} x = \cot^{-1} \dfrac{1}{x}$ for $x > 0$

55. $\tan^{-1} x = \cot^{-1} \dfrac{1}{x} - \pi$ for $x < 0$

56. $2 \tan^{-1} x = \tan^{-1} \dfrac{2x}{1 - x^2}$ for $-1 < x < 1$

57. $2 \tan^{-1} x = \sin^{-1} \dfrac{2x}{1 + x^2}$ for $-1 \le x \le 1$

58. $\cos^{-1} x = \frac{1}{2} \cos^{-1} (2x^2 - 1)$ for $0 \le x \le 1$

59. $2 \sin^{-1} x = \sin^{-1} (2x \sqrt{1 - x^2})$ if $-\dfrac{1}{\sqrt{2}} \le x \le \dfrac{1}{\sqrt{2}}$

60. Graph f if (a) $f(x) = \arctan (x + 1)$; (b) $f(x) = 1 + \arctan x$; (c) $f(x) = \arctan 2x$.

61. Let y be a real number in $(0, \pi)$. Find all real numbers x for which $\cos^{-1} \dfrac{x + 1}{x - 1} = y$.

62. Let a, b, c, and d be any numbers with $a \ne 0$ and $b \ne 0$. Find all numbers x for which $a \sin (bx + c) = d$.

Answer Exercises 63–70 true or false.

63. $\sin^{-1} x = \sin^{-1} (-x)$ for $-1 \le x \le 1$

64. $\sin^{-1} x = -\sin^{-1} (-x)$ for $-1 \le x \le 1$

65. $\cos^{-1} x = \cos^{-1} (-x)$ for $-1 \le x \le 1$

66. $\cos^{-1} x = -\cos^{-1}(-x)$ for $-1 \leq x \leq 1$

67. $\tan^{-1} x = \tan^{-1}(-x)$ for all numbers x

68. $\tan^{-1} x = -\tan^{-1}(-x)$ for all numbers x

69. $(\sin^{-1} x)^2 + (\cos^{-1} x)^2 = 1$ for $-1 \leq x \leq 1$

70. $\tan^{-1} x = \dfrac{\sin^{-1} x}{\cos^{-1} x}$ for $-1 \leq x \leq 1$

71. Use tables and graphs to find (approximately) all positive numbers t for which (a) $\arctan t = t/2$ and (b) $\arctan t = t^3$.

9-2 THE LAWS OF SINES AND COSINES. SOLVING TRIANGLES

This section is concerned with the information that is needed to determine a triangle. We say that a triangle is determined when the lengths of its three sides are known. However, the three angles of a triangle do not determine it, since similar triangles need not be congruent.

On the other hand, a triangle *is* determined when two of its angles and a side, or two sides and the angle between them, are known. In the first case, the remaining two sides may be found using the *law of sines*, and in the second case, the remaining side may be found using the *law of cosines*.

The **law of sines** states that

$$\frac{\sin \alpha}{a} = \frac{\sin \beta}{b} = \frac{\sin \gamma}{c}$$

where α, β, and γ are the three angles of a triangle, and a, b, and c are the lengths of the sides opposite these three angles, respectively.

The **law of cosines** states that

$$c^2 = a^2 + b^2 - 2ab \cos \gamma$$

where a, b, and c are the lengths of the sides of a triangle, and γ is the angle opposite the side of length c.

To verify the law of sines, consider the triangles in Figure 9-13. Since $d/a = \sin \beta$ and $d/b = \sin \alpha$, we have $d = a \sin \beta = b \sin \alpha$, and hence $(\sin \alpha)/a = (\sin \beta)/b$; and since this argument applies to *any* two of the three angles, the law of the sines follows.

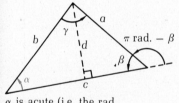

α is acute (i.e. the rad. measure of α is $< \frac{\pi}{2}$)

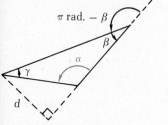

α is obtuse (i.e. the rad. measure of α is $> \frac{\pi}{2}$)

Figure 9-13

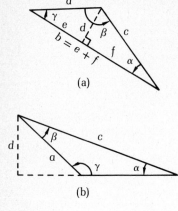

(a)

(b)

Figure 9-14

To verify the law of cosines for the case $0 < \gamma \leq \pi/2$, consider the triangle in Figure 9-14(a). Here we have $d/a = \sin \gamma$, $e/a = \cos \gamma$, and hence $d = a \sin \gamma$, $e = a \cos \gamma$, and $f = b - e = b - a \cos \gamma$. Since c, d, and f are the sides of a right triangle, we obtain, therefore,

$$
\begin{aligned}
c^2 &= d^2 + f^2 \\
&= (a \sin \gamma)^2 + (b - a \cos \gamma)^2 \\
&= a^2 \sin^2 \gamma + b^2 - 2ba \cos \gamma + a^2 \cos^2 \gamma \\
&= a^2(\sin^2 \gamma + \cos^2 \gamma) + b^2 - 2ab \cos \gamma \\
&= a^2 + b^2 - 2ab \cos \gamma,
\end{aligned}
$$

which is the law of cosines. A similar argument holds for the case $\pi/2 < \gamma < \pi$ [see Figure 9-14(b)].

Example 1

If, in Figure 9-14(a), $\alpha = 44°10'$ (where $60' = 1°$), $\beta = 61°20'$, and $c = 5$ feet, find the lengths a and b.

Solution

$\gamma = 180° - 44°10' - 61°20' = 74°30'$. From the tables, $\sin \alpha \approx 0.6967$, $\sin \beta \approx 0.8774$, and $\sin \gamma \approx 0.9636$. Hence, since $\dfrac{\sin \alpha}{a} = \dfrac{\sin \gamma}{c}$, we see that $a = \dfrac{c \sin \alpha}{\sin \gamma}$; that is,

$$ a \approx \frac{5 \times .6967}{.9636} \approx 3.6151 \text{ feet} $$

$$ b \approx \frac{5 \times .8774}{.9636} \approx 4.5527 \text{ feet} $$

Example 2

Each of two men sights a stationary object that is directly above a straight line path one-half mile long between the men. The angles of elevation are, respectively, $64°12'$ and $48°31'$ (see Figure 9-15). How far is the object from each of the men? How far is the object above the ground?

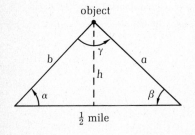

Figure 9-15

Solution

From Figure 9-15, the distances of the object to the men are, respectively, b and a, and the distance of the object from the ground is h. If $\alpha = 64°12'$, $\beta = 48°31'$, then the remaining angle in the triangle so determined is

$$ 180° - (64°12' + 48°31') = 180° - 112°43' = 67°17' $$

From the tables, $\sin \alpha \approx .9003$, $\sin \beta \approx .7492$ and $\sin \gamma \approx .9224$.

Thus, since

$$\frac{a}{\sin \alpha} = \frac{\frac{1}{2}}{\sin \gamma} \quad \text{and} \quad \frac{b}{\sin \beta} = \frac{\frac{1}{2}}{\sin \gamma}$$

we have

$$a = \frac{1}{2} \cdot \frac{.9003}{.9224} \approx .488 \text{ (mile)}$$

and

$$b = \frac{1}{2} \cdot \frac{.7492}{.9224} \approx .406 \text{ (mile)}$$

Finally,

$$h = b \sin \alpha \approx (.406)(.9003) \approx .366 \text{ mile}$$

□

Given real numbers a and b and an angle α, is there a triangle having two legs of lengths a and b, with α the angle opposite the side of length a? Figure 9-16(a) illustrates why there is no such triangle if $a < b \sin \alpha$. In fact, note that by the law of sines, if there were an angle β opposite the side of length b, then $\sin \beta$ would be equal to $\frac{b \sin \alpha}{a}$, and hence, since $a < b \sin \alpha$, it would follow that $\sin \beta > 1$, which is impossible.

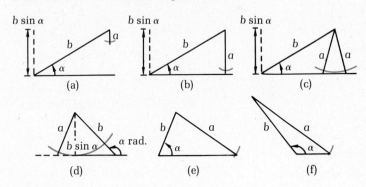

Figure 9-16

There are six possible cases covered by the following list:

1. $a < b \sin \alpha$: no such triangle   [Figure 9-16(a)]
2. $a = b \sin \alpha$: exactly one triangle   [Figure 9-16(b)]
3. $b \sin \alpha < a < b$ and $\alpha < 90°$:
exactly two triangles   [Figure 9-16(c)]
4. $b \sin \alpha < a < b$ and $\alpha > 90°$:
no such triangle   [Figure 9-16(d)]
5. $b < a$ and $\alpha < 90°$: exactly one triangle   [Figure 9-16(e)]
6. $b < a$ and $\alpha > 90°$: exactly one triangle   [Figure 9-16(f)]

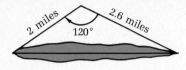

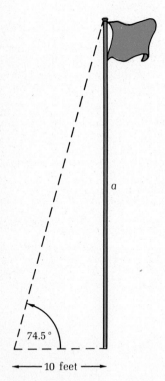

Figure 9-17

Figure 9-18

Example 3

At a point that is 2 miles from one end of a lake and 2.6 miles from the other end, the lake subtends an angle of 120° (see Figure 9–17). Find the length of the lake.

Solution

If c is the length of the lake, then, from the law of cosines,

$$c^2 = 4 + (2.6)^2 - (2)(2)(2.6) \cos \frac{2\pi}{3} \quad \text{(miles)}$$

Now,

$$\cos \frac{2\pi}{3} = -\cos \left(\pi - \frac{2\pi}{3} \right) = -\cos \frac{\pi}{3} = -\frac{1}{2}$$

so

$$c = \sqrt{15.96} \text{ miles}$$

□

Remark   It is a theorem from plane geometry that no three real numbers a, b, and c can be the lengths of the sides of a triangle, unless the sum of any two is greater than the third. This shows up in the law of cosines, for if $a + b < c$, then $(a + b)^2 < c^2$, or $a^2 + b^2 + 2ab < c^2$, or $a^2 + b^2 - c^2 < -2ab$; thus, if there were a triangle with angle γ opposite the side of length c, then, from the law of cosines,

$$\cos \gamma = \frac{a^2 + b^2 - c^2}{2ab} < -1$$

and, consequently, $\cos \gamma < -1$, which is impossible.

The following are more examples of problems in which tables of trigonometric functions are used.

Example 4

From a point on the ground the distance to the base of a flagpole is 10 feet. The angle of elevation (see Figure 9-18) is 74.5°. Find the height of the flagpole.

Solution

Let a be the height of the flagpole in feet. Then (see Figure 9-18) $\frac{a}{10} = \tan 74.5° \approx 3.61$. Hence, $a \approx 10 \ (3.61) \approx 36.1 \ \text{(feet)}$.

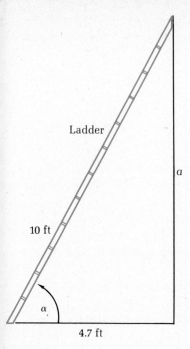

Ladder

a

10 ft

α

4.7 ft

Figure 9-19

Example 5

A ladder 10 feet long leans against a wall. The bottom of the ladder is 4.7 feet from the wall. Find the angle the ladder makes with the floor and the height of the top of the ladder. See Figure 9-19.

Solution

Referring to Figure 9-19, we must find the radian measure of α, the angle the ladder makes with the floor, and the height of the top of the ladder *a*. We see that $\cos \alpha = 4.7/10 = .47$, and since the radian measure of α is between 0 and $\pi/2$, we see from the table that $\alpha = 61°58'$. The number *a* can be found in a number of ways; for example, using the Theorem of Pythagoras or by observing that $a \approx 10 \sin 61°58'$.

Exercises 9-2

1. If two sides of a triangle have lengths 100 inches and 30 inches, respectively, and if the angle between them is 30°, find the length of the remaining side.

2. Find the distance *PQ* across a lake if a line *PR* of length 56 yards is laid off and angles *QPR* and *PRQ* are 80.3° and 70.4°.

3. In a parallelogram *ABCD*, suppose that side *AB* has length 60 inches, side *AD* has length 100 inches, and angle *BAD* is 60°. Find the length of the diagonal *AC* and angle *CAD*.

In Exercises 4–7, we are given the triangle in the accompanying figure. Find:

4. *c*   if $a = 3$, $b = 5$, $\gamma = 60°$

5. *b*   if $a = 15$, $c = 11$, $\beta = 35°40'$

6. *a*   if $b = 12$, $c = 8$, $\alpha = 120°$

7. *c*   if $a = 120$, $b = 150$, $\gamma = 140°15'$

Let A, B, and C be the vertices of a triangle with angles α, β, γ and length of sides a, b, and c. In Exercises 8–11, find a, b, given that:

8. $\cos \alpha = \frac{1}{2}$, $\cos \beta = \frac{1}{3}$, $c = 10$

9. $\cos \alpha = \frac{3}{5}$, $\cos \beta = \frac{7}{10}$, $c = 5$

10. $\alpha = 20°$, $\beta = 50°$, $c = 10$

11. $\alpha = 62°$, $\beta = 18°$, $c = 5$

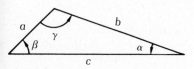

a   *b*

γ

β   *c*   α

Exercises 4–7

In Exercises 12–16, refer to the figure for Exercises 4–7 and find the lengths of the remaining sides and the remaining angles, given that:

12. $\beta = 50°$, $\gamma = 49.7°$, $c = 8$

13. $\gamma = 95°$, $\beta = 43.4°$, $b = 40$

14. $a = 16$, $\alpha = 30°$, $\beta = 65°$

15. $c = 120$, $\beta = 75°\,50'$, $\gamma = 35°\,15'$

16. $b = 18$, $\beta = 64°\,3'$, $\gamma = 18°\,25'$

17. Find the smallest and largest angles of the triangle whose sides are 5, 10, and 12. (*Hint:* Use the law of cosines to find the cosines of the angles.)

18. In a triangle ABC, find the smallest angle if $a = 4$, $b = 5$, $c = 6$.

19. In a triangle ABC, find the largest angle if $a = 5$, $b = 8$, $c = 10$.

20. In a triangle ABC, find (angle) β if $a = 14$, $b = 48$, $c = 50$.

21. Two ships, A and B, start from the same place at the same time. Ship A sails 15° west of north at 10 miles per hour; ship B sails 48.2° east of north at 12 miles per hour. At the end of one hour, how far apart are the ships?

22. A town is on a straight freeway. A major shopping center is 30 miles from the town on a road that makes an angle of $\pi/6$ radians with the freeway. Is it possible to build a 20-mile road which will connect the shopping center with the freeway?

9-3 POLAR COORDINATES

We shall now consider another system of coordinates. If a point (x, y) lies on a circle of radius r (whose center is the origin), then

$$x = r \cos t \quad \text{and} \quad y = r \sin t$$

where t is the radian measure of the angle between the horizontal axis and the segment joining (x, y) to the origin. See Figure 9-20.

The real numbers r and t determine the point (x, y) and are called the **polar coordinates** of this point. To avoid confusion with the *rectangular coordinates* x and y, the point (x, y) in polar coordinates is designated by $(r, t)_P$. Note that

$$r = \sqrt{x^2 + y^2} \quad \text{and} \quad \tan t = \frac{y}{x}$$

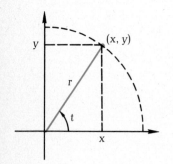

Figure 9-20

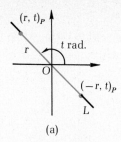

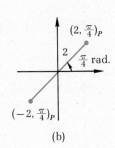

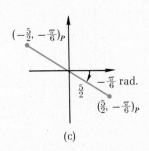

Figure 9-27

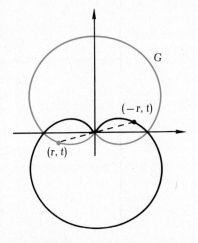

Figure 9-28

Example 8

Let a be a positive real number. Show that the graph of the set of all points $(r, t)_P$ for which $r = a \sin t$ is the circle of radius $a/2$ and center $(a/2, \pi/2)_P$ [i.e., with center $(0, a/2)$ in rectangular coordinates]. See Figure 9-29(a).

Solution

If the point $(r, t)_P$ on the graph has rectangular coordinates x, y, then, since $r = \sqrt{x^2 + y^2}$ and $\dfrac{y}{r} = \sin t$, we obtain from the fact that $r = a \sin t$

$$\sqrt{x^2 + y^2} = a \frac{y}{\sqrt{x^2 + y^2}}$$

that is, $x^2 + y^2 = ay$, or, completing the square,

$$x^2 + \left(y - \frac{a}{2}\right)^2 = \left(\frac{a}{2}\right)^2$$

Thus, the graph is the circle of radius $\dfrac{a}{2}$ and center $\left(0, \dfrac{a}{2}\right)$.

$\square$

By similar arguments we could show that the graph of the set of all points $(r, t)_P$ for which $r = a \cos t$, $r \geq 0$, is the circle with radius $a/2$ and center $\left(\dfrac{a}{2}, 0\right)$. See Figure 9-29(b).

In Example 8, we changed an equation that was given in polar coordinates to an equation expressed in rectangular coordinates. Thus, the set of all points with polar coordinates r, t, with $r \geq 0$, for which $r = \sin t$, is the set of all points with rectangular coordinates x, y for which $x^2 + (y - \frac{1}{2})^2 = \frac{1}{4}$. In the following examples, we shall change equations given in rectangular coordinates to equations expressed in polar coordinates.

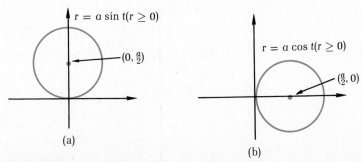

Figure 9-29

Example 9

The graph of the set of all (x, y) for which $x^2 - y^2 = 1$ is a hyperbola. Find the set of all points $(r, t)_P$ which are on this hyperbola.

Solution

$(r, t)_P$ has rectangular coordinates x, y, with $x = r \cos t$ and $y = r \sin t$, and therefore is on the graph if and only if

$$r^2 \cos^2 t - r^2 \sin^2 t = 1$$

or, equivalently,

$$r^2(\cos^2 t - \sin^2 t) = 1$$
$$r^2 \cos 2t = 1 \quad (\text{since } \cos^2 t - \sin^2 t = \cos 2t)$$

or, finally,

$$r^2 = \sec 2t$$

Example 10

Find the equation in polar coordinates for the circle whose equation (in rectangular coordinates) is $(x - 1)^2 + (y + 2)^2 = 1$.

Solution

With $x = r \cos t$ and $y = r \sin t$, we see that $(r, t)_P$ is on the circle if and only if

$$[(r \cos t) - 1]^2 + [(r \sin t) + 2]^2 = 1$$
$$r^2 \cos^2 t - 2r \cos t + 1 + r^2 \sin^2 t + 4r \sin t + 4 = 1$$
$$r^2(\cos^2 t + \sin^2 t) - 2r(\cos t - 2 \sin t) + 4 = 0$$

or, finally,

$$r^2 - 2r(\cos t - 2 \sin t) + 4 = 0$$

Exercises 9-3

In Exercises 1–5, find the Cartesian coordinates of each of the given points.

1. $\left(2, \dfrac{\pi}{6}\right)_P$

2. $\left(3, \dfrac{3\pi}{2}\right)_P$

3. $(4, \pi)_P$

4. $(4, 0)_P$

5. $\left(5, \dfrac{3\pi}{4}\right)_P$

In Exercises 6–10, find polar coordinates of each of the given points.

6. $(1, -1)$ **7.** $(-1, -\sqrt{3})$

8. $(-5, 0)$ **9.** $(0, 0)$

10. $(-\pi, 0)$

In Exercises 11–14, graph the given sets of points.

11. $(3, s)_P$,   s any number

12. $(3, s)_P$, $0 \le s \le \pi$

13. $(q, 0)_P$,   for any real number q

14. $\left(q, \dfrac{\pi}{2}\right)_P$, $q > 1$

Sketch the graph for each of the equations in Exercises 15–18 [i.e., sketch the set of all $(r, t)_P$].

15. $r = 1 - \cos t$ **16.** $r = a \cos t$, $a > 0$, $0 \le t \le \pi/2$

17. $r = 2 - \sin t$ **18.** $r = \sin 3t$, $0 \le t \le \dfrac{2\pi}{3}$

Express Exercises 19–23 in rectangular coordinates.

19. $r = -4 \cos t$ **20.** $r = 3 \csc t$

21. $r = \dfrac{6}{2 - \sin t}$ **22.** $r^2 = \sin t$

Express Exercises 23–27 in polar coordinates.

23. $x(x^2 + y^2) = 2ay^2$ **24.** $x^2 + y^2 - 2ay = 0$

25. $y^2 = 4ax + 4a^2$

26. $x^2 + y^2 = ay + a\sqrt{x^2 + y^2}$

27. $(x^2 + y^2)^2 = 2a^2(x^2 - y^2)$

9-4 COMPLEX NUMBERS (CONTINUED)

We have seen, as a by-product of our study of trigonometric functions, how *polar coordinates* are used to locate points in the plane. In this section, a closely related use of trigonometry will be applied to a further investigation of complex numbers.

Note, first, that if $a + bi$ is a nonzero complex number, then

$$a + bi = \sqrt{a^2 + b^2}\left(\frac{a}{\sqrt{a^2 + b^2}} + i\frac{b}{\sqrt{a^2 + b^2}}\right)$$

From Figure 9-30 we see that if θ is the angle measured from the positive side of the horizontal axis to the point (a, b), then

$$\frac{a}{\sqrt{a^2 + b^2}} = \cos \theta \quad \text{and} \quad \frac{b}{\sqrt{a^2 + b^2}} = \sin \theta$$

The number θ ("theta") is called the **argument** (or **direction**) of $a + bi$ and is frequently denoted by **arg** $(a + bi)$.

Example 1

(a) $\arg (1) = 0$ (b) $\arg (i) = \dfrac{\pi}{2}$ (c) $\arg (-1) = \pi$

(d) $\arg (-i) = \dfrac{3\pi}{2}$ (e) $\arg (3) = 0$ (f) $\arg (1 + i) = \dfrac{\pi}{4}$

Example 2

$$1 + \sqrt{3}i = 2 \left(\frac{1}{2} + \frac{\sqrt{3}}{2} i \right)$$

Here $|1 + \sqrt{3}i| = \sqrt{1 + 3} = 2$, $\dfrac{1}{2} = \cos \theta$, and $\dfrac{\sqrt{3}}{2} = \sin \theta$ where $\theta = \arg (1 + \sqrt{3}i) = 60° = \pi/3$ radians. See Figure 9-31. $\square$

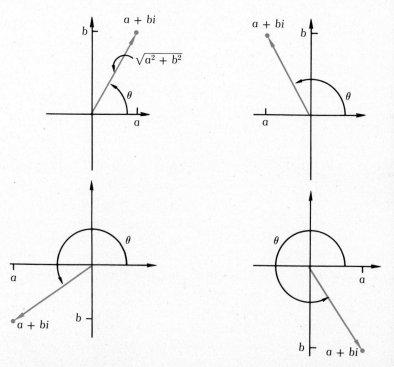

Figure 9-30

Thus, every complex number z is the product of its *absolute value* (or *length*), $|z|$, and the *unit* complex number (that is, the complex number of length 1) whose direction is the same as that of z:

$$z = |z| (\cos \theta + i \sin \theta)$$

[where $\theta = \arg (z)$]. In particular, for every complex number v on the unit circle, we have $v = \cos\theta + i \sin\theta$, where $\theta = \arg (v)$; and conversely, for every angle θ, $\cos\theta + i \sin\theta$ is a complex number on the unit circle (since $\sqrt{\cos^2\theta + \sin^2\theta} = 1$).

This leads to the function u, defined by

$$u(x) = \cos x + i \sin x$$

where, as in the case of trigonometric functions, the domain of u may be taken to be the set of all angles or the set of real numbers.*

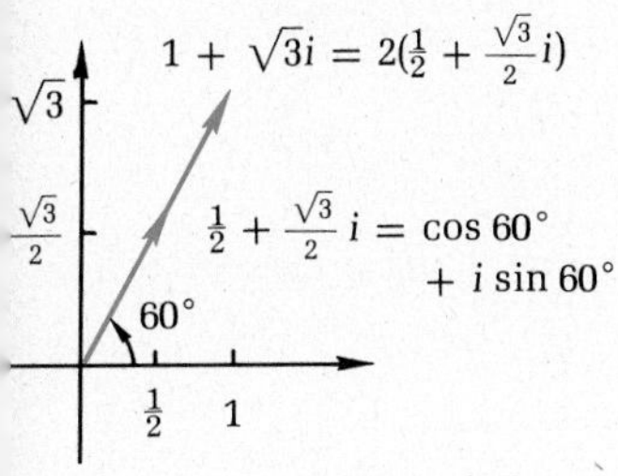

□

We shall now investigate a few interesting properties of u. As has been already noted, for every angle (or real number) x, $u(x)$ is a complex number on the unit circle; that is, $|u(x)| = 1$. Furthermore, if y is also an angle (or a real number), then†

$$\begin{aligned}
u(x) \cdot u(y) &= (\cos x + i \sin x) \cdot (\cos y + i \sin y) \\
&= (\cos x \cos y - \sin x \sin y) \\
&\qquad + i(\sin x \cos y + \cos x \sin y) \\
&= \cos (x + y) + i \sin (x + y) \\
&= u(x + y)
\end{aligned}$$

In other words, *the product of two unit complex numbers is the unit complex number whose argument is the sum of the arguments of the two factors.* From a geometric standpoint, this means that to multiply two unit complex numbers amounts to *rotating* one of them by the argument of the other. See Figure 9-32.

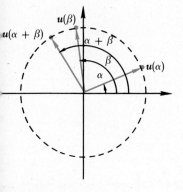

*The function u is sometimes referred to as "cis," which is an abbreviation of $\cos + i \sin$.
†Note the similarity between the behavior of u and that of the exponential functions.

Figure 9-31

Figure 9-32

It follows in general that if z_1 and z_2 are complex numbers, and θ_1 and θ_2 are their arguments, respectively, then

$$
\begin{aligned}
z_1 \cdot z_2 &= |z_1| \cdot u(\theta_1) \cdot |z_2| \cdot u(\theta_2) \\
&= |z_1| \cdot |z_2| \cdot u(\theta_1) \cdot u(\theta_2) \\
&= |z_1| \cdot |z_2| \cdot u(\theta_1 + \theta_2)
\end{aligned}
$$

that is, *the product of any two complex numbers is the complex number whose length is the product of the two lengths and whose argument is the sum of the two arguments.*

Example 3

We saw in Example 2 that

$$
\frac{1}{2} + \frac{\sqrt{3}}{2} i = \cos 60° + i \sin 60° = u(60°)
$$

Similarly,

$$
\frac{\sqrt{3}}{2} + \frac{1}{2} i = \cos 30° + i \sin 30° = u(30°)
$$

Thus,

$$
\left(\frac{1}{2} + \frac{\sqrt{3}}{2} i \right)\left(\frac{\sqrt{3}}{2} + \frac{1}{2} i \right) = u(60°) \cdot u(30°) = u(90°) = i
$$

or, directly:

$$
\left(\frac{1}{2} + \frac{\sqrt{3}}{2} i \right) \cdot \left(\frac{\sqrt{3}}{2} + \frac{1}{2} i \right)
$$

$$
= \left(\frac{\sqrt{3}}{4} - \frac{\sqrt{3}}{4} \right) + \left(\frac{3}{4} i + \frac{1}{4} i \right) = i
$$

Example 4

$$
2u(46°) \cdot 3u(104°) = 6u(150°)
$$

$$
= 6 \left(-\frac{\sqrt{3}}{2} + i \frac{1}{2} \right) = -3\sqrt{3} + 3i
$$

Example 5

$$
(1 + i)^2 = \left[\sqrt{2} \left(\frac{1}{\sqrt{2}} + \frac{1}{\sqrt{2}} i \right) \right]^2 = \left[\sqrt{2} \left(\cos \frac{\pi}{4} + i \sin \frac{\pi}{4} \right) \right]^2
$$

$$
= 2 \left[u \left(\frac{\pi}{4} \right) \right]^2 = 2u \left(\frac{\pi}{2} \right) = 2i
$$

or

$$
(1 + i)^2 = 1 + 2i + i^2 = 2i
$$

□

Since multiplication is commutative and associative, it follows that for complex numbers z_1, z_2, $\ldots$, z_n, with arguments θ_1, θ_2, $\ldots$, θ_n, we have

$$z_1 \cdot z_2 \cdots z_n = |z_1| \cdot |z_2| \cdots |z_n| \cdot u(\theta_1 + \theta_2 + \cdots + \theta_n)^*$$

In particular, if z is a complex number with argument θ, then

$$z^n = |z|^n \cdot u(n\theta) = |z|^n (\cos n\theta + i \sin n\theta)$$

Also note that $u(-\theta) \cdot u(\theta) = u(0°) = 1$, so that $u(-\theta)$ is the reciprocal of $u(\theta)$. Furthermore,

$$u(-\theta) = \cos(-\theta) + i \sin(-\theta) = \cos\theta - i \sin\theta = \overline{u(\theta)}$$

Thus,

$$\frac{1}{u(\theta)} = u(-\theta) = \overline{u(\theta)}$$

Example 6

$(1 + i)^8 = (\sqrt{2})^8 \cdot [u(45°)]^8 = 2^4 \cdot u(360°) = 16$. (Note that we have just shown that $1 + i$ is an eighth root of 16.)

Example 7

Find the reciprocal of $1 + i$.

Solution

With arg $(1 + i)$ in radians,

$$\frac{1}{1 + i} = \frac{1}{\sqrt{2} \cdot u(\pi/4)} = \frac{1}{\sqrt{2}} u(-\pi/4)$$

$$= \frac{1}{\sqrt{2}}\left(\frac{1}{\sqrt{2}} - i\frac{1}{\sqrt{2}}\right) = \frac{1 - i}{2}$$

Finally, observe that for any angle θ, any natural number n, and any whole number k with $0 \leq k < n$, we have

$$\left[u\left(\frac{\theta + k \cdot 360°}{n}\right)\right]^n = u(\theta + k \cdot 360°) = u(\theta)$$

and so each of the n complex numbers

$$u\left(\frac{\theta}{n}\right), \quad u\left(\frac{\theta + 360°}{n}\right), \ldots, u\left(\frac{\theta + (k - 1) \cdot 360°}{n}\right)$$

*This is known as *De Moivre's Theorem*.

is an nth root of $u(\theta)$. We shall leave it for the reader to verify (a) that these n roots are distinct, and (b) that there are no others. Thus, *if z is a complex number, θ its argument, and n a natural number, then z has precisely the following nth roots:*

$$\sqrt[n]{|z|}\ u\left(\frac{\theta + k \cdot 360°}{n}\right)$$

or

$$\sqrt[n]{|z|}\ u\left(\frac{\theta + 2\pi k}{n}\right), \quad k = 0, 1, \cdots, n-1$$

Example 8

Find (a) the three third roots of 1, and (b) the six sixth roots of 64.

Solution

(a) Since $1 = u(0°)$, the third roots of 1 are

$$u(0°) = 1, \quad u\left(\frac{360°}{3}\right) = u(120°) = -\frac{1}{2} + \frac{\sqrt{3}}{2}i,$$

and

$$u\left(\frac{720°}{3}\right) = u(240°) = -\frac{1}{2} - \frac{\sqrt{3}}{2}i$$

(b) Since $1 = u(0)$, the sixth roots of 64 are

$$2, \quad 2u(\pi/3) = 1 + \sqrt{3}i, \quad 2u(2\pi/3) = -1 + \sqrt{3}i,$$
$$2u(\pi) = -2, \quad 2u(4\pi/3) = -1 - \sqrt{3}i,$$

and

$$2u(5\pi/3) = 1 - \sqrt{3}i$$

Example 9

Find the fourth roots of $1 + \sqrt{3}i$.

Solution

We have

$$1 + \sqrt{3}i = 2\left(\frac{1}{2} + \frac{\sqrt{3}}{2}i\right) = 2u(60°)$$

Hence, the fourth roots of $1 + \sqrt{3}i$ are:

$$\sqrt[4]{2}u(15°) = \sqrt[4]{2}(\cos 15° + i \sin 15°)$$
$$\sqrt[4]{2}u(15° + 90°) = \sqrt[4]{2}u(105°) = \sqrt[4]{2}(\cos 105° + i \sin 105°)$$
$$\sqrt[4]{2}u(15° + 180°) = \sqrt[4]{2}u(195°) = \sqrt[4]{2}(\cos 195° + i \sin 195°)$$

and

$$\sqrt[4]{2}u(15° + 270°) = \sqrt[4]{2}u(285°) = \sqrt[4]{2}(\cos 285° + i \sin 285°)$$

Example 10

Find all complex numbers x for which $x^3 + 8i = 0$.

Solution

The required numbers are clearly the cube roots of $-8i$. Since $-i = u(270°)$, the cube roots of $-8i$ are

$$2u(90°) = 2i,$$

$$2u(210°) = 2\left(-\frac{\sqrt{3}}{2} - \frac{1}{2}i\right) = -\sqrt{3} - i,$$

and

$$2u(330°) = 2\left(\frac{1}{2} - \frac{\sqrt{3}}{2}i\right) = 1 - \sqrt{3}i$$

Example 11

Let $f(z) = z^3 + 8i$ for any complex number z. Then, f is a complex polynomial function,* and, from Example 10, we see that the zeros of f are $2i$, $-\sqrt{3} - i$ and $1 - \sqrt{3}i$.

Exercises 9-4

In Exercises 1–5, find the length $|z|$ and the argument arg (z) of each of the given complex numbers.

1. π

2. $3 + 4i$

3. $3 - 4i$

4. $-\sqrt{2} - \sqrt{2}i$

5. $2\pi i$

In Exercises 6–10, find real numbers a and b for which $z = a + bi$, given that:

6. $|z| = \pi$,   arg $(z) = \pi$

7. $|z| = \sqrt{6}$,   arg $(z) = \dfrac{\pi}{4}$

8. $|z| = 1$,   arg $(z) = 2\pi$

9. $|z| = 2$,   arg $(z) = \dfrac{3\pi}{2}$

10. $|z| = k$,   arg $(z) = (2n + 1)\pi$, n an integer, k a real number

*See Section 5-6.

In Exercises 11–15, write each of the given complex numbers as $r \cdot u(\theta)$ for the appropriate numbers r and θ (in radians), and then perform the indicated computation.

11. $z_1 = 1 - \sqrt{3}i$, $z_2 = \sqrt{3} + i$;   compute $z_1 \cdot z_2$.

12. $z_1 = 1 + \sqrt{3}i$, $z_2 = \sqrt{3} - i$;   compute $\dfrac{z_1}{z_2}$.

13. $z = -\sqrt{3} - i$;   compute $\dfrac{1}{z}$.

14. $z = -1 + \sqrt{3}i$;   compute z^2.

15. $z_1 = -1 + \sqrt{3}i$, $z_2 = -1 - \sqrt{3}i$;   compute $z_1 \cdot z_2$.

16. Find all the fifth roots of unity.

17. Find the four fourth roots of -16. (Express answer in radians.)

18. Find all zeros of the function f given by $f(z) = z^4 + 1$, z complex.

19. Find the two square roots of $(2\sqrt{3} - 2i)$. (Express answer in degrees.)

20. Find all complex numbers z for which $z^6 + 1 = \sqrt{3}i$.

21. Find all complex numbers z for which $2z^2 + 2iz - 5 = 0$. (*Hint:* Quadratic equation.)

22. Prove that $z^n = |z|^n (\cos n\theta + i \sin n\theta)$. (*Hint:* Use mathematical induction.)

23. Let z_1 and z_2 be nonzero complex numbers. What relationships must exist between arg (z_1) and arg (z_2) such that the quotient z_1/z_2 will be (a) complex and not real? (b) real?

24. Show that if z_1 is any complex nonreal (i.e., imaginary) nth root of unity, then

$$1 + z_1 + z_1^2 + \cdots + z_1^{n-1} = 0$$

(*Hint:* Consider the sum of the appropriate geometric progression.)

Miscellaneous Problems

For Problems 1 through 5: Let the sides of a triangle have lengths a, b, and c, and suppose that α is the angle opposite the side of length a. Let s = (a + b + c)/2.

1. Show that $1 + \cos \alpha = \dfrac{(b + c + a)(b + c - a)}{2bc}$. (*Hint:* Consider the law of cosines.)

2. Show that $1 - \cos \alpha = \dfrac{(a - b + c)(a + b - c)}{2bc}$.

3. Show that $\cos \dfrac{\alpha}{2} = \sqrt{\dfrac{s(s - a)}{bc}}$. (*Hint:* Consider the half-angle formula for cos.)

4. Show that $\sin \dfrac{\alpha}{2} = \sqrt{\dfrac{(s - b)(s - c)}{bc}}$.

5. Use Problems 3 and 4 to show that the area of any triangle is equal to $\sqrt{s(s - a)(s - b)(s - c)}$.

6. Three circles of radii 4, 5, and 6 are pairwise tangent to each other. A triangle is formed by connecting the centers of the circles. Find the angles of this triangle. (*Hint:* Find the lengths of the sides of the triangle.)

7. Show that $\tan (\sin^{-1} x) = \dfrac{x \sqrt{1 - x^2}}{1 - x^2}$.

8. Show that $\sin (2 \sin^{-1} x) = 2x \sqrt{1 - x^2}$.

9. Let $z = \cos \theta + i \sin \theta$ for some number θ. Show that all of the powers of z (i.e., $z, z^2, z^3, z^4, \ldots$) are points on the unit circle whose equation is $x^2 + y^2 = 1$.

10. Show that the distance between the points $(r, t)_P$ and $(r', t')_P$ is

$$\sqrt{r^2 + r'^2 - 2rr' \cos (t - t')}$$

11. Find an equation in polar form for the set of points $(r, t)_P$ the product of whose distance from the fixed points $(a, \pi)_P$ and $(a, 0)_P$ is the constant a^2. (*Hint:* See Problem 9.)

12. Find the four complex numbers z for which $z^4 + 4 = 0$, and use these to obtain the factorization

$$z^4 + 4 = (z^2 + 2z + 2)(z^2 - 2z + 2)$$

13. Find the six complex numbers z for which $z^6 - 64 = 0$, and use them to obtain the factorization

$$z^6 - 64 = (z^2 + 4)(z^2 + 2z + 4)(z^2 - 2z + 4)$$

4

LINEAR SYSTEMS AND AN INTRODUCTION TO ANALYTIC GEOMETRY

Part Four introduces three topics that occur in the study of calculus.

Chapter 10 studies vectors, first without coordinates, then with coordinates.

Chapter 11 introduces the conics and concludes with an introduction to three-dimensional coordinate systems and some graphs of spheres and planes.

Chapter 12 studies the problem of solving linear systems and develops from this the basic matrix concepts of row operation, rank, determinant, and inverses as well as algebraic operations on matrices. The chapter includes an introduction to linear programming.

Although these three chapters are essentially independent, the definition of matrix multiplication in Chapter 12, Section 6, makes use of "dot product," introduced in Chapter 10, Section 2. Also, the introduction of planes in a three-dimensional coordinate system in Chapter 11, Section 7, provides a geometric background for some of the linear systems in Chapter 12.

10 Vectors

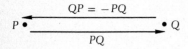

QP = −PQ

Figure 10-1

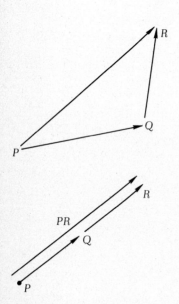

Figure 10-2

10-1 VECTORS: INTRODUCTION

Vectors provide a means for the description of phenomena that have the properties of direction and magnitude, as we shall see in the following pages.

Given two points P and Q, we shall denote by $\boldsymbol{PQ}$ the *directed line segment* from P to Q, and by $|\boldsymbol{PQ}|$, the length of PQ.

Note that QP is the directed line segment from Q to P and, therefore, has the same *length* as PQ (i.e., $|QP| = |PQ|$) but is opposite in *direction*. This latter fact is denoted by

$$QP = -PQ$$

See Figure 10-1.

The physical effect of successive displacements of an object suggests a rather natural definition of addition of directed line segments: If an object is moved from a point P to a point Q, and then from Q to a point R, the result is the same as if the object had been moved directly from P to R. Hence, we define the directed line segment PR to be the **sum** of PQ and QR:

$$PQ + QR = PR$$

for any three points P, Q, and R. See Figure 10-2.

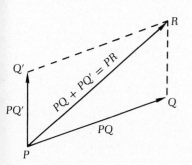

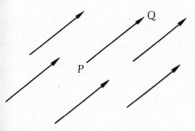

Figure 10-3

Figure 10-4

Many physical quantities can be represented by directed line segments; for example, PQ might represent

1. a displacement of an object from P to Q;
2. a force (that is, a push or pull) acting at P in the direction of Q and of magnitude $|PQ|$;
3. the velocity of a particle at P moving in the direction of Q with speed $|PQ|$.

Let P, Q, and Q' be noncollinear points. It is a fact that the combined effect of two forces PQ and PQ', acting at the same point P, is the same as that of the single force PR, where P, Q, Q', and R form the vertices of a parallelogram with diagonal PR. Hence, we may also define $PQ + PQ'$ to be PR (see Figure 10-3). Note that this definition is equivalent to the preceding one.

Now, a quantity such as wind velocity has magnitude and direction, but is not associated with any particular point. It can be represented not only by, say, PQ, but by any other directed line segment having the same length and the same direction as PQ. See Figure 10-4.

We are thus led to isolating what is common to all of these directed line segments, and we define $\overrightarrow{PQ}$, *the vector determined by PQ, to be the length and direction of PQ.*

We shall see that addition of *vectors* subsumes and unites the addition of chained displacements and the addition of forces at a common point. This is another reason for studying them.

There are alternative, but essentially equivalent, definitions,* but whichever one is taken, the following statement defines **equality** for vectors, and *that* is the important thing:

$\overrightarrow{PQ} = \overrightarrow{P'Q'}$ if and only if PQ and $P'Q'$ have the same *length* and the same *direction*.

In fact, this property by itself implicitly defines what a vector is, for all mathematical purposes. Informally, *it is helpful to think of a vector as a movable directed line segment that can be shifted freely but not turned.*

Now, let $\boldsymbol{u}$ and $\boldsymbol{v}$ be any vectors.† Choose any point P. Then, there is exactly one point Q for which $\boldsymbol{u} = \overrightarrow{PQ}$ and exactly

*For example, the vector determined by a directed line segment PQ may be defined as the set of all directed line segments having the same length and direction as PQ.
†Henceforth, any letter in boldface shall denote a vector.

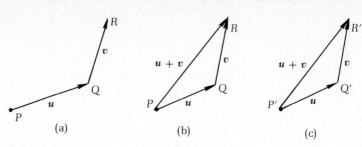

Figure 10-5

one point R for which $v = \overrightarrow{QR}$ [see Figure 10-5(a)]. By definition, the **sum** of the vectors u and v is the vector $\overrightarrow{PR}$; we denote this by

$$u + v = \overrightarrow{PR}$$

[see Figure 10-5(b)]. Observe that $u + v$ is unambiguously defined, because we get the same vector no matter what point we start with. Choosing, say, P' instead of P [see Figure 10-5(c)], we get $u = \overrightarrow{P'Q'}$, $v = \overrightarrow{Q'R'}$, and $u + v = \overrightarrow{P'R'}$; clearly $\overrightarrow{PR} = \overrightarrow{P'R'}$, and so we obtain the same sum [compare Figures 10-5(b) and 10-5(c)].

Suppose now that u and v do not have the same direction, and let P, Q, and Q' be three points for which $u = \overrightarrow{PQ}$ and $v = \overrightarrow{PQ'}$. Then, as before, the diagonal PR of the completed parallelogram is the sum of PQ and PQ'. Hence, $u + v = \overrightarrow{PR}$. See Figure 10-6. Note that PQ' is just a shift of QR, so that $\overrightarrow{PQ'} = \overrightarrow{QR}$; hence,

$$\overrightarrow{PQ} + \overrightarrow{PQ'} = \overrightarrow{PR} = \overrightarrow{QR} = u + v$$

In Figure 10-7 we construct various vectors u and v, and then $u + v$. For any vectors u and v, we see that

$$u + v = v + u$$

It is also possible to show that $(u + v) + w = u + (v + w)$; more will be said about this in the next section.

The following definitions are useful.

$\mathbf{0} = $ the **zero vector**, the vector of length 0.

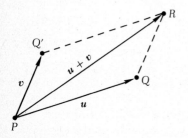

Figure 10-6

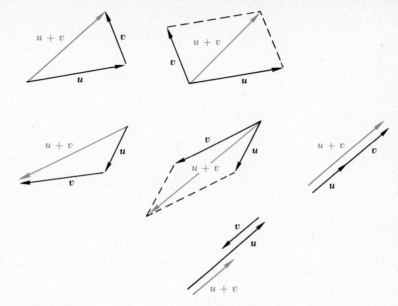

Figure 10-7

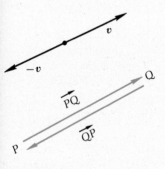

Figure 10-8

Observe that unlike any other vector, **0** has *no* direction. Also by definition,

$-v$ = the vector having the same length as v, but opposite direction.

Hence, if $v = \overrightarrow{PQ}$, then $-v = \overrightarrow{QP}$. See Figure 10-8.

If $v = \overrightarrow{PQ}$, then the **length,** or **magnitude,** of v is defined to be the length of PQ, and is denoted by $|v|$. Thus,

$$|v| = |PQ|$$

A vector whose length is 1 is called a **unit vector.**
Observe that

$$v + 0 = v$$

and

$$v + (-v) = 0$$

Vector subtraction is defined as follows:

$$u - v = u + (-v)$$

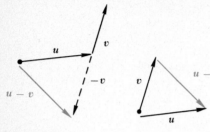

Figure 10-9

See Figure 10-9. Thus, $-v$ is the *additive inverse* of v, and $\mathbf{0}$ is the *additive identity* for the set of vectors.

Now, if a is a real number, then, by definition, the vector av is a stretch of the vector v by a factor of $|a|$. That is, av is the vector of length $|a| \cdot |v|$, and direction the same as that of v if $a > 0$, and opposite to that of v if $a < 0$. We call av a **multiple of v**. See Figure 10-10. Note that $(-1)v = -v$. Also note that *if $v \neq \mathbf{0}$, then the vector* $\dfrac{1}{|v|} \cdot v$ *is a unit vector with the same direction as v.* It has the same direction as v because $\dfrac{1}{|v|} > 0$; it is a unit vector because its length is $\dfrac{1}{|v|} \cdot |v| = 1$. Since every nonzero vector v can be written as

$$v = 1 \cdot v = \frac{|v|}{|v|} \cdot v = |v| \left(\frac{1}{|v|} \cdot v \right)$$

we see that *every nonzero vector v is the product of its magnitude $|v|$ and the unit vector* $\dfrac{1}{|v|} \cdot v$ *in the same direction as v.*

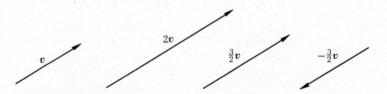

Figure 10-10

Example 1

A force of 3 pounds directed south and a force of 4 pounds directed west are applied to an object. Find the resultant of these forces.

Solution

The forces are given by the vectors u and v in Figure 10-11. (Here, $\frac{1}{4}$ inch = 1 pound of force); thus, $|u| = 4$ and $|v| = 3$. The resultant of these forces is the vector $u + v$. By the Theorem of Pythagoras, the magnitude of the resultant force is $|u + v| = \sqrt{3^2 + 4^2} = 5$ (pounds). The angle $u + v$ makes with the vector u is arc cos $\frac{4}{5}$. (See Figure 10-11.)

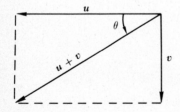

Figure 10-11

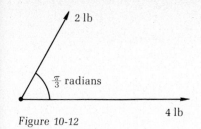

Figure 10-12

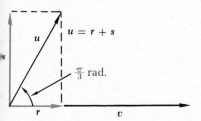

Figure 10-13

Example 2

A force of 2 pounds and a force of 4 pounds are applied to a point as shown in Figure 10-12 (note that the directions of the forces are 60°, or $\pi/3$ radians, apart). Find the magnitude of the resultant force and the angle that force makes with the given forces.

Solution

Let the forces of 2 pounds and 4 pounds be given by vectors u and v, respectively. Thus, $|u| = 2$ and $|v| = 4$. We shall first express u as a sum of two vectors, one along the line of action of v, call it r, and one perpendicular to the line of action of v, call it s, as in Figure 10-13. Note that

$$|r| = |u| \cos \frac{\pi}{3} = 2 \cdot \frac{1}{2} = 1$$

and

$$|s| = |u| \sin \frac{\pi}{3} = 2 \cdot \frac{\sqrt{3}}{2} = \sqrt{3}$$

The resultant of the vectors u and v is the vector $u + v$. Since $u = r + s$, we see that

$$u + v = (r + s) + v = s + (r + v)$$

Since r and v are collinear and have the same direction,

$$|r + v| = |r| + |v| = 5$$

Thus, the resultant $u + v$ is the sum of two perpendicular vectors, s and $r + v$, of magnitudes $\sqrt{3}$ and 5, respectively. See Figure 10-14. Consequently, the magnitude of the resultant force is $\sqrt{(\sqrt{3})^2 + 5^2}$, that is, $2\sqrt{7}$ pounds. The angle θ that the resultant makes with the 4-pound force (or, equivalently, $r + v$) is arc tan $\dfrac{\sqrt{3}}{5}$ radians, which is approximately 19°. The angle that the resultant makes with u is approximately 60° − 19° = 41°.

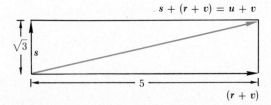

Figure 10-14

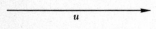

Exercise 1

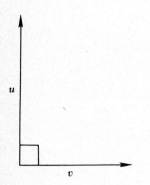

Exercise 2

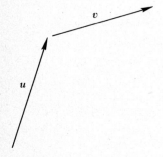

Exercise 3

Exercises 10-1

1. Let the vector u be given as in the accompanying figure. Construct each of the following vectors.

$$2u - u$$
$$\tfrac{1}{2}u - 3u$$
$$-\tfrac{7}{2}u$$

2. Let the vectors u and v be given as in the accompanying figure. Construct each of the following vectors:

$u + v$	$u - v$
$v - u$	$-2u + v$
$3u + 2v$	$-2u - 3v$

3. Let the vectors u and v be given as in the accompanying figure. Construct the vectors

$u + v$	$u - v$
$v - u$	$-u - v$
$2u$	$2u + 3v$

4. Let $u, v, w, r,$ and s be as in the accompanying figure. Find $u + v + w + r + s.$

5. Let $u, v,$ and w be given as in the accompanying figure. Show that $u + (v + w) = (u + v) + w.$

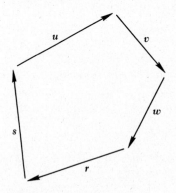

Exercise 4

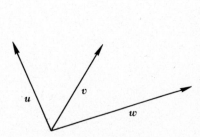

Exercise 5

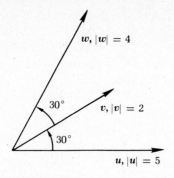

Exercise 14(a)

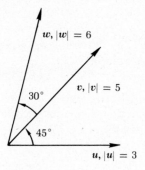

Exercise 14(b)

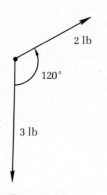

Exercise 21

In Exercises 6–13, the magnitudes of vectors *u* and *v* are given. Assuming *u* is a horizontal vector and θ is the angle from *u* to *v*, find $|u + v|$ and the angle between *u* and *u + v*.

6. $|u| = 3, |v| = 2, \theta = 30°$

7. $|u| = 5, |v| = 2, \theta = 45°$

8. $|u| = 4, |v| = 3, \theta = 26°$

9. $|u| = 6, |v| = 3, \theta = 55°$

10. $|u| = 2, |v| = 7, \theta = 45°$

11. $|u| = 2, |v| = 4, \theta = 20°$

12. $|u| = 4, |v| = 5, \theta = 120°$

13. $|u| = 5, |v| = 2, \theta = 135°$

14. (a) Let *u, v,* and *w* be given as in the figure Exercise 14(a). Construct *u + v + w* and find $|u + v + w|$.

Repeat if *u, v,* and *w* are as in the figure Exercise 14(b).

In Exercises 15–17, a force of 3 pounds directed north is applied to an object, and another force of 5 pounds directed east is applied to the same object.

15. Construct a vector representation of this (let $\frac{1}{2}$ inch represent 1 pound of force). Construct the **resultant** of these forces (that is, the sum of the vectors that represent the forces).

16. What is the magnitude of the resultant force?

17. What is the angle between the resultant force and the 5-pound force?

18. Repeat Exercises 15–17, now assuming forces of 6 pounds directed west and 2 pounds directed south from the object. (In Exercise 18, find the angle between the resultant and the 6-pound force.)

19. A force of 40 pounds directed north is applied to an object. What must be the magnitude of a force directed east if the magnitude of the resultant force is to be 80 pounds? What angle does the resultant force make with the easterly direction?

20. A force of 100 pounds directed west is applied to an object. What must be the magnitude of a second force applied to the same object if its direction is 45° north of the 100-pound force and the magnitude of their resultant force is 180 pounds?

◊**21.** Two forces are applied to an object as shown in the accompanying figure. Find a single third force which will keep the object stationary.

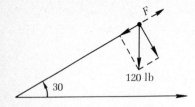

Exercise 22

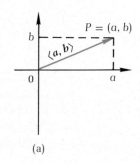

(a)

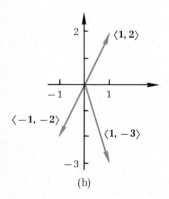

(b)

Figure 10-15

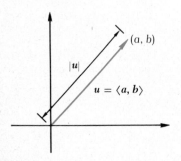

Figure 10-16

◇**22.** A weight of 120 pounds is on a smooth (that is, frictionless) inclined plane that makes an angle of 30° with the horizontal plane (see the accompanying figure). What force F along the inclined plane must be applied to the weight so that the weight will be stationary? (*Hint:* Express the force of 120 pounds as a sum of forces, one in the direction of the inclined plane and one perpendicular to that direction.)

10-2 VECTORS WITH COORDINATES

In the previous section, the vectors we investigated were "coordinate-free"; that is, they did not depend on the choice of a particular coordinate system. In this section, we shall view vectors as objects in a coordinate plane. Though we lose some of the generality of the previous section, we make up for it by gaining the convenience of a coordinate system.

Let u be a vector in a coordinate plane with origin O. Then, $u = \overrightarrow{OP}$ for a unique point $P = (a, b)$; that is, every vector in the plane determines, and is determined by, an ordered pair of real numbers. From now on, by definition,

$$\langle a, b \rangle \text{ is the vector } \overrightarrow{OP} \text{ with } P = (a, b)$$

See Figure 10-15(a).

Example 1

$\langle 1, 2 \rangle$ is the vector $\overrightarrow{OP}$ with $P = (1, 2)$; $\langle 1, -3 \rangle$ is the vector $\overrightarrow{OP}$ with $P = (1, -3)$; $\langle -1, -2 \rangle$ is the vector $\overrightarrow{OP}$ with $P = (-1, -2)$. See Figure 10-15(b).

The **length** of the vector $\langle a, b \rangle$ is simply the distance between the points $(0, 0)$ and (a, b). Thus,

$$|u| = |\langle a, b \rangle| = \sqrt{a^2 + b^2}$$

See Figure 10-16.

Example 2

For $u = \langle -3, 4 \rangle$, we have $|u| = \sqrt{3^2 + 4^2} = 5$; if $v = \langle -3, \pi \rangle$, then $|v| = \sqrt{9 + \pi^2}$.

Example 3

For any number t, let $w = \langle \cos t, \sin t \rangle$, then

$$|w| = \sqrt{\cos^2 t + \sin^2 t} = 1,$$

and thus w is a unit vector.

□

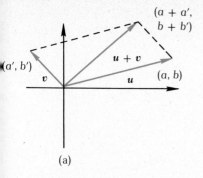

(a)

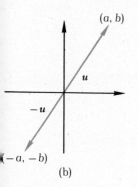

(b)

Figure 10-17

In the previous section, we defined vector addition and subtraction, and a multiple of a vector. The following theorem translates these concepts to vectors in a coordinate plane.

Theorem

If $u = \langle a, b \rangle$, and $v = \langle a', b' \rangle$, then

1. $u + v = \langle a, b \rangle + \langle a', b' \rangle = \langle a + a', b + b' \rangle$
2. $-u = -\langle a, b \rangle = \langle -a, -b \rangle$
3. $u - v = \langle a - a', b - b' \rangle$
4. If c is a number, then $cu = c\langle a, b \rangle = \langle ca, cb \rangle$
5. $\mathbf{0} = \langle 0, 0 \rangle$

Proof The proofs of these assertions are exercises in plane geometry. For example, to show (1), it suffices to show that the fourth vertex P of the parallelogram determined by the points $(0, 0)$, (a, b), $(a'b')$, and P, with OP a diagonal, is $P = (a + a', b + b')$. See Figure 10-17(a). A proof of (2) is indicated by Figure 10-17(b). The rest are left as exercises.

Example 4

(a) If $u = \langle -1, 3 \rangle$ and $v = \langle 3, -2 \rangle$, then

$$u + v = \langle -1, 3 \rangle + \langle 3, -2 \rangle = \langle -1 + 3, 3 - 2 \rangle = \langle 2, 1 \rangle$$

(b) $\langle 3, -2 \rangle + \langle -1, 3 \rangle = \langle 2, 1 \rangle$

(c) $\langle -5, 17 \rangle + \mathbf{0} = \langle -5, 17 \rangle + \langle 0, 0 \rangle = \langle -5, 17 \rangle$

(d) If $u = \langle -1, 4 \rangle$, then

$$-3u = -3\langle -1, 4 \rangle = \langle (-3)(-1), (-3)(4) \rangle = \langle 3, -12 \rangle$$

(e) $0\langle -87, 141 \rangle = \langle 0, 0 \rangle = \mathbf{0}$

(f) If $u = \langle -1, 3 \rangle$ and $v = \langle 1, -1 \rangle$, then

$$3u - 2v = \langle -3, 9 \rangle + \langle -2, 2 \rangle = \langle -5, 11 \rangle$$

□

Many of the properties of vector addition and subtraction are similar to corresponding properties of ordinary addition and subtraction (of numbers). The following list summarizes many of these basic properties. Many of these properties were already observed in the previous section. For any vectors u, v, and w,

1. $u + v = v + u$ (vector addition is *commutative*)
2. $(u + v) + w = u + (v + w)$ (vector addition is *associative*)
3. $u + 0 = u = 0 + u$ (**0** is an *additive identity*)
4. $u + (-u) = u - u = 0$ ($-u$ is the additive inverse of u)
5. If c and d are real numbers, then $c(du) = (cd)u$
6. If c and d are real numbers, then $(c + d)u = cu + du$
7. If c is a real number, then $c(u + v) = cu + cv$
8. $1 \cdot u = u$
9. $0 \cdot u = 0$
10. $c \cdot 0 = 0$ for any real number c

We shall verify (4) and (5) and leave the rest as exercises. Let $u = \langle a, b \rangle$. To prove (4), note that

$$u + (-u) = \langle a, b \rangle + \langle -a, -b \rangle$$
$$= \langle a - a, b - b \rangle = \langle 0, 0 \rangle = 0$$

To prove (5), note that

$$c(du) = c\langle da, db \rangle = \langle c(da), c(db) \rangle = \langle (cd)a, (cd)b \rangle$$
$$= (cd)\langle a, b \rangle = (cd)u$$

Note that any nonzero vector $\langle a, b \rangle$ is the sum of the vectors $\langle a, 0 \rangle$ and $\langle 0, b \rangle$. These are called, respectively, the horizontal and vertical **projections** of $\langle a, b \rangle$. The relationship between a vector and its horizontal and vertical projections is indicated in Figure 10-18. Thus, given a vector v, *its horizontal and vertical components are*

$$|v| \cos \theta \quad \text{and} \quad |v| \sin \theta$$

respectively, where θ is the angle v makes with the positive horizontal axis. Thus,

$$v = \langle |v| \cos \theta, |v| \sin \theta \rangle$$
$$= |v|\langle \cos \theta, \sin \theta \rangle$$

If $v = \langle a, b \rangle \neq 0$, then, referring to Figure 10-18, we see that

$$\cos \theta = \frac{a}{\sqrt{a^2 + b^2}} \quad \text{and} \quad \sin \theta = \frac{b}{\sqrt{a^2 + b^2}}$$

and thus

$$v = \langle a, b \rangle = |(a, b)| \left(\frac{a}{\sqrt{a^2 + b^2}}, \frac{b}{\sqrt{a^2 + b^2}} \right)$$
$$= \sqrt{a^2 + b^2} \left(\frac{a}{\sqrt{a^2 + b^2}}, \frac{b}{\sqrt{a^2 + b^2}} \right)$$

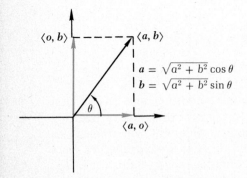

Figure 10-18

This is again the observation that every nonzero vector v is the product of its length and the unit vector in the same direction as v $\left[\text{i.e., that } v = |v|\left(\dfrac{1}{|v|} \cdot v\right)\right]$.

Example 5

Express the vector $\langle 3, -2 \rangle$ as the product of a real number and a unit vector with the same direction as $\langle 3, -2 \rangle$.

Solution

Since $|\langle 3, -2 \rangle| = \sqrt{13}$, we see that

$$\langle 3, -2 \rangle = \sqrt{13}\left\langle \frac{3}{\sqrt{13}}, -\frac{2}{\sqrt{13}} \right\rangle$$

and $\left\langle \dfrac{3}{\sqrt{13}}, -\dfrac{2}{\sqrt{13}} \right\rangle$ is the unit vector in the same direction as $\langle 3, -2 \rangle$. Equivalently, if θ is the angle the vector $\langle 3, -2 \rangle$ makes with the positive horizontal axis, then

$$\begin{aligned}\langle 3, -2 \rangle &= |\langle 3, -2 \rangle|\langle \cos\theta, \sin\theta \rangle \\ &= \sqrt{13}\,\langle \cos\theta, \sin\theta \rangle\end{aligned}$$

where $\langle \cos\theta, \sin\theta \rangle = \left\langle \dfrac{3}{\sqrt{13}}, -\dfrac{2}{\sqrt{13}} \right\rangle$ is the unit vector in the direction of w.

□

From the Pythagorean Theorem, we know that two nonzero vectors u and v are perpendicular (i.e., $u \perp v$) if and only if $|u|^2 + |v|^2 = |u + v|^2$ (see Figure 10-19).

The proof of the following theorem is a consequence of this observation.

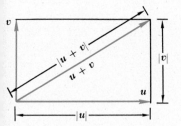

Figure 10-19

Theorem

If $u = \langle a, b \rangle$ and $v = \langle a', b' \rangle$ are nonzero vectors, then u and v are perpendicular if and only if $aa' + bb' = 0$; that is,

$$\langle a, b \rangle \perp \langle a', b' \rangle \quad \text{if and only if} \quad aa' + bb' = 0$$

Proof The (nonzero) vectors u and v are perpendicular if and only if $|u|^2 + |v|^2 = |u + v|^2$; that is,

$$(a^2 + b^2) + (a'^2 + b'^2) = (a + a')^2 + (b + b')^2$$

or, equivalently,

$$a^2 + b^2 + a'^2 + b'^2 = a^2 + 2aa' + a'^2 + b^2 + 2bb' + b'^2$$

that is,

$$aa' + bb' = 0$$

which proves the theorem.

The *number* $aa' + bb'$ is called the **inner product*** of the vectors $\boldsymbol{u} = \langle a, b \rangle$ and $\boldsymbol{v} = \langle a', b' \rangle$ and is denoted by $\boldsymbol{u} \cdot \boldsymbol{v}$. Thus,

$$\boldsymbol{u} \cdot \boldsymbol{v} = \langle a, b \rangle \cdot \langle a', b' \rangle = \boldsymbol{aa' + bb'}$$

Example 6

Let $\boldsymbol{u} = \langle 1, 3 \rangle$ and $\boldsymbol{v} = \langle -6, 2 \rangle$. Then,

$$\boldsymbol{u} \cdot \boldsymbol{v} = (1)(-6) + (3)(2) = 0$$

Thus, the vectors $\boldsymbol{u}$ and $\boldsymbol{v}$ are perpendicular.

Example 7

Let $\boldsymbol{u} = \langle -1, 2 \rangle$ and $\boldsymbol{v} = \langle 2, 4 \rangle$. Then,

$$\boldsymbol{u} \cdot \boldsymbol{v} = (-1)(2) + (2)(4) = 6$$

Thus, $\boldsymbol{u}$ and $\boldsymbol{v}$ are **not** perpendicular vectors.
□

Notice that if $\boldsymbol{u} = \langle a, b \rangle$, then

$$\boldsymbol{u} \cdot \boldsymbol{u} = \boldsymbol{a}^2 + \boldsymbol{b}^2 = |\boldsymbol{u}|^2$$

From the law of cosines, we have the following important result.

Theorem

Let θ denote the angle of least positive measure between the nonzero vectors $\boldsymbol{u}$ and $\boldsymbol{v}$. See Figure 10-20. Then,

$$\cos \theta = \frac{\boldsymbol{u} \cdot \boldsymbol{v}}{|\boldsymbol{u}| \cdot |\boldsymbol{v}|}$$

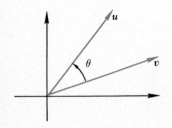

Figure 10-20

*Other names for this are *dot product* and *scalar product*.

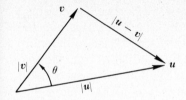

Figure 10-21

Proof Refer to Figure 10-21. From the law of cosines, we have

$$|u - v|^2 = |u|^2 + |v|^2 - 2|u| \cdot |v| \cos \theta$$

Let $u = \langle a, b \rangle$ and $v = \langle a', b' \rangle$, then $u - v = \langle a - a', b - b' \rangle$, and therefore

$$
\begin{aligned}
|u - v|^2 &= (a - a')^2 + (b - b')^2 \\
&= a^2 + b^2 + a'^2 + b'^2 - 2aa' - 2bb' \\
&= |u|^2 + |v|^2 - 2u \cdot v
\end{aligned}
$$

Thus,

$$|u|^2 + |v|^2 - 2|u| \cdot |v| \cos \theta = |u|^2 + |v|^2 - 2u \cdot v$$

and, hence,

$$\cos \theta = \frac{u \cdot v}{|u| \cdot |v|}$$

Example 8

For $u = \langle 1, 3 \rangle$, $v = \langle -1, 2 \rangle$, and, as before, θ the angle of least positive measure between u and v, we have

$$\cos \theta = \frac{u \cdot v}{|u| \cdot |v|} = \frac{(1)(-1) + (3)(2)}{(\sqrt{1^2 + 3^2})(\sqrt{(-1)^2 + 2^2})} = \frac{5}{\sqrt{50}}$$

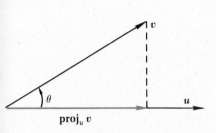

$\cos \theta > 0$

$\cos \theta < 0$

Figure 10-22

Suppose we want to project a vector v onto a vector u (see Figure 10-22), thus obtaining a vector called *the vector projection of v on u*, and denoted by $\mathbf{proj}_u v$. Note (see Figure 10-22) that

$$|\mathbf{proj}_u v| = |v| \cdot |\cos \theta| = |v| \cdot \frac{|u \cdot v|}{|u| \cdot |v|} = \frac{|u \cdot v|}{|u|}$$

Observe that $\mathbf{proj}_u v$ and u have the same direction if $\cos \theta > 0$ (that is, if $u \cdot v > 0$) and the opposite direction if $\cos \theta < 0$ (that is, if $u \cdot v < 0$). Since $\frac{1}{|u|} \cdot u$ is a unit vector in the direction of u, it is also a unit vector in the direction of $\mathbf{proj}_u v$ if $u \cdot v > 0$ and in the opposite direction if $u \cdot v < 0$. Thus, in either case,

$$\mathbf{proj}_u v = \frac{u \cdot v}{|u|} \cdot \frac{1}{|u|} \cdot u = \frac{u \cdot v}{|u|^2} \cdot u$$

That is,

$$\mathbf{proj}_u v = \frac{u \cdot v}{|u|^2} \cdot u$$

□

The coordinate treatment of vectors can be easily generalized to higher dimensions. Vectors in the plane are ordered pairs of real numbers. We simply replace pairs of real numbers by triples, quadruples, or, in general, by n-tuples of real numbers (where n is a positive integer), and define the operations on vectors by analogy with the operations on plane vectors. Specifically, the **vector space of (real) n-tuples,** often called **real n-spaces,** is the set of all n-tuples $(x_1, x_2, \ldots, x_n)$ of real numbers with the following definitions.

If $x_1, \ldots, x_n, y_1, \ldots, y_n$ and c are real numbers, then

$$\langle x_1, x_2, \ldots, x_n \rangle + \langle y_1, y_2, \ldots, y_n \rangle$$
$$= \langle x_1 + y_1, x_2 + y_2, \ldots, x_n + y_n \rangle$$

and

$$c \langle x_1, \ldots, x_n \rangle = \langle cx_1, \ldots, cx_n \rangle$$

The **zero vector (0)** is $\langle 0, 0, \ldots, 0 \rangle$. If u and v are vectors, then, as before,

$$u - v = u + (-v)$$

Thus,

$$\langle x_1, \ldots, x_n \rangle - \langle y_1, \ldots, y_n \rangle = \langle x_1 - y_1, \ldots, x_n - y_n \rangle$$

and hence

$$u - u = 0$$

If $n = 3$ (that is, in 3-dimensional space), the vector (x_1, x_2, x_3) can be interpreted as the directed line segment from $(0, 0, 0)$ to (x_1, x_2, x_3); thus, these vectors can be thought of as directed line segments in 3-dimensional space.

All the properties of plane vectors carry over to real n-space. Thus, vector addition is commutative and associative, 0 is an additive identity, $-u$ is the additive inverse of u, and so forth. Inner product and length are defined as follows:

If $u = \langle x_1, \ldots, x_n \rangle$ and $v = \langle y_1, \ldots, y_n \rangle$, then, by definition,

$$u \cdot v = x_1 y_1 + x_2 y_2 + \cdots + x_n y_n$$

and

$$|u|^2 = u \cdot u = x_1^2 + x_2^2 + \cdots + x_n^2$$

Example 9

$u = \langle 3, -7, -5, 2, 8, -3 \rangle$ is a vector in 6-space. If v is the vector $\langle -3, -1, 2, 0, -5, 0 \rangle$, then

$$u + v = \langle 0, -8, -3, 2, 3, -3 \rangle$$

$$-u = \langle -3, 7, 5, -2, -8, 3 \rangle$$

$$3v = \langle -9, -3, 6, 0, -15, 0 \rangle$$

$$3v - u = \langle -12, 4, 11, -2, -23, 3 \rangle$$

$$u \cdot v = -9 + 7 - 10 + 0 - 40 + 0 = -52$$

and

$$|v| = \sqrt{9 + 1 + 4 + 25} = \sqrt{39}$$

□

Two nonzero vectors are said to be **orthogonal,** or **perpendicular,** if their inner product is zero. Thus, $\langle -1, 0, 2, -3 \rangle$ is orthogonal to $\langle 1, 3, -1, -1 \rangle$. Note that we have now defined perpendicularity between vectors in terms of inner product; for $n = 2$ or 3, this definition agrees with the usual meaning of perpendicularity.

We shall use the concept of vector in the study of matrices.

Exercises 10-2

Let $u = \langle -3, 2 \rangle$ and $v = \langle -5, -4 \rangle$. In Exercises 1-4, find the indicated sum.

1. $u + 3v$ 　　　　　　　　　　**2.** $-2u + 4v$

3. $\frac{1}{2}(u + v)$ 　　　　　　　　**4.** $-u + 0$

In Exercises 5-12, let $u = \langle -2, 3 \rangle$ and let $v = \langle 3, -4 \rangle$. Find:

5. $|u + v|$ 　　　　　　　　　　**6.** $|-4u|$

7. $|2u + 5v|$ 　　　　　　　　　**8.** $|-2u + v|$

9. $|-4u - 5v|$ 　　　　　　　　**10.** $u \cdot v$

11. $(u + v) \cdot (2u + 5v)$ 　　　　**12.** $(-2u + v) \cdot (-2u + v)$

In Exercises 13-16, let $u = \langle -1, 3 \rangle$ and $v = \langle 3, -4 \rangle$. Find numbers a and b for which:

13. $3u + \langle a, b \rangle = -v$ 　　　　**14.** $-2u + 3\langle a, b \rangle = 3v$

15. $\frac{1}{2}\langle u + v \rangle - \frac{1}{2}[\langle a, b \rangle + \langle 1, 3 \rangle] = \left\langle -\frac{1}{2}, 1 \right\rangle$

16. $3u + 2\langle a, b \rangle = -v + 3\langle a - 1, b + 2 \rangle$

17. Find real numbers a and b for which $2\langle a, -b \rangle - 3\langle 2b, a \rangle = \langle -2, -8 \rangle$.

18. Find real numbers a and b for which $2\langle -1, 3 \rangle + 3\langle a, b \rangle = \langle 2, 3 \rangle - 2\langle b, a \rangle$.

11 Analytic Geometry

11-1 INTRODUCTION

The problem of finding the graph of a given equation has already been treated. We now consider the converse problem of finding an equation for a given curve in the plane. The curves to be considered in this chapter are conics—circles, ellipses, parabolas, and hyperbolas. All of these curves can be obtained by intersecting the plane with right circular cones. (See Figure 11-1.)

Each type of conic can also be defined in terms of distances in the plane, and these definitions will be used to obtain their equations. For example, a circle (which is a special case of an ellipse) is the set of all points equidistant from a given point and has the equation $(x - a)^2 + (y - b)^2 = r^2$, where (a, b) is the given point and r is the given distance.

Some examples of conic sections are the orbits of planets and the paths of atomic particles. Conics are also used in the design of lenses and mirrors as well as in the study of projectiles.

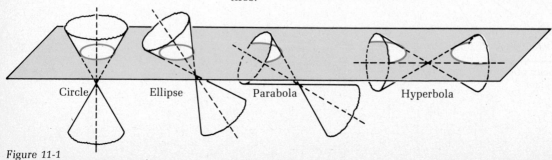

Circle   Ellipse   Parabola   Hyperbola

Figure 11-1

Miscellaneous Problems

1. Suppose u and v are nonzero vectors. Show that if we let $c = u \cdot v / |v|^2$, then $u - cv$ will be orthogonal to v.

2. Suppose u and v are any nonzero vectors. Show that the vectors $|v|u + |u|v$ and $|v|u - |u|v$ are orthogonal.

3. Suppose u, v, w, and x are vectors. Show that
(a) $(u + v) \cdot w = u \cdot w + v \cdot w$
(b) $(u + v) \cdot (w + x) = u \cdot w + v \cdot w + u \cdot x + v \cdot x$

4. Suppose u and v are any vectors. Show that

$$|u + v| \leq |u| + |v|$$

(*Hint:* Use the fact that $|u|^2 = u \cdot u$.)

5. Show that if u and v are any vectors, then

$$|u + v|^2 + |u - v|^2 = 2|u|^2 + 2|v|^2$$

This is called the **parallelogram law;** can you explain why?

6. Let u and v be nonzero orthogonal vectors. Show that

$$|u + v|^2 = |u|^2 + |v|^2$$

This is called the **Pythagorean theorem;** explain why.

7. Show that the vectors u and v have the same or opposite direction if and only if there are nonzero real numbers a and b for which $au + bv = \mathbf{0}$.

8. A plane traveling north encounters a 100 mile per hour wind from the east. The cruising speed of the plane is 500 miles per hour. In what direction should the plane point if it wants to continue traveling north? How fast will it be traveling? See the accompanying figure.

9. A bullet is fired with muzzle velocity 2000 feet/sec from an airplane flying horizontally 500 miles per hour. The bullet is fired horizontally 30° away from the path in which the plane is traveling. Find the magnitude and the direction of the resultant (or actual) velocity of the bullet. See the accompanying figure.

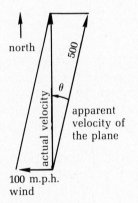

Problem 8

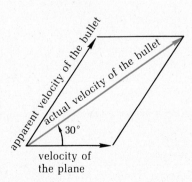

Problem 9

11 Analytic Geometry

11-1 INTRODUCTION

The problem of finding the graph of a given equation has already been treated. We now consider the converse problem of finding an equation for a given curve in the plane. The curves to be considered in this chapter are conics—circles, ellipses, parabolas, and hyperbolas. All of these curves can be obtained by intersecting the plane with right circular cones. (See Figure 11-1.)

Each type of conic can also be defined in terms of distances in the plane, and these definitions will be used to obtain their equations. For example, a circle (which is a special case of an ellipse) is the set of all points equidistant from a given point and has the equation $(x - a)^2 + (y - b)^2 = r^2$, where (a, b) is the given point and r is the given distance.

Some examples of conic sections are the orbits of planets and the paths of atomic particles. Conics are also used in the design of lenses and mirrors as well as in the study of projectiles.

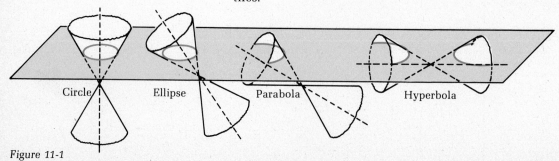

Circle   Ellipse   Parabola   Hyperbola

Figure 11-1

Example 9

$u = \langle 3, -7, -5, 2, 8, -3 \rangle$ is a vector in 6-space. If v is the vector $\langle -3, -1, 2, 0, -5, 0 \rangle$, then

$$u + v = \langle 0, -8, -3, 2, 3, -3 \rangle$$
$$-u = \langle -3, 7, 5, -2, -8, 3 \rangle$$
$$3v = \langle -9, -3, 6, 0, -15, 0 \rangle$$
$$3v - u = \langle -12, 4, 11, -2, -23, 3 \rangle$$
$$u \cdot v = -9 + 7 - 10 + 0 - 40 + 0 = -52$$

and

$$|v| = \sqrt{9 + 1 + 4 + 25} = \sqrt{39}$$

□

Two nonzero vectors are said to be **orthogonal,** or **perpendicular,** if their inner product is zero. Thus, $\langle -1, 0, 2, -3 \rangle$ is orthogonal to $\langle 1, 3, -1, -1 \rangle$. Note that we have now defined perpendicularity between vectors in terms of inner product; for $n = 2$ or 3, this definition agrees with the usual meaning of perpendicularity.

We shall use the concept of vector in the study of matrices.

Exercises 10-2

Let $u = \langle -3, 2 \rangle$ and $v = \langle -5, -4 \rangle$. In Exercises 1-4, find the indicated sum.

1. $u + 3v$
2. $-2u + 4v$
3. $\frac{1}{2}(u + v)$
4. $-u + 0$

In Exercises 5-12, let $u = \langle -2, 3 \rangle$ and let $v = \langle 3, -4 \rangle$. Find:

5. $|u + v|$
6. $|-4u|$
7. $|2u + 5v|$
8. $|-2u + v|$
9. $|-4u - 5v|$
10. $u \cdot v$
11. $(u + v) \cdot (2u + 5v)$
12. $(-2u + v) \cdot (-2u + v)$

In Exercises 13-16, let $u = \langle -1, 3 \rangle$ and $v = \langle 3, -4 \rangle$. Find numbers a and b for which:

13. $3u + \langle a, b \rangle = -v$
14. $-2u + 3\langle a, b \rangle = 3v$
15. $\frac{1}{2}\langle u + v \rangle - \frac{1}{2}[\langle a, b \rangle + \langle 1, 3 \rangle] = \left\langle -\frac{1}{2}, 1 \right\rangle$
16. $3u + 2\langle a, b \rangle = -v + 3\langle a - 1, b + 2 \rangle$
17. Find real numbers a and b for which $2\langle a, -b \rangle - 3\langle 2b, a \rangle = \langle -2, -8 \rangle$.
18. Find real numbers a and b for which $2\langle -1, 3 \rangle + 3\langle a, b \rangle = \langle 2, 3 \rangle - 2\langle b, a \rangle$.

19. Find real numbers a and b for which $2\langle a^2, b \rangle - 3\langle a, b^2 \rangle = \langle -1, -5 \rangle$.

20. Find all real numbers a for which $\langle a, -2 \rangle \cdot \langle a, a \rangle = 3$.

21. Find the vector $\langle a, b \rangle$ for which $\langle 4, 5 \rangle \cdot \langle a, b \rangle = 6$ and $\langle 2, -3 \rangle \cdot \langle a, b \rangle = -8$.

22. Find the vector $\langle a, b \rangle$ for which $\langle 3, -1 \rangle \cdot \langle a, b \rangle = 4$ and $\langle -1, 2 \rangle \cdot \langle a, b \rangle = -1$.

In Exercises 23–28, find the angle of least positive measure between the given vectors u and v.

23. $u = \langle -1, 0 \rangle, \quad v = \langle 0, -3 \rangle$

24. $u = \langle 1, 1 \rangle, \quad v = \langle 2, 2 \rangle$

25. $u = \langle 2, -3 \rangle, \quad v = \langle -3, 2 \rangle$

26. $u = \langle -2, 3 \rangle, \quad v = \langle -3, -2 \rangle$

27. $u = \left\langle \dfrac{1}{2}, -1 \right\rangle, \quad v = \langle 2\sqrt{2}, 1 \rangle$

28. $u = \langle \sqrt{3}, \sqrt{2} \rangle, \quad v = \langle 1, -1 \rangle$

29. Let $u = \left\langle -5, \dfrac{1}{2} \right\rangle$. Find the two vectors of unit length that are perpendicular to u.

30. Repeat Exercise 29 with $u = \left\langle \dfrac{3}{2}, \dfrac{1}{8} \right\rangle$.

31. Let $F_1 = \langle 2, 1 \rangle$ and $F_2 = \langle 3, 2 \rangle$ be two forces in the plane. Find the magnitude of the resultant force $F_1 + F_2$, and the angle between the resultant force and F_1.

32. Repeat Exercise 31 with $F_1 = \langle -1, 2 \rangle$ and $F_2 = \langle 4, 1 \rangle$.

33. Repeat Exercise 31 with $F_1 = \langle -2, 1 \rangle$ and $F_2 = \langle -3, -4 \rangle$.

34. Repeat Exercise 31 with $F_1 = \langle 4, -1 \rangle$ and $F_2 = \langle -5, 1 \rangle$.

35. Let $u = \langle 1, 4 \rangle$ and $v = \langle 6, 3 \rangle$. Find $\mathbf{proj}_v\, u$.

36. Let $u = \langle a, b \rangle$ and $v = \langle a', b' \rangle$. Find $\mathbf{proj}_v\, u$, the vector projection of u on v. Then find $|\mathbf{proj}_v\, u|$.

37. Verify that

$$u + v = v + u$$
$$(u + v) + w = u + (v + w)$$

and

$$c(u + v) = cu + cv \quad \text{(where c is a real number)}$$

In Exercises 38–42, let $u = \langle 2, 0, -1, -2 \rangle$ and $v = \langle -1, 1, 2, -2 \rangle$. Find:

38. $2u - 3v$

39. $|u + 2v|$

40. $u \cdot v$

41. $(-3u) \cdot (2v)$

42. $(2u - 3v) \cdot (u + 2v)$

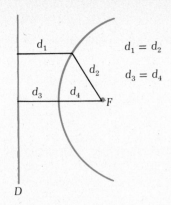

Figure 11-2

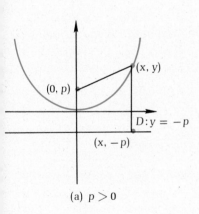

(a) $p > 0$

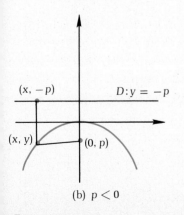

(b) $p < 0$

Figure 11-4

11-2 PARABOLAS

A **parabola** can be defined as the set of all points equidistant from a given point F, called the **focus**, and a given line D, called the **directrix** (see Figure 11-2). The point on the parabola that lies on the perpendicular from the focus to the directrix is called the **vertex**.

We shall now obtain an equation for a parabola with focus $F = (p, 0)$ and whose directrix D is the vertical line given by $x = -p$. Figures 11-3(a) and 11-3(b) give the shapes of the

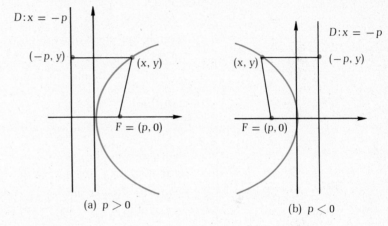

(a) $p > 0$ (b) $p < 0$

Figure 11-3

parabola for $p > 0$ and $p < 0$, respectively. In either case, we have

$$|x + p| = \sqrt{(x - p)^2 + y^2}$$

for any point (x, y) on the parabola. Squaring and simplifying, we obtain

$$y^2 = 4px$$

as the equation of the parabola with focus $(p, 0)$ and directrix given by $x = -p$.

We leave it to the reader to verify that if the focus is the point $(0, p)$ on the vertical axis, and the directrix is given by $y = -p$, then an equation of the parabola is

$$x^2 = 4py$$

Its shapes are given in Figures 11-4(a) and 11-4(b) for $p > 0$ and $p < 0$, respectively.

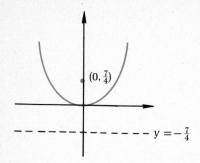

Figure 11-5

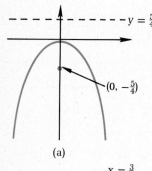

(a)

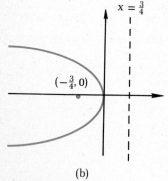

(b)

Figure 11-6

Example 1

Find the focus and directrix of the parabola given by $x^2 = 7y$.

Solution

The equation is equivalent to

$$x^2 = 4 \cdot \tfrac{7}{4}y$$

Thus, the focus is $(0, \tfrac{7}{4})$ and the directrix is the line given by $y = -\tfrac{7}{4}$. See Figure 11-5.

Example 2

Sketch the parabolas given by
(a) $x^2 = -5y$ (b) $y^2 + 3x = 0$
In each case locate the focus and the directrix.

Solution

See Figures 11-6(a) and 11-6(b).

Exercises 11-2

In Exercises 1–10, sketch the parabola of the given equation. Find the vertex, the focus, and the equation of the directrix.

1. $x^2 = 6y$ **2.** $3x^2 = -y$

3. $-x^2 + 7y = 0$ **4.** $3y + 5x^2 = 0$

5. $-3y + 5x^2 = 0$ **6.** $y^2 = -2x$

7. $y^2 = \dfrac{x}{2}$ **8.** $2y^2 + 3x = 0$

9. $-2y^2 + 3x = 0$ **10.** $\tfrac{3}{5}y^2 + \tfrac{4}{7}x = 0$

In Exercises 11–14, find an equation of the parabola whose focus and directrix are given.

11. Focus $(\sqrt{2}, 0)$, directrix $x = -\sqrt{2}$

12. Focus $(-\tfrac{3}{2}, 0)$, directrix $x = \tfrac{3}{2}$

13. Focus $(0, \tfrac{1}{8})$, directrix $y = -\tfrac{1}{8}$

14. Focus $(0, -\pi)$, directrix $y = \pi$

In Exercises 15–17, sketch the region given by:

15. $3y + 2x^2 \geq 0$ **16.** $-3y^2 + 2x < 0$

17. $2x - 5y^2 < 0$

18. Sketch the graph of the equation

$$(3x^2 + y)(x^2 - 6y)(y^2 - x) = 0$$

19. Find an equation for the parabola whose vertex is $(0, 0)$ and which passes through the points $(4, 5)$ and $(-4, 5)$.

20. Find an equation for the parabola whose vertex is the origin and which passes through the points $(-10, \pm 2)$.

21. Find the points of intersection of the parabola whose focus is $(\frac{3}{4}, 0)$ and whose directrix has equation $x = -\frac{3}{4}$, and the circle whose center is $(0, 0)$ and whose radius is 2.

22. For a parabola, show that every chord that is parallel to the directrix is bisected by the line that passes through the vertex and the focus of the parabola.

11-3 THE ELLIPSE AND THE HYPERBOLA

An **ellipse** can be defined as the set of all points P the sum of whose distances from two given points F_1 and F_2, called the **foci**, is a given positive number, say k.* (See Figure 11-7.) It will simplify things later if we let

$$a = \frac{k}{2}; \quad \text{that is,} \quad k = 2a$$

If the foci are $(c, 0)$ and $(-c, 0)$, with $c > 0$ (see Figure 11-8), then for any point (x, y) on the ellipse,

$$\sqrt{(x - c)^2 + y^2} + \sqrt{[x - (-c)]^2 + y^2} = 2a$$

that is,

$$\sqrt{(x + c)^2 + y^2} = 2a - \sqrt{(x - c)^2 + y^2}$$

Squaring and simplifying, we obtain

$$a\sqrt{(x - c)^2 + y^2} = a^2 - cx$$

Squaring and simplifying again, we have

$$(a^2 - c^2)x^2 + a^2y^2 = a^2(a^2 - c^2)$$

The number $a^2 - c^2$ is positive.† Letting $b = \sqrt{a^2 - c^2}$, we obtain

$$b^2x^2 + a^2y^2 = a^2b^2$$

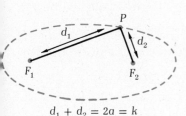

$$d_1 + d_2 = 2a = k$$

Figure 11-7

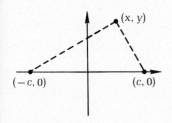

Figure 11-8

---

*Imagine a string of length k with its ends tacked to the loci F_1 and F_2 and then pulled taut with a pencil. If the pencil is moved keeping the string taut, the curve traced out is an ellipse.

†To see this, observe that $2c$ is the distance between the foci, while $2a$ is the sum of the lengths of the other two sides of the triangle determined by the foci and any point of the ellipse. Hence, $2a > 2c$; that is, $a > c$, and hence $a^2 > c^2$.

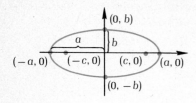

Figure 11-9

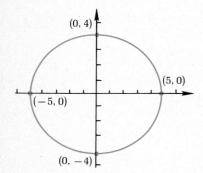

Figure 11-10

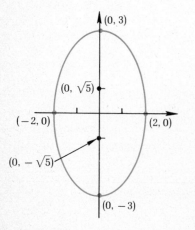

Figure 11-11

or, equivalently,*

$$\frac{x^2}{a^2} + \frac{y^2}{b^2} = 1$$

where $b^2 = a^2 - c^2$, and hence $a > b$.

Note that if $x = 0$, then $y = \pm b$; and if $y = 0$, then $x = \pm a$. The points $(\pm a, 0)$ and $(0, \pm b)$ are the **vertices** of the ellipse. See Figure 11-9.

If the foci are chosen to be $(0, c)$ and $(0, -c)$ [instead of $(c, 0)$ and $(-c, 0)$], then, by a similar argument, an equation for the ellipse is

$$\frac{y^2}{a^2} + \frac{x^2}{b^2} = 1$$

where $a > b$ and $b^2 = a^2 - c^2$.

Example 1

Find an equation for the ellipse which passes through $(5, 0)$ and has foci $(3, 0)$ and $(-3, 0)$.

Solution

In this case, with $a = 5$ and $c = 3$, we have

$$b^2 = a^2 - c^2 = 25 - 9 = 16$$

Hence,

$$\frac{x^2}{25} + \frac{y^2}{16} = 1$$

is an equation for the ellipse. See Figure 11-10.

Example 2

Sketch the set of all points (x, y) for which $\dfrac{x^2}{4} + \dfrac{y^2}{9} = 1$.

Solution

The graph is an ellipse. If $x = 0$, then $y = \pm 3$, and if $y = 0$, then $x = \pm 2$. See Figure 11-11. With $a^2 = 9$, $b^2 = 4$, and c the second coordinate of the focus, we see that $4 = 9 - c^2$; therefore $c = \sqrt{5}$, and the foci are $(0, \pm\sqrt{5})$.

□

*Although we arrived at this equation by squaring, it can be shown that the graph of this equation contains only points on the given ellipse.

The **eccentricity** of an ellipse is defined to be the number c/a. Since $a > c > 0$, we see that $0 < c/a < 1$.

If the eccentricity is close to 0, then the ellipse is almost a circle. To see this recall that $b^2 = a^2 - c^2$; hence,

$$\frac{b^2}{a^2} = 1 - \frac{c^2}{a^2} = 1 - \left(\frac{c}{a}\right)^2$$

Thus, if c/a is close to 0, then b^2/a^2 is close to 1, and so $b \approx a$.

If the eccentricity is close to 1, then b is small compared to a and the ellipse has the shape indicated in Figure 11-12. For example, the ellipse in Figure 11-10 has eccentricity $\frac{c}{a} = \frac{3}{5} = .6$, while the ellipse in Figure 11-11 has eccentricity $\frac{c}{a} = \frac{\sqrt{5}}{3} \approx .74$, which is greater than .6. We there-

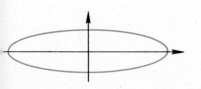

Figure 11-12

fore expect the second ellipse to be flatter than the first. An inspection of the graphs show that this is indeed the case. □

A **hyperbola** can be defined as the set of all points the *difference* of whose distances from two given points (again called the *foci*) is in absolute value a given positive number $2a$. Let the foci be the points $(0, c)$ and $(0, -c)$ where $c > 0$ and $c > a$. Then, an equation for the hyperbola is

$$\frac{y^2}{a^2} - \frac{x^2}{b^2} = 1 \qquad \text{where } c^2 - a^2 = b^2$$

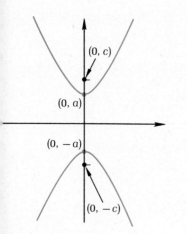

Figure 11-13

(See Figure 11-13.) Notice that if $x = 0$, $y = \pm a$. The points $(0, \pm a)$ are the **vertices** of this hyperbola.

If the foci are the points $(c, 0)$ and $(-c, 0)$, we obtain

$$\frac{x^2}{a^2} - \frac{y^2}{b^2} = 1 \qquad \text{where } c^2 - a^2 = b^2$$

In this case the vertices are $(\pm a, 0)$. (See Figure 11-14.)

The derivation of the equation for a hyperbola is analogous to the derivation of that for an ellipse; we leave the details as an exercise.

There are two lines of special interest that are associated with hyperbolas. Let us consider the hyperbola given by

$$\frac{x^2}{a^2} - \frac{y^2}{b^2} = 1$$

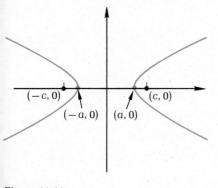

Figure 11-14

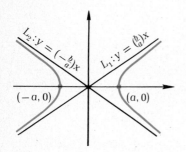

Figure 11-15

The two lines given by

$$\frac{x}{a} - \frac{y}{b} = 0 \quad \text{and} \quad \frac{x}{a} + \frac{y}{b} = 0$$

or, equivalently,

$$y = \frac{bx}{a} \quad \text{and} \quad y = -\frac{bx}{a}$$

are **asymptotes** for the hyperbola; that is, as $|x|$ gets larger and larger, the hyperbola gets closer and closer to these lines. See Figure 11-15.

This is suggested by the following observation: if we confine our attention, say, to the branch of the hyperbola given by

$$y = \frac{b}{a}\sqrt{x^2 - a^2} = \frac{bx}{a}\sqrt{1 - \frac{a^2}{x^2}} \quad \text{with } x \geq a > 0$$

we see that for x large, $\frac{a^2}{x^2}$ is small, so $\sqrt{1 - \frac{a^2}{x^2}}$ is close to 1. This suggests that the straight line given by $y = \frac{bx}{a}$ may be an asymptote. To see that this is actually the case we now show that the hyperbola approaches the line $y = \frac{bx}{a}$ as x increases:

$$\lim_{x \to \infty}\left(\frac{bx}{a} - \frac{b}{a}\sqrt{x^2 - a^2}\right) = \frac{b}{a}\lim_{x \to \infty}(x - \sqrt{x^2 - a^2})$$

$$= \frac{b}{a}\lim_{x \to \infty}\left[(x - \sqrt{x^2 - a^2})\frac{x + \sqrt{x^2 - a^2}}{x + \sqrt{x^2 - a^2}}\right]$$

$$= ab\lim_{x \to \infty}\frac{1}{x + \sqrt{x^2 - a^2}} \quad \text{(Why?)}$$

$$= 0$$

which is what we wanted to show.

These asymptotes are useful for graphing hyperbolas. A frequent method is to locate the vertices, draw the asymptotes, and then construct the hyperbola so that it passes through the vertices and "tends to" the asymptotes. If the hyperbola is of the form $\frac{y^2}{a^2} - \frac{x^2}{b^2} = 1$, then the corresponding asymptotes are given by

$$y = \frac{ax}{b} \quad \text{and} \quad y = -\frac{ax}{b}$$

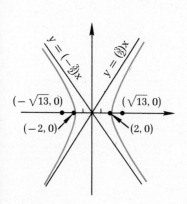

Figure 11-16

Example 3

Find the vertices and the foci of the hyperbola given by $\dfrac{x^2}{4} - \dfrac{y^2}{9} = 1$. Graph the hyperbola.

Solution

Note that if $y = 0$, then $x = 2$ or $x = -2$. Hence, $(2, 0)$ and $(-2, 0)$ are the vertices. Thus, the vertices are on the horizontal axis and so are the foci. With $a^2 = 4$ and $b^2 = 9$, we see that $c^2 = a^2 + b^2 = 4 + 9 = 13$. Therefore, $c = \sqrt{13}$ or $-\sqrt{13}$, and $(\sqrt{13}, 0)$ and $(-\sqrt{13}, 0)$ are the foci of the hyperbola. Since $a = 2$ and $b = 3$, we see that the asymptotes are the lines given by $y = \dfrac{3x}{2}$ and $y = -\dfrac{3x}{2}$. The graph of this hyperbola is given in Figure 11-16.

Example 4

Find the vertices and the foci of the hyperbola given by $y^2 - \dfrac{x^2}{8} = 1$.

Solution

The foci of this hyperbola are on the y axis. We see that if $x = 0$, then $y = 1$ or -1. Thus, $(0, 1)$ and $(0, -1)$ are the vertices of the hyperbola. With $a^2 = 1$ and $b^2 = 8$, we see that $c^2 = a^2 + b^2 = 1 + 8 = 9$ and, consequently, $c = 3$ or -3. Hence, $(0, 3)$ and $(0, -3)$ are the foci. With $a = 1$ and $b = \sqrt{8} = 2\sqrt{2}$, we see that the asymptotes to this hyperbola are given by

$$ y = \frac{x}{2\sqrt{2}} = \frac{x}{2\sqrt{2}} \cdot \frac{\sqrt{2}}{\sqrt{2}} = \frac{\sqrt{2}x}{4} $$

(notice that $\dfrac{\sqrt{2}}{4}$ is approximately equal to 0.35), and

$$ y = -\frac{x}{2\sqrt{2}} = -\frac{\sqrt{2}x}{4} $$

This hyperbola is graphed in Figure 11-17.

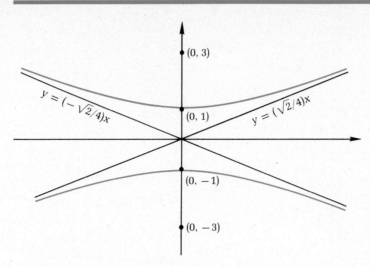

$y = (-\sqrt{2}/4)x$

$(0, 3)$

$(0, 1)$

$y = (\sqrt{2}/4)x$

$(0, -1)$

$(0, -3)$

Figure 11-17

Exercises 11-3

In Exercises 1–6, sketch the graph for each of the given equations. Find the vertices, the foci, and in the case of hyperbolas, the asymptotes.

1. $\dfrac{x^2}{4} + \dfrac{y^2}{3} = 1$

2. $\dfrac{x^2}{25} + \dfrac{y^2}{24} = 1$

3. $4x^2 - 9y^2 = 36$

4. $\dfrac{x^2}{25} + \dfrac{y^2}{9} = 1$

5. $9y^2 - 4x^2 = 1$

6. $4x^2 - 9y^2 = 1$

In Exercises 7–10, describe the graph of the function f given by:

7. $f(x) = 2\sqrt{4 - x^2}$ (Hint: First graph f^2.)

8. $f(x) = -2\sqrt{1 - 4x^2}$

9. $f(x) = -2\sqrt{4 + x^2}$ 10. $f(x) = 2\sqrt{x^2 - 4}$

In Exercises 11–13, find the point of intersection (if any) of the line L and ellipse E whose equations are, respectively:

11. L: $x + 2y = 7$; E: $x^2 + 4y^2 = 25$

12. L: $10y + 3x = 25$; E: $4x^2 + 25y^2 = 100$

13. L: $y = 5x + 15$; E: $\dfrac{x^2}{4} + \dfrac{y^2}{16} = 1$

14. Find an equation for the ellipse whose vertices are (a) $(\pm 6, 0)$ and $(0, \pm 5)$; (b) $(\pm 2, 0)$ and $(0, \pm 1)$.

15. Find an equation for the ellipse whose foci are $(\pm 4, 0)$ and two of whose vertices are $(\pm 6, 0)$.

16. Let a be a real number, and let C be the set of all (x, y) for which

$$\frac{x^2}{7 - a} + \frac{y^2}{4 - a} = 1$$

Show that if $a < 4$, then C is an ellipse, and that if $4 < a < 7$, then C is a hyperbola.

11-4 TRANSLATION

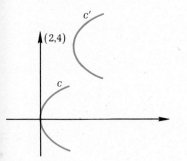

Figure 11-18

We have already seen that $x^2 + y^2 = r^2$ is an equation for a circle C with center at the origin. If this circle is shifted h units horizontally and k units vertically then the center is shifted to (h, k) and consequently $(x - h)^2 + (y - k)^2 = r^2$ is an equation for the shifted circle C'.

Similarly, if the parabola C given by $y^2 = 6x$ is shifted 2 units horizontally and 4 units vertically, then $(y - 4)^2 = 6(x - 2)$ is an equation for the shifted parabola C'. (See Figure 11-18.) To see this, observe that for any point (c, d), $d^2 = 6c$ if and only if $(d' - 4)^2 = 6(c' - 2)$, where (c', d') is the shifted point $(c + 2, d + 4)$.

In general, if the conic sections given by

$$y^2 = 4px, \quad x^2 = 4py, \quad \frac{x^2}{a^2} + \frac{y^2}{b^2} = 1, \quad \frac{x^2}{a^2} - \frac{y^2}{b^2} = 1,$$

and

$$\frac{y^2}{a^2} - \frac{x^2}{b^2} = 1$$

are translated so that the point at the origin (that is, the *center*) is moved to (h, k), then the resulting conics have, respectively, the equations:

$(y - k)^2 = 4p(x - h)$ Parabola;   vertex (h, k), focus$(p + h, k)$, directrix $x = -p + h$

$(x - h)^2 = 4p(y - k)$ Parabola; vertex (h, k), focus $(h, p + k)$,   directrix $y = -p + k$

$\dfrac{(x - h)^2}{a^2} + \dfrac{(y - k)^2}{b^2} = 1$ Ellipse; vertices $(h \pm a, k)$ and $(h, k \pm b)$, center (h, k), foci $(h \pm c, k)$ if $c^2 = a^2 - b^2$, and $(h, k \pm c)$ if $c^2 = b^2 - a^2$

$$\frac{(x - h)^2}{a^2} - \frac{(y - k)^2}{b^2} = 1$$

Hyperbola; foci $(h \pm c, k)$, where $c^2 = a^2 + b^2$, vertices $(h \pm a, k)$

$$\frac{(y - k)^2}{a^2} - \frac{(x - h)^2}{b^2} = 1$$

Hyperbola; foci $(h, k \pm c)$, where $c^2 = a^2 + b^2$, vertices: $(h, k \pm a)$

Example 1

Find an equation for and sketch the ellipse with center at $(2, 3)$, a focus at $(2, 5)$, and a vertex at $(2, 6)$.

Solution

Here, with $c = 5 - 3 = 2$ and $b = 6 - 3 = 3$, we have $a^2 = b^2 - c^2 = 5$. Hence,

$$\frac{(x - 2)^2}{5} + \frac{(y - 3)^2}{9} = 1$$

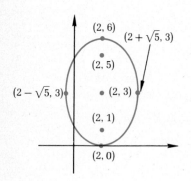

Figure 11-19

is an equation of this ellipse. The other focus is at $(2, 3 - 2) = (2, 1)$, and the other vertices are at $(2, 3 - 3) = (2, 0), (2 - \sqrt{5}, 3)$, and $(2 + \sqrt{5}, 3)$. See Figure 11-19.

Example 2

Identify and sketch the graph of the equation

$$-x/2 = y^2 + 4y + 8.$$

Solution

For any point (x, y) on the graph, we have

$$\begin{aligned}
-x &= 2(y^2 + 4y) + 16 \\
&= 2(y^2 + 4y + 4 - 4) + 16 \quad \text{[Completing} \\
&= 2(y + 2)^2 + 8 \quad\quad\quad\quad \text{the square]}
\end{aligned}$$

and hence $-(x + 8) = 2(y + 2)^2$; that is,

$$(y + 2)^2 = -\tfrac{1}{2}(x + 8) = 4(-\tfrac{1}{8})(x + 8)$$

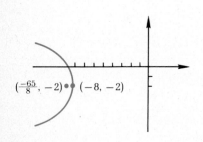

Figure 11-20

Hence, the graph is a parabola with vertex $(-8, -2)$; and since $p = -\tfrac{1}{8}$, we see that the focus is $(-\tfrac{65}{8}, -2)$. See Figure 11-20.

Example 3

Identify and describe the graph of $x^2 + 2x - y^2 - 4y = 7$.

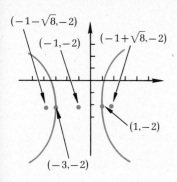

$(-1-\sqrt{8},-2)$

$(-1,-2)$ $(-1+\sqrt{8},-2)$

$(1,-2)$

$(-3,-2)$

Figure 11-21

Solution

For any point (x, y) on the graph we have

$$x^2 + 2x - y^2 - 4y = (x^2 + 2x + 1) - (y^2 + 4y + 4) + 3$$
$$= (x + 1)^2 - (y + 2)^2 + 3 = 7$$

or, equivalently,

$$\frac{(x + 1)^2}{4} - \frac{(y + 2)^2}{4} = 1$$

Hence, the graph is a hyperbola with $a = b = 2$, $c^2 = a^2 + b^2 = 8$, center at $(-1, -2)$, vertices at $(-1 \pm 2, -2)$, and foci at $(-1 \pm \sqrt{8}, -2)$. See Figure 11-21.

Exercises 11-4

In Exercises 1–12, identify and sketch the graph of the given equation, and find the vertex (or vertices.)

1. $4(y + 3)^2 - 9(x - 2)^2 = 1$

2. $9(x - 2)^2 - 4(y + 3)^2 = 36$

3. $y^2 + 4x = 8$

4. $y^2 + 8x - 2y = 7$

5. $25x^2 - 4y^2 - 150x - 16y + 109 = 0$

6. $16y^2 - 9x^2 - 64y - 54x - 161 = 0$

7. $3y^2 = 6y - 5x - 13$

8. $x^2 - 12x + 12y + 48 = 0$

9. $9x^2 + 25y^2 - 90x - 150y + 225 = 0$

10. $9x^2 + 4y^2 - 36x + 16y + 16 = 0$

11. $2y^2 + 5x - 3y + 4 = 0$

12. $16x^2 + y^2 - 32x = -(4y + 16)$

◇**13.** Find an equation for the parabola whose vertex is $(-1, -1)$ and which passes through the points $(2, -2)$ and $(-4, -2)$.

◇**14.** Find an equation for the parabola whose vertex is $(0, 8)$ and which intersects the horizontal axis at $(-3, 0)$ and $(3, 0)$.

◇**15.** Sketch the graph whose equation is $|x - y^2| = 1$.

◇**16.** Sketch the graph whose equation is

$$(x^2 - 2x + 4y^2 + 8y + 4)(x^2 - 2x - 4y^2 - 8y - 4) = 0$$

11-5 ROTATION

A companion to the problem of *translation* is that of **rotation.**
If the point (c, d) is rotated through an angle θ to the point (c', d'), then

$$c' = c \cos \theta - d \sin \theta$$

and

$$d' = c \sin \theta + d \cos \theta$$

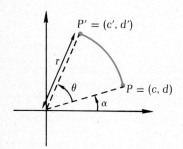

Figure 11-22

To see this, let r be the distance from the origin to (c, d), and let α be the angle (c, d) makes with the horizontal axis as shown in Figure 11-22. Then

$$c = r \cos \alpha \quad \text{and} \quad d = r \sin \alpha$$

and

$$c' = r \cos (\alpha + \theta)$$
$$= r \cos \alpha \cos \theta - r \sin \alpha \sin \theta$$

from which we see that

$$c' = c \cos \theta - d \sin \theta$$

Similarly,

$$d' = c \sin \theta + d \cos \theta$$

Solving this system for c and d, we obtain

$$c = c' \cos \theta + d' \sin \theta$$

and

$$d = -c' \sin \theta + d' \cos \theta$$

Example 1

If the point $(3, 4)$ is rotated by $30°$, we obtain the point (x', y'), where

$$x' = 3 \cos 30° - 4 \sin 30° = \frac{3\sqrt{3}}{2} - \frac{4}{2} = \frac{3\sqrt{3} - 4}{2}$$

and

$$y' = 3 \sin 30° + 4 \cos 30° = \frac{3}{2} + \frac{4\sqrt{3}}{2} = \frac{3 + 4\sqrt{3}}{2}$$

that is, the result is the point $\left(\dfrac{3\sqrt{3} - 4}{2}, \dfrac{3 + 4\sqrt{3}}{2} \right)$.

□

Using the equations relating the coordinates of a point to the coordinates of its image under rotation, we have the following result:

Let C be the graph of the equation $f(x, y) = 0$,* let C' be the curve obtained by rotating C through the angle θ, and let $g(x, y) = f(x \cos \theta + y \sin \theta, -x \sin \theta + y \cos \theta)$. Then $g(x, y) = 0$ is an equation for C'.

To see this, observe that $f(c, d) = f(c' \cos \theta + d' \sin \theta, -c' \sin \theta + d' \cos \theta) = g(c', d')$ so (c, d) is on C if and only if $g(c', d') = 0$; that is, the graph of $g(x, y) = 0$ is C'.

Example 2

Let C be the graph of $x^2 - y^2 - 1 = 0$. Let C' be the graph obtained by rotating C through $45°$. Then an equation for C' is

$$(x \cos 45° + y \sin 45°)^2$$
$$- (-x \sin 45° + y \cos 45°)^2 - 1 = 0$$

which can be simplified to

$$\left(\frac{x}{\sqrt{2}} + \frac{y}{\sqrt{2}} \right)^2 - \left(-\frac{x}{\sqrt{2}} + \frac{y}{\sqrt{2}} \right)^2 - 1 = 0$$

or

$$2xy - 1 = 0$$

This example shows that $2xy - 1 = 0$ is an equation for a hyperbola. In general, the curve given by $Ax^2 + By^2 + Cxy + Dx + Ey + F = 0$ for any real numbers A, B, C, D, E, F is a conic.† The following example illustrates how to identify such curves.

Example 3

Identify and sketch the curve C given by $x^2 + xy + y^2 - 4 = 0$.

---

*The notation $f(x, y) = 0$ is used here to indicate equations such as $x^2 + y^2 - 1 = 0$, $y^2 + 6x = 0$, etc. It is also the case that f is a function whose domain consists of points (x, y).

†If we call lines, points and the empty set conics.

Solution

Let C' be the curve obtained by rotating C through the angle θ. We will choose θ so that the equation for C' has no xy term, that is, C' will be one of the conics in standard position. The equation for C' is

$$(x \cos \theta + y \sin \theta)^2 + (x \cos \theta + y \sin \theta)$$
$$(-x \sin \theta + y \cos \theta) + (-x \sin \theta + y \cos \theta)^2 - 4 = 0$$

which reduces to

$$(1 - \sin \theta \cos \theta)x^2 + (1 + \sin \theta \cos \theta)y^2$$
$$+ (\cos^2 \theta - \sin^2 \theta)xy - 4 = 0$$

We want $\cos^2 \theta - \sin^2 \theta = 0$, so we choose $\theta = 45°$, and the equation for C' becomes

$$\left(1 - \frac{1}{\sqrt{2}} \frac{1}{\sqrt{2}}\right) x^2 + \left(1 + \frac{1}{\sqrt{2}} \frac{1}{\sqrt{2}}\right) y^2 - 4 = 0$$

which reduces to

$$\frac{x^2}{8} + \frac{y^2}{\frac{8}{3}} = 1$$

Therefore C' is an ellipse, and C is the ellipse obtained by rotating C' through $-45°$. See Figure 11-23.

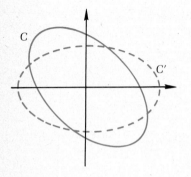

Figure 11-23

Exercises 11-5

Rotate each of the points in Exercises 1–4 30°, 45°, and 60°.

1. $(1, 3)$ **2.** $(-1, 3)$

3. $(1, -3)$ **4.** $(-1, -3)$

5. Rotate the point $(-1, 2)$ 105° by first rotating it 45° and then 60°.

6. Rotate the point $(2, 3)$ 75° by first rotating it 45° and then 30°.

7. Rotate the point $(1, 2)$ 15° by first rotating it 60° and then $-45°$.

In Exercises 8–10, find the equation of the line obtained by rotating

8. The line whose equation is $y = 5x$, 30° (with the origin fixed).

9. The line whose equation is $3y - 2x = 1$, 60° (with the origin fixed).

10. The line whose equation is $y - x = 1$, 90° (with the origin fixed).

11. Show that the transformation $x' = -y$, $y' = x$ corresponds to rotating the coordinate axes 90° (with the origin fixed).

In Exercises 12–16, identify and sketch the curves given by each of the following equations.

12. $3x^2 - 2xy + 3y^2 - 2 = 0$

13. $5x^2 - 6xy + 5y^2 - 32 = 0$

14. $x^2 - 2xy + y^2 - 8x - 8y = 0$

15. $2x^2 + 2y^2 + 3\sqrt{3}\,xy - 2 = 0$

16. $x^2 + y^2 + 3xy - 1 = 0$

11-6 INEQUALITIES

For illustration, consider the pair of inequalities

$$x^2 + 4y^2 - 1 < 0 \quad \text{and} \quad x^2 + 4y^2 - 1 > 0$$

Notice that the graphs of these inequalities are disjoint sets in a plane, and these sets taken together comprise the entire plane except for their common border, which is the ellipse given by $x^2 + 4y^2 = 1$. In fact, one of these graphs is the "inside" of the ellipse, and the other is the "outside." This can be generalized as follows.

The graphs of $f(x, y) < 0$ and $f(x, y) > 0$ are disjoint and comprise the whole plane except for their common border, which is given by the equation $f(x, y) = 0$.

Example 1

Sketch the graph given by $\dfrac{x^2}{4} + \dfrac{y^2}{9} - 1 < 0$.

Solution

This graph is either the inside or the outside of the ellipse given by $\dfrac{x^2}{4} + \dfrac{y^2}{9} - 1 = 0$. We can decide which by classifying just one known point! The origin $(0, 0)$ is inside the ellipse and is on the graph since $\dfrac{0^2}{4} + \dfrac{0^2}{9} - 1 < 0$; hence the graph of $\dfrac{x^2}{4} + \dfrac{y^2}{9} - 1 < 0$ is the inside of the ellipse.

Example 2

Sketch the set of all points (x, y) for which $y^2 + 3x \geq 0$.

Solution

This graph is either the "inside" or the "outside" of the parabola given by $y^2 + 3x = 0$, together with the parabola itself. Again, we can decide by considering just one point. The focus $\left(-\dfrac{3}{4}, 0\right)$ is inside the parabola and is *not* on the graph since $0^2 + 3\left(-\dfrac{3}{4}\right) < 0$; hence, the graph of $y^2 + 3x \geq 0$ is the "outside" of the parabola plus the parabola itself.

Exercises 11-6

In Exercises 1–14, sketch the set of all points (x, y) for which

1. $y^2 \leq 4x$ **2.** $y^2 \geq 4x + 2$

3. $(x - 1)^2 \leq y$ **4.** $(x + 2)^2 \leq y + 1$

5. $(x - 1)^2 + (y + 2)^2 \geq 1$ **6.** $(x + 3)^2 + (y - 4)^2 > 0$

7. $\dfrac{x^2}{25} + \dfrac{y^2}{16} < 1$ **8.** $\dfrac{x^2}{9} + \dfrac{y^2}{7} \geq 1$

9. $\dfrac{x^2}{4} - \dfrac{y^2}{49} \geq 1$ **10.** $\dfrac{x^2}{9} - y^2 < 1$

11. $\dfrac{(x - 1)^2}{25} + \dfrac{(y + 2)^2}{16} < 1$ **12.** $\dfrac{(x + 1)^2}{9} - (y - 1)^2 < 1$

13. $16x^2 + 25y^2 - 32x + 100y - 284 < 0$

14. $x^2 - y^2 + 2(x + y) - 1 > 0$

11-7 DISTANCE. SPHERES. PLANES

The reader is already familiar with the fact that the solution of a system of linear equations such as

$$3x + 2y = 1$$
$$4x + 3y = 0$$

is simply the point of intersection of the lines given by these equations, that is, the point $(3, -4)$, as the reader should verify (see Figure 11-24).

In general, the solution of the system

$$ax + by = k$$
$$cx + dy = l$$

is the intersection of the lines L_1 and L_2 given by these two equations; that is, x and y are numbers for which $ax + by = k$

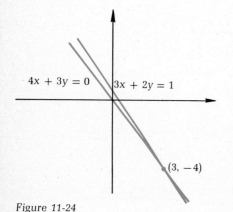

Figure 11-24

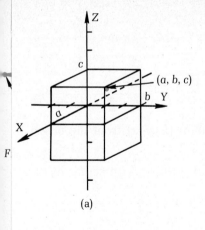

(a)

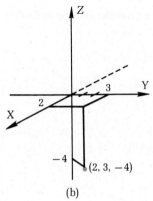

(b)

Figure 11-25

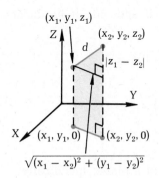

Figure 11-26

and $cx + dy = l$ if and only if (x, y) is a point on both the lines L_1 and L_2. In view of this, we see that there are three possibilities:

1. The two lines are parallel, in which case the corresponding system has *no* solution.

2. The lines are the same, in which case there are infinitely many solutions to the corresponding system.

3. The lines are not parallel (that is, the lines intersect), in which case the corresponding system has exactly one solution.

Thus, the familiar two-dimensional rectangular coordinate system provides a geometric interpretation for *linear equations in two unknowns*. Similarly, as we shall see, a three-dimensional rectangular coordinate system provides a geometric interpretation for linear equations in three unknowns.

Consider three mutually perpendicular number lines, called the X axis, Y axis, and Z axis respectively, whose origins coincide. Then the "point" (a, b, c), where a, b, and c are any numbers, is the point c units above the point $P = (a, b)$ in the XY coordinate plane if $c > 0$, and $|c|$ units below it if $c < 0$. [See Figure 11-25(a).] For example, the point $(2, 3, -4)$ is 4 units directly *below* the point $P = (2, 3)$ in the XY coordinate plane. [See Figure 11-25(b).] Similarly, the point $(2, 3, 4)$ is 4 units directly *above* the point P.

We shall now show that the *distance d between two points* $P_1 = (x_1, y_1, z_1)$ and $P_2 = (x_2, y_2, z_2)$ is

$$\sqrt{(x_1 - x_2)^2 + (y_1 - y_2)^2 + (z_1 - z_2)^2}$$

To see this, we use the Theorem of Pythagoras twice. First, the distance from $(x_1, y_1, 0)$ to $(x_2, y_2, 0)$

$$\sqrt{(x_1 - x_2)^2 + (y_1 - y_2)^2}$$

Thus (see Figure 11-26), d is the length of the hypotenuse of the right triangle whose other sides have lengths

$$\sqrt{(x_1 - x_2)^2 + (y_1 - y_2)^2} \quad \text{and} \quad |z_1 - z_2|$$

Thus,

$$d^2 = |z_1 - z_2|^2 + (\sqrt{(x_1 - x_2)^2 + (y_1 - y_2)^2})^2$$
$$= (z_1 - z_2)^2 + (x_1 - x_2)^2 + (y_1 - y_2)^2$$

from which the desired result follows.

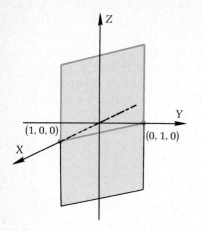

Figure 11-29

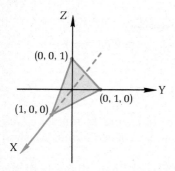

Figure 11-30

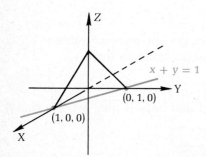

Figure 11-31

or, equivalently,

$$2x - 2y - 8z = -6$$

which is an equation for S.

Example 4

The set of all points (x, y, z) for which $x + y = 1$ is a plane. In fact, it is the plane perpendicular to the XY plane and passing through the line in the XY plane having equation $x + y = 1$. See Figure 11-29.

Example 5

Sketch the graph of the equation $x + y + z = 1$.

Solution

The graph is a plane; call it S. Since three points determine a plane, it is sufficient to find three points on S. For example, if $x = y = 0$, then $z = 1$, and so $(0, 0, 1)$ is on the graph. Similarly, $(0, 1, 0)$ and $(1, 0, 0)$ are on S (see Figure 11-30). Notice that the intersection of S and the XY plane is the set of all points (x, y, z) for which

$$x + y + z = 1$$

and

$$z = 0$$

and hence the intersection is the line in the XY plane given by the equation $x + y = 1$ (see Figure 11-31). Thus, the *system*

$$x + y + z = 1$$
$$z = 0$$

has infinitely many solutions; namely, $(x, y, 0)$ where x and y are any numbers for which $x + y = 1$.

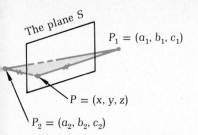

The plane S

$P_1 = (a_1, b_1, c_1)$

$P = (x, y, z)$

$P_2 = (a_2, b_2, c_2)$

Figure 11-28

We can show this as follows. Suppose S is a plane. Consider a line segment perpendicular to S with endpoints equidistant from S, and call these points $P_1 = (a_1, b_1, c_1)$ and $P_2 = (a_2, b_2, c_2)$. See Figure 11-28. Then any point $P = (x, y, z)$ on S is equidistant from P_1 and P_2, and a point not on S cannot be equidistant from P_1 and P_2. Thus, if $P = (x, y, z)$ is on S, then

$$\overline{|PP_1|}^2 = \overline{|PP_2|}^2$$

or, equivalently,

$$(x - a_1)^2 + (y - b_1)^2 + (z - c_1)^2$$
$$= (x - a_2)^2 + (y - b_2)^2 + (z - c_2)^2$$

That is, after squaring and simplifying,

$$ax + by + cz = k$$

where

$$a = 2(a_2 - a_1), \quad b = 2(b_2 - b_1), \quad c = 2(c_2 - c_1),$$

and

$$k = (a_2^2 + b_2^2 + c_2^2) - (a_1^2 + b_1^2 + c_1^2)$$

Conversely, *for any numbers a, b, c, and k, the set of all points (x, y, z) for which ax + by + cz = k is a plane.* We shall not prove this.

Note that every point in the XY coordinate plane is of the form $(x, y, 0)$, and thus $z = 0$ *is an equation for the XY coordinate plane.* Similarly, $x = 0$ *and* $y = 0$ *are equations for the YZ and XZ coordinate planes, respectively.*

Example 3

Let S be the plane which bisects the line segment joining $P_1 = (1, 1, 3)$ and $P_2 = (2, 0, -1)$. Find an equation for S.

Solution

A point $P = (x, y, z)$ is on S if and only if $\overline{|PP_1|} = \overline{|PP_2|}$; that is, if and only if

$$\sqrt{(x - 1)^2 + (y - 1)^2 + (z - 3)^2}$$
$$= \sqrt{(x - 2)^2 + (y - 0)^2 + (z + 1)^2}$$

or

$$x^2 - 2x + 1 + y^2 - 2y + 1 + z^2 - 6z + 9$$
$$= x^2 - 4x + 4 + y^2 + z^2 + 2z + 1$$

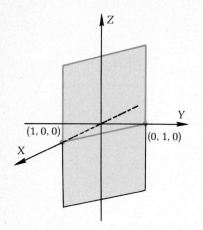

Figure 11-29

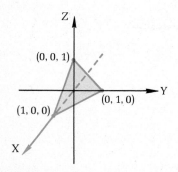

Figure 11-30

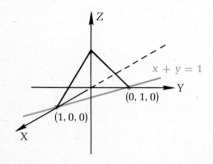

Figure 11-31

or, equivalently,

$$2x - 2y - 8z = -6$$

which is an equation for S.

Example 4

The set of all points (x, y, z) for which $x + y = 1$ is a plane. In fact, it is the plane perpendicular to the XY plane and passing through the line in the XY plane having equation $x + y = 1$. See Figure 11-29.

Example 5

Sketch the graph of the equation $x + y + z = 1$.

Solution

The graph is a plane; call it S. Since three points determine a plane, it is sufficient to find three points on S. For example, if $x = y = 0$, then $z = 1$, and so $(0, 0, 1)$ is on the graph. Similarly, $(0, 1, 0)$ and $(1, 0, 0)$ are on S (see Figure 11-30). Notice that the intersection of S and the XY plane is the set of all points (x, y, z) for which

$$x + y + z = 1$$

and

$$z = 0$$

and hence the intersection is the line in the XY plane given by the equation $x + y = 1$ (see Figure 11-31). Thus, the *system*

$$x + y + z = 1$$
$$z = 0$$

has infinitely many solutions; namely, $(x, y, 0)$ where x and y are any numbers for which $x + y = 1$.

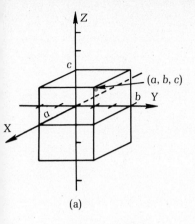

(a)

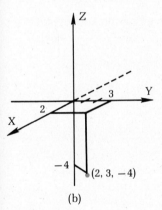

(b)

Figure 11-25

and $cx + dy = l$ if and only if (x, y) is a point on both the lines L_1 and L_2. In view of this, we see that there are three possibilities:

1. The two lines are parallel, in which case the corresponding system has *no* solution.
2. The lines are the same, in which case there are infinitely many solutions to the corresponding system.
3. The lines are not parallel (that is, the lines intersect), in which case the corresponding system has exactly one solution.

Thus, the familiar two-dimensional rectangular coordinate system provides a geometric interpretation for *linear equations* in two unknowns. Similarly, as we shall see, a three-dimensional rectangular coordinate system provides a geometric interpretation for linear equations in three unknowns.

Consider three mutually perpendicular number lines, called the X axis, Y axis, and Z axis respectively, whose origins coincide. Then the "point" (a, b, c), where a, b, and c are any numbers, is the point c units above the point $P = (a, b)$ in the XY coordinate plane if $c > 0$, and $|c|$ units below it if $c < 0$. [See Figure 11-25(a).] For example, the point $(2, 3, -4)$ is 4 units directly *below* the point $P = (2, 3)$ in the XY coordinate plane. [See Figure 11-25(b).] Similarly, the point $(2, 3, 4)$ is 4 units directly *above* the point P.

We shall now show that the *distance d between two points* $P_1 = (x_1, y_1, z_1)$ and $P_2 = (x_2, y_2, z_2)$ is

$$\sqrt{(x_1 - x_2)^2 + (y_1 - y_2)^2 + (z_1 - z_2)^2}$$

To see this, we use the Theorem of Pythagoras twice. First, the distance from $(x_1, y_1, 0)$ to $(x_2, y_2, 0)$

$$\sqrt{(x_1 - x_2)^2 + (y_1 - y_2)^2}$$

Thus (see Figure 11-26), d is the length of the hypotenuse of the right triangle whose other sides have lengths

$$\sqrt{(x_1 - x_2)^2 + (y_1 - y_2)^2} \quad \text{and} \quad |z_1 - z_2|$$

Thus,

$$d^2 = |z_1 - z_2|^2 + (\sqrt{(x_1 - x_2)^2 + (y_1 - y_2)^2})^2$$
$$= (z_1 - z_2)^2 + (x_1 - x_2)^2 + (y_1 - y_2)^2$$

from which the desired result follows.

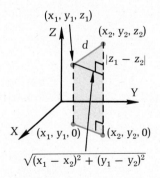

Figure 11-26

Example 1

The distance between the points $(-1, 2, -3)$ and $(1, -3, -4)$ is

$$\sqrt{[1-(-1)]^2 + (-3-2)^2 + [-4-(-3)]^2}$$
$$= \sqrt{2^2 + 5^2 + 1^2} = \sqrt{30}$$

In particular, the distance of a point (x, y, z) from the origin $(0, 0, 0)$ is $\sqrt{x^2 + y^2 + z^2}$.

Hence, (x, y, z) is on the sphere of radius r with center at the origin if and only if

$$x^2 + y^2 + z^2 = r^2$$

In other words, $x^2 + y^2 + z^2 = r^2$ is an equation for the sphere of radius r with center at the origin.

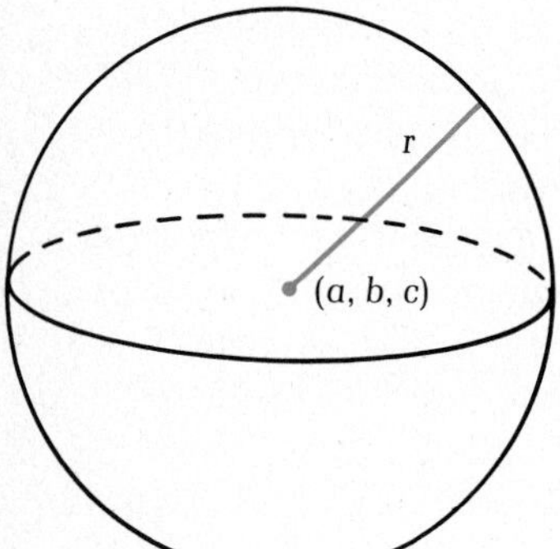

Similarly, a point (x, y, z) is on the sphere of radius r with center at (a, b, c) if and only if

$$(x-a)^2 + (y-b)^2 + (z-c)^2 = r^2$$

In other words, $(x-a)^2 + (y-b)^2 + (z-c)^2 = r^2$ is an equation for the sphere of radius r with center at (a, b, c). (See Figure 11-27.)

Figure 11-27

Example 2

Find an equation for the sphere of radius 2 with center at $(1, 0, -4)$.

Solution

A point (x, y, z) is on this sphere if and only if its distance from $(1, 0, -4)$ is 2; that is,

$$\sqrt{(x-1)^2 + (y-0)^2 + (z-(-4))^2} = 2$$

Hence,

$$(x-1)^2 + y^2 + (z+4)^2 = 4$$

is an equation for the sphere.

□

We shall next obtain an equation of a plane; specifically:

if S is a plane, then there are numbers a, b, c, and k for which the equation of S is

$$ax + by + cz = k$$

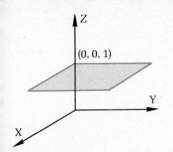

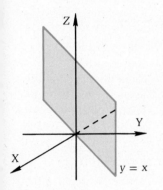

Figure 11-32

Figure 11-33

Example 6

Graph the set of all points (x, y, z) for which z = 1.

Solution

This is the set of all points 1 unit above the XY plane. See Figure 11-32.

Example 7

Graph the set of all points (x, y, z) for which y = x.

Solution

A point in the XY plane is on the graph if and only if it is on the line in the XY plane given by the equation y = x. Any point *directly above or below* this line is also on the graph because for any such point (x, y, z) we still have y = x. See Figure 11-33.

We can see then that if we are considering the set of all points (x, y, z) for which

$$a_1x + b_1y + c_1z = k_1$$
$$a_2x + b_2y + c_2z = k_2$$

and

$$a_3x + b_3y + c_3z = k_3$$

we are actually considering the intersection of the three corresponding planes. Thus, we obtain the following possibilities:

1. The system has no solution. Some of the possibilities for the related planes are given in Figure 11-34.
2. The system has infinitely many solutions that lie on a line. In this case, the planes intersect in a line. See Figure 11-35.

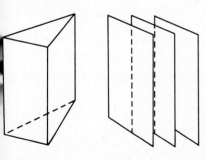

Figure 11-34

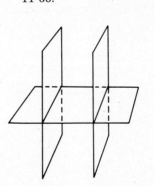

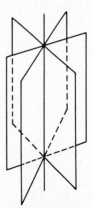

Figure 11-35

3. The system has exactly one solution. In this case, the planes have exactly one point in common.

4. The system has infinitely many solutions, and these solutions lie on a plane. In this case, the equations are all equations for the same plane.

Example 8

If x, y, and z are numbers for which $2x + 3y - z = 2$, then $2x + 3y - z \neq 3$. Thus, the system

$$2x + 3y - z = 2$$
$$2x + 3y - z = 3$$

has no solution. Equivalently, the planes given by these equations have no point in common (that is, they are parallel).
□

In the next chapter we shall investigate procedures for determining the intersection of three planes; that is, we shall investigate the solutions of systems of equations (of planes).

Exercises 11-7

In Exercises 1–4, write an equation for the sphere with:

1. center $(0, 0, 0)$ and radius 9
2. center $(0, 0, 0)$ and radius 5
3. center $(1, 0, -3)$ and radius 4
4. center $(1, -1, 1)$ and radius 7

In Exercises 5 and 6, write an equation for the plane which bisects the line segments joining:

5. $(1, 3, 1)$ and $(2, -1, 5)$
6. $(1, 0, 1)$ and $(1, 1, 0)$

In Exercises 7–13, sketch the set of all points (x, y, z) for which:

7. $x = 1$
8. $x + y + 3z = 3$
9. $x + y + z = 1$ and $x + y = 1$
10. $x^2 + y^2 + z^2 = 4$ and $z = 1$
11. $z = 1$ and $y = x^2$
12. $x^2 + 2x + y^2 + z^2 = 4$
13. $x^2 - 6x + y^2 - 4y + z^2 - 2z = -13$ and $z = 1$

Miscellaneous Problems

1. Johann Kepler (1571–1630), a famous German astronomer, showed that the earth travels around the sun in an elliptic path with the sun at one focus. The *major diameter,* that is, the chord passing through the foci, is the chord that joins the summer solstice (on June 21) and the winter solstice (on December 21). Assuming that the length of this major diameter (i.e., the number $2a$) is equal to 184,000,000 miles, and that $c = 1,000,000$, find an equation for the elliptic path.

2. Show that $ax^2 + cy^2 + dx + ey + f = 0$ is an equation of a parabola when $a = 0$, $c \neq 0$ and $d \neq 0$, or in case $c = 0$, $a \neq 0$, $e \neq 0$.

3. Consider an ellipse whose equation is $\dfrac{x^2}{a^2} + \dfrac{y^2}{b^2} = 1$, with $a > b$. If e is the eccentricity of this ellipse, we define the *directrices* of this ellipse as the lines given by $x = \pm a/e$. Find an equation for the ellipse for which $e = 2/3$, if one of the directrices is given by $x = 9$ and the corresponding focus is the point $(4, 0)$.

4. Show that the hyperbola given by $x^2 - y^2 = 4$ and the ellipse given by $x^2 + 9y^2 = 9$ have the same foci.

5. Suppose a circle has equation $x^2 + y^2 = r^2$, and is rotated through an angle Θ. Show that a point (x', y') is on the new curve if and only if $x'^2 + y'^2 = r^2$. Thus, a circle remains a circle under *any* rotation.

The center of an ellipse or hyperbola is the point midway between its vertices and on the line connecting the vertices. For Problems 6 through 8, describe the graph of the equation, telling if it is an ellipse, hyperbola, etc., and giving such information as the center, the foci, and, for the hyperbola, the asymptotes.

6. $6x^2 + 9y^2 - 24x - 54y + 51 = 0$

7. $4y^2 - 9x^2 + 16y + 18x = 29$

8. $8x^2 + 2y^2 + 48x - 32y = -168$

9. It can be shown that the graph of the set of points (x, y, z) for which

$$\frac{x + 1}{2} = \frac{y}{3} = z - 1$$

is a line in space. Where will this line intersect the plane given by $2x + y - z = 3$? (*Hint:* Let $t = \frac{x + 1}{2} = \frac{y}{3} = z - 1$. Thus, $x = -1 + 2t$, with corresponding expressions for y and z. At a point of intersection, this triple (x, y, z) must be the same on the line and the plane.)

10. Show that the line in space given by

$$\frac{x - 1}{2} = \frac{y + 1}{3} = \frac{z - 2}{4}$$

and the plane given by $x - 2y + z = 6$ are parallel. (*Hint:* See the hint for Problem 9.)

11. Consider the ellipse given by

$$\frac{x^2}{11} + \frac{y^2}{7} = 1$$

Find an equation for the hyperbola whose vertices are the foci of the given ellipse and whose foci are the vertices of the ellipse that lie on the horizontal axis.

12. The arch of a toy bridge is in the shape of the top half of an ellipse. The horizontal span of the bridge is 20 inches, and the height above the center of the bridge is 1 inch. Assuming the center of the bridge is the point $(0, 0)$, find an equation for the bridge. How high is the arch 5 inches from the center of the bridge?

12 Linear Systems, Matrices, and Linear Programming

12-1 INTRODUCTION. ECHELON FORMS

We shall now consider the problem of finding all solutions of a system of linear equations. We begin with some special cases and later show that it is always possible to reduce the general problem to one of such special cases.

Example 1

Find all ordered triples (x, y, z) of real numbers for which

$$x + z = 5 \quad \text{and} \quad y + z = 10$$

Solution

Suppose we try $z = 1$; then we must have $x = 4$ and $y = 9$. In fact, $(4, 9, 1)$ *is* a solution. Similarly, if we try $z = 7$, then $x = -2$ and $y = 3$, and, in fact, $(-2, 3, 7)$ is also a solution. In this way we can obtain as many solutions as we please, since we are free to let z be any real number; then $(5 - z, 10 - z, z)$ is a solution. Furthermore, for any solution (x, y, z), we have $x = 5 - z$ and $y = 10 - z$; thus, the solutions are

$$(5 - z, 10 - z, z)$$

where z is any real number. (Note that there are infinitely many solutions.)

The foregoing example has the special feature that any choice of the third coordinate (that is, coordinate z) determines the other two coordinates (coordinates x and y).

We describe this feature by saying that the system has one degree of freedom; that is, there is no constraint on our choice of z, but once a real number z is chosen, the numbers x and y are determined.* The number x is determined by the first equation in which y does not appear, and the number y is determined by the second equation in which x does not appear.

Example 2

Find all ordered quadruples (x_1, x_2, x_3, x_4) of real numbers for which

$$x_1 + 3x_3 - 4x_4 = 5$$

and

$$x_2 + 6x_3 - 8x_4 = 10$$

Solution

The solutions are $(5 - 3x_3 + 4x_4, 10 - 6x_3 + 8x_4, x_3, x_4)$, where x_3 and x_4 are any real numbers. In this case, x_3 and x_4 are "free" and x_1 and x_2 appear in exactly one equation. For example, if we let $x_3 = x_4 = 0$, then $x_1 = 5$, $x_2 = 10$, and $(5, 10, 0, 0)$ is a solution. Similarly, if we let $x_3 = 1$ and $x_4 = 2$, then $x_1 = 5 - 3 \cdot 1 + 4 \cdot 2 = 10$, $x_2 = 10 - 6 \cdot 1 + 8 \cdot 2 = 20$, and so $(10, 20, 1, 2)$ is also a solution. In general, we can let $x_1 = 5 - 3x_3 + 4x_4$ and $x_2 = 10 - 6x_3 + 8x_4$, where x_3 and x_4 are any real numbers. Thus, this system has "two degrees of freedom."

Remark   Since the particular letters used in the last example do not matter, the solutions may be given by

$$(5 - 3r + 4s, 10 - 6r + 8s, r, s)$$

where r and s are any real numbers.

□

In each of the preceding examples, the system being considered has infinitely many solutions. The next example illustrates a system which has exactly one solution.

---

*Note, however, that in the above example either x or y could also be a "free" coordinate; that is, once any of the three coordinates is chosen, this choice determines the other two.

Example 3

Find all ordered triples (x, y, z) of real numbers for which

$$-2x + y + 4z = -1$$
$$2y - z = 5$$
$$z = -1$$

Solution

Let (x, y, z) be a solution. Since $z = -1$, we see from the second equation that

$$2y = z + 5 = -1 + 5 = 4$$

and so $y = 2$. Finally, from the first equation,

$$-2x = -y - 4z - 1 = -2 + 4 - 1 = 1$$

and so $x = -\frac{1}{2}$. Thus, the solution to this system is $(-\frac{1}{2}, 2, -1)$.

The system in the preceding example is said to be in echelon (or triangular) form. A system of m linear equations is in **echelon** (or **triangular**) form, if for each integer k between 1 and m the kth equation of the system does not involve the first $k - 1$ unknowns or coordinates. Thus, the system

$$a_{11}x_1 + a_{12}x_2 + a_{13}x_3 + \cdots + a_{1n}x_n = b_1$$
$$a_{22}x_2 + a_{23}x_3 + \cdots + a_{2n}x_n = b_2$$
$$a_{33}x_3 + \cdots + a_{3n}x_n = b_3$$
$$\vdots \qquad \vdots \qquad \vdots$$
$$a_{mm}x_m + \cdots + a_{mn}x_n = b_m$$

is in echelon, or triangular, form. Here, $m \leq n$.

□

Now, two linear systems are *equivalent* if they have precisely the same solutions. The next example illustrates how a system can be reduced to an equivalent system in echelon form, from which, as in Example 3, the solutions to the system can be easily read off.

For the sake of verbal economy, let us agree that from now on (x_1, x_2) is an ordered pair of real numbers, (x_1, x_2, x_3) is an ordered triple of real numbers, and so on.

Example 4

Find all (x_1, x_2, x_3) for which

$$2x_1 - x_2 - 4x_3 = 1$$
$$4x_1 - 9x_3 = 7$$

and

$$2x_2 - 2x_3 = 6$$

Solution

We shall proceed by writing down a sequence of systems equivalent to the one that has been given and in which the last system is in echelon form. Thus, the object is to eliminate the first coordinate (i.e., x_1) from the second equation, and the first two coordinates from the third equation. To help us keep track of the computations, we shall label the equations, respectively, A_1, B_1, C_1 in the first system, A_2, B_2, C_2 in the second system, and so on, and indicate to the right of each equation "where it came from." That each system in the sequence is equivalent to the preceding one follows from the fact that multiplication, addition, and subtraction all "preserve equality."

Thus, (x_1, x_2, x_3) is a solution of the system

$$A_1: \quad 2x_1 - x_2 - 4x_3 = 1$$
$$B_1: \quad 4x_1 \qquad - 9x_3 = 7$$
$$C_1: \qquad\qquad 2x_2 - 2x_3 = 6$$

if and only if it is a solution of the system

$$A_2: \quad 2x_1 - x_2 - 4x_3 = 1 \qquad (A_2 = A_1)$$
$$B_2: \qquad\quad 2x_2 - x_3 = 5 \qquad (B_2 = B_1 - 2A_1)^*$$
$$C_2: \qquad\quad 2x_2 - 2x_3 = 6 \qquad (C_2 = C_1)$$

and, in turn, the system

$$A_3: \quad 2x_1 - x_2 - 4x_3 = 1 \qquad (A_3 = A_2)$$
$$B_3: \qquad\quad 2x_2 - x_3 = 5 \qquad (B_3 = B_2)$$
$$C_3: \qquad\qquad\quad x_3 = -1 \qquad (C_3 = B_2 - C_2)$$

the last being precisely the system of Example 3. Thus, the solution is $(-\frac{1}{2}, 2, -1)$; that is, $x_1 = -\frac{1}{2}$, $x_2 = 2$, and $x_3 = -1$.
□

Note that to "solve" a system means nothing more than to reduce it to an equivalent system in which $x_1, x_2, \ldots,$ are

---

*Note: "$B_2 = B_1 - 2A_1$" simply *means* that $2x_2 - x_3 = 4x_1 - 9x_3 - 2(2x_1 - x_2 - 4x_3)$ and $5 = 7 - 2 \cdot 1$.

exhibited as explicitly as possible. For the above examples, such a system is

$$x_1 \qquad\qquad = -\tfrac{1}{2}$$
$$\qquad x_2 \qquad = \quad 2$$
$$\qquad\qquad x_3 = -1$$

The echelon form is but a convenient intermediate form from which this final system is easily obtained.

Example 5

Solve the system

$$2x_1 + x_2 + \ x_3 = 5$$
$$3x_1 + x_2 - 4x_3 = 7$$

Solution

The following sequence of equivalent systems leads to a system in echelon form.

$$\begin{cases} 3(2x_1 + x_2 + \ x_3) = 3 \cdot 5 \\ 2(3x_1 + x_2 - 4x_3) = 2 \cdot 7 \end{cases} \quad [multiplication]$$

$$\begin{cases} 6x_1 + 3x_2 + 3x_3 = 15 \\ 6x_1 + 3x_2 + 3x_3 - (6x_1 + 2x_2 - 8x_3) \\ \qquad\qquad\qquad\qquad = 15 - 14 \end{cases} \quad [subtraction]$$

$$\begin{cases} 2x_1 + x_2 + \ x_3 = 5 \\ \qquad x_2 + 11x_3 = 1 \end{cases}$$

The last system is in echelon form. Note that it has one degree of freedom and (since subtraction preserves equality) is equivalent to

$$2x_1 - 10x_3 = 4$$
$$x_2 + 11x_3 = 1$$

Thus, if t is any real number, letting $x_3 = t$, we obtain

$$x_1 = 2 + 5t$$

and

$$x_2 = 1 - 11t$$

Hence, for any real number t

$$(2 + 5t, 1 - 11t, t)$$

is a solution.

Example 6

Find all (x_1, x_2, x_3) for which

$$2x_1 + 3x_2 - x_3 = 1$$

and

$$2x_1 + 3x_2 - x_3 = 4$$

Solution

If (x_1, x_2, x_3) is a triple for which $2x_1 + 3x_2 - x_3 = 1$, then $2x_1 + 3x_2 - x_3 \neq 4$. Thus, there are *no* solutions for this system.

Exercises 12-1

In Exercises 1–26, find all triples of real numbers (x, y, z) which satisfy the given conditions.

1. $x - z = 10$
   $y + z = -4$

2. $2x + y = -3$
   $z + y = 4$

3. $3x - 2y = 4$
   $-x + z = 1$

4. $x - z = 10$
   $-y - z = 10$

5. $x - y + z = 4$
   $y - z = -1$

6. $2x + y - 3z = 8$
   $-y + 3z = 4$

7. $x + y - 2z = 3$
   $2y + z = 4$

8. $3x + 2y - z = 2$
   $-2y + z = -2$

9. $2x - y - 2z = 1$
   $-x + y + 3z = -1$

10. $-3x + y + 3z = 4$
    $2x + 2y - z = -1$

11. $-x - y + z = 8$
    $2x + 2y - 2z = 15$

12. $3x + 4y - 8z = 10$
    $2x - 2y + 4z = 5$

13. $x + y - 2z = 7$
    $2x - 3y - 2z = 0$
    $x - 2y - 3z = 3$

14. $x + y + z = 3$
    $-2x + y - z = 4$
    $x + 2y + z = 5$

15. $3x + 2y - z = -1$
    $-2x + y - 2z = -1$
    $x + y - z = 0$

16. $x - y + 2z = 3$
    $-2x + 3y + z = 1$
    $-2x + 2y - 4z = 5$

17. $z + 2y = 0$
    $z - 5x = 1$
    $x + 3y = -2$

18. $-x + y - z = 0$
    $-x + 3y + z = 0$
    $x + 2y + 4z = 0$

19. $x + y + 2z = 0$
    $2x + 2y - z = 0$
    $x + y + z = 0$

20. $x - y - z = 0$
    $-x + 3y + 5z = 0$
    $3x + y + 6z = 0$

21. $\quad x + 3y - 2z = 0$
$\quad -5x - 3y + 7z = 0$
$\quad\ \ x -\ \ y -\ \ z = 0$

22. $\quad 2x -\ \ y - 3z = 0$
$\quad -6x -\ \ y + 7z = 0$
$\quad\ \ 2x + 3y -\ \ z = 0$

23. $\ x +\ \ y - 2z = 0$
$\ 2x - 3y - 2z = 0$
$\ \ x - 2y - 3z = 0$

24. $\quad\ \ x +\ \ y + z = 0$
$\quad -2x +\ \ y - z = 4$
$\quad\ \ \ x + 2y + z = 5$

25. $3x - 2y - 13z = 0$
$\ x + 4y +\ \ 5z = 0$
$\ 2x +\ \ y -\ \ 4z = 0$

26. $3x - 7y + 3z =\ \ \ 7$
$2x - 3y +\ \ z =\ \ \ 3$
$\ x +\ \ y -\ \ z = -2$

In Exercises 27–34, find all (x, y, z, w) which satisfy the given conditions.

27. $\quad\ \ x -\ \ y + 2z +\ \ w = -2$
$\quad\ 2x\qquad\ \ + 5z +\ \ w = -3$
$\quad\ \ x -\ \ y + 3z + 3w = -3$
$\quad -2x + 2y - 4z +\ \ w =\ \ \ 7$

28. $\quad 2x -\ \ y + 3z -\ \ w =\ \ \ \ \ 9$
$\quad\ \ x -\ \ 4y +\ \ z\qquad\ \ =\ \ \ 11$
$\quad\ 3x\qquad\quad - 5z + 2w =\ \ \ 13$
$\quad -4x + 23y - 3z -\ \ w = -56$

29. $\ x +\ \ y +\ \ 3z + 2w =\ \ \ 6$
$\ x - 2y +\ \ \ z - 3w =\ \ \ 7$
$\ 2x + 2y - 11z - 3w = -6$
$\ x +\ \ y -\ \ 2z +\ \ w =\ \ \ 9$

30. $\quad\ x + 2y -\ \ z + 2w =\ \ \ 1$
$\quad\ 2x - 5y + 5z -\ \ w = -4$
$\quad -4x - 9y + 5z - 4w = -7$
$\quad\ \ x +\ \ y + 2z + 5w = -1$

31. $\quad\ \ x -\ \ y + 2z +\ \ w =\ \ \ 3$
$\quad -2x + 3y - 5z - 3w = -5$
$\quad\ -x +\ \ y -\ \ z +\ \ w = -3$
$\quad -3x + 3y -\ \ z + 7w = -9$

32. $\quad -x -\ \ y +\ \ z - 2w = -1$
$\quad\ 2x -\ \ y -\ \ z -\ \ w = -1$
$\quad -3x + 3y -\ \ z + 2w =\ \ \ 3$
$\quad\ -x + 2y - 2z +\ \ w =\ \ \ 2$

33. $3x\qquad\quad +\ \ z - 2w =\ \ \ 1$
$3x +\ \ y + 2z -\ \ w =\ \ \ 2$
$\quad -\ \ y +\ \ z -\ \ w =\ \ \ 1$
$\ x - \frac{2}{3}y - \frac{1}{3}z - \frac{4}{3}w = -\frac{1}{3}$

34.
$$x + 2y - 2z - 3w = -3$$
$$2x - y + z + w = 2$$
$$3x + y - z - 2w = -1$$
$$x + 7y - 7z - 8w = -13$$

35. Find all (x, y, z) for which
$$x - y - z = 0$$
$$3x + y - z = 0$$
$$2x + 4y + z = 0$$
$$9x + 13y + 2z = 0$$

36. Find all (x, y, z) for which
$$x - y + z = -2$$
$$-4x + y - 3z = 5$$
$$-2x - z = 6$$
$$4x - 3y + 2z = 0$$
$$2x - 3y + z = 6$$

37. Find all (x, y, z, w) for which
$$3x - y + 2z - 5w = 1$$
$$-2x + 3y - z + 2w = -3$$
$$-x - 5y - w = 2$$

38. Find all (x, y, z, w) for which
$$x + 2y - z - w = -2$$
$$-2x - 2y + z + 2w = -1$$
$$-x + 4y - 2z + w = -10$$

39. Find all (x, y, z) for which
$$x - 2y + 3z = 4$$
$$x - y + z = 3$$
$$2x + y - 4z = 3$$
$$-3x + 4y - z = -2$$
$$-7x + 2y + 7z = -8$$

40. Find (x, y, z, w) for which
$$-x + 3y - z + 2w = -13$$
$$-4x + y + w = -20$$
$$-5x + 2z + 3w = -28$$
$$x + 4y - 2z + 8w = -18$$
$$x - 14y + 5z - 9w = 45$$

41. Find (x, y, z, w) for which

$$
\begin{aligned}
2x + 3y + z - w &= 1 \\
3x - y - 2z + 3w &= 2 \\
x + y - 2z - 2w &= 3 \\
-x + 2y \phantom{{}+2z} - w &= 4 \\
-2x \phantom{{}+2y} + 3z + 6w &= 2 \\
- 3y + z + 5w &= -3
\end{aligned}
$$

42. Find (x, y, z, r, s) for which

$$
\begin{aligned}
-3x + y + 2z + r + 4s &= 6 \\
-2x + 2y + 3z + 2r - s &= 5 \\
x - y + z - 2r - 2s &= -4 \\
-x - 2y + z + r - 2s &= 9 \\
-2x + 2y + 2z + 2r + s &= 5
\end{aligned}
$$

43. Find (x, y, z, r, s) for which

$$
\begin{aligned}
2x + 6y + 2z - r + 2s &= -3 \\
- 6y + 8z + 2r - 2s &= 1 \\
2x + y + z \phantom{{}+2r} - s &= -1 \\
6x + 4y - 2z + r \phantom{{}+2s} &= 4 \\
- 8y + 6z - r + 4s &= 3
\end{aligned}
$$

12-2 ECHELON MATRICES*

We can now formulate the general problem as follows: Find all ordered n-tuples $(x_1, \ldots, x_n)$ of real numbers for which

$$
\begin{aligned}
a_{11}x_1 + a_{12}x_2 + \cdots + a_{1n}x_n &= c_1 \\
a_{21}x_1 + a_{22}x_2 + \cdots + a_{2n}x_n &= c_2 \\
& \vdots \\
a_{m1}x_1 + a_{m2}x_2 + \cdots + a_{mn}x_n &= c_m
\end{aligned}
$$

where $a_{11}, a_{12}, \ldots, a_{mn}$, and $c_1, c_2, \ldots, c_m$ are real numbers.

*For a more complete discussion of matrices, see Section 12-6.

This leads us to consider the following rectangular array of real numbers:

$$\begin{pmatrix} a_{11} & a_{12} & \cdots & a_{1n} \\ a_{21} & a_{22} & \cdots & a_{2n} \\ \cdot & \cdot & & \cdot \\ \cdot & \cdot & & \cdot \\ \cdot & \cdot & & \cdot \\ a_{m1} & a_{m2} & \cdots & a_{mn} \end{pmatrix}$$

Such an array is called an **m by n matrix.**[*] The numbers a_{11}, a_{12}, a_{21} and so on, are called the **elements,** or **entries,** of the matrix. The ordered array $(a_{11}, a_{22}, a_{33}, \ldots, a_{nn})$ is called the **main diagonal** of the matrix.

We shall represent matrices by capital letters. For example,

$$B = \begin{pmatrix} 1 & 2 & 3 \\ -4 & 0 & 5 \end{pmatrix}$$

is a 2 by 3 (written 2×3) matrix and

$$C = \begin{pmatrix} 4 & 7 & 0 \\ 0 & -1 & \frac{1}{2} \\ \pi & \sqrt{2} & 4 \end{pmatrix}$$

is a 3×3 matrix. The main diagonal of C is $(4, -1, 4)$.

As we shall see in the next section, the following definition is related to the reduction of linear systems.

Definition

Two matrices A and B are said to be **equivalent,** written $A \approx B$, if A can be transformed into B by a sequence of operations, each operation being one of the following three types:

1. Adding a multiple of one row to some other row.
2. Interchanging two rows.
3. Multiplying a row by a nonzero real number.

These operations are called **elementary operations.**

Remark To *multiply a row of a matrix by a number* means to multiply each entry in that row by the given number; to *add a row of a matrix to another row* means to add the corresponding elements of those rows.

[*]Because the matrix has m rows and n columns.

In each of the following examples a matrix is transformed into an equivalent matrix having all zeros below the main diagonal.

Example 1

$$\begin{pmatrix} 1 & 2 \\ 3 & 4 \end{pmatrix} \approx \begin{pmatrix} 1 & 2 \\ 0 & -2 \end{pmatrix} \quad \text{[add } -3 \text{ times row (1) to row (2)]}$$

Example 2

$$\begin{pmatrix} 2 & 1 & -1 \\ 3 & 2 & 1 \\ -4 & -3 & 2 \end{pmatrix} \approx \begin{pmatrix} 2 & 1 & -1 \\ 6 & 4 & 2 \\ -4 & -3 & 2 \end{pmatrix} \quad \begin{array}{l}\textit{[multiplying row (2)}\\ \textit{by 2]}\end{array}$$

$$\approx \begin{pmatrix} 2 & 1 & -1 \\ 0 & 1 & 5 \\ -4 & -3 & 2 \end{pmatrix} \quad \begin{array}{l}\textit{[adding } -3 \textit{ times}\\ \textit{row (1) to row (2)]}\end{array}$$

$$\approx \begin{pmatrix} 2 & 1 & -1 \\ 0 & 1 & 5 \\ 0 & -1 & 0 \end{pmatrix} \quad \begin{array}{l}\textit{[adding 2 times}\\ \textit{row (1) to row (3)]}\end{array}$$

$$\approx \begin{pmatrix} 2 & 1 & -1 \\ 0 & 1 & 5 \\ 0 & 0 & 5 \end{pmatrix} \quad \begin{array}{l}\textit{[adding row (2)}\\ \textit{to row (3)]}\end{array}$$

Hence,

$$\begin{pmatrix} 2 & 1 & -1 \\ 3 & 2 & 1 \\ -4 & -3 & 2 \end{pmatrix} \approx \begin{pmatrix} 2 & 1 & -1 \\ 0 & 1 & 5 \\ 0 & 0 & 5 \end{pmatrix}$$

In general, *a matrix B is said to be an* **echelon matrix** *if B has all zeros below the main diagonal.*

Example 3

The matrices

$$\begin{pmatrix} 1 & 4 & 0 & 5 \\ 0 & 1 & 5 & 3 \end{pmatrix} \quad \text{and} \quad \begin{pmatrix} 2 & -1 & \sqrt{2} & 0 \\ 0 & 4 & 0 & 1 \\ 0 & 0 & \pi & 7 \end{pmatrix}$$

are in echelon form.

□

In the previous two examples, the given matrices were replaced by equivalent matrices that are in echelon form.

Note that *it is always possible to transform a matrix to an echelon matrix using elementary transformations.* This fact enables us to solve linear systems by considering matrices whose entries are the numbers that appear in the system, as the following examples illustrate.

Example 4

Find all (x_1, x_2) for which

$$2x_1 - x_2 = 3$$

and

$$-x_1 + 3x_2 = -4$$

Solution

We shall see that the usual steps taken in solving this problem correspond to elementary operations on the matrix

$$\begin{pmatrix} 2 & -1 & 3 \\ -1 & 3 & -4 \end{pmatrix}$$

This correspondence is indicated below. Note that (x_1, x_2) is a solution if and only if

the system	corresponding matrices
$\begin{aligned} 2x_1 - x_2 &= 3 \\ -x_1 + 3x_2 &= -4 \end{aligned}$	$\begin{pmatrix} 2 & -1 & 3 \\ -1 & 3 & -4 \end{pmatrix}$
	[*multiply row* (2) *by* 2]
$\begin{aligned} 2x_1 - x_2 &= 3 \\ -2x_1 + 6x_2 &= -8 \end{aligned}$	$\begin{pmatrix} 2 & -1 & 3 \\ -2 & 6 & -8 \end{pmatrix}$
	[*add row* (1) *to row* (2)]
$\begin{aligned} 2x_1 - x_2 &= 3 \\ 5x_2 &= -5 \end{aligned}$	$\begin{pmatrix} 2 & -1 & 3 \\ 0 & 5 & -5 \end{pmatrix}$

Now, we see that (x_1, x_2) is a solution if and only if

$$x_2 = -1 \quad \text{and} \quad x_1 = \frac{3 + x_2}{2} = \frac{3 - 1}{2} = 1$$

Thus, there is exactly one solution: $(1, -1)$.

Example 5

Find all (x_1, x_2, x_3) for which

$$\begin{aligned} 3x_1 \quad\quad - 2x_3 &= -1 \\ x_1 - x_2 + 2x_3 &= 6 \\ 2x_1 + x_2 - x_3 &= -1 \end{aligned}$$

Solution

Let

$$B = \begin{pmatrix} 3 & 0 & -2 & -1 \\ 1 & -1 & 2 & 6 \\ 2 & 1 & -1 & -1 \end{pmatrix}$$

We see that

$$B \approx \begin{pmatrix} 1 & -1 & 2 & 6 \\ 3 & 0 & -2 & -1 \\ 2 & 1 & -1 & -1 \end{pmatrix} \quad \text{[\textit{interchange rows (1) and (2)}]}$$

$$\approx \begin{pmatrix} 1 & -1 & 2 & 6 \\ 0 & 3 & -8 & -19 \\ 2 & 1 & -1 & -1 \end{pmatrix} \quad \begin{array}{l} \textit{[adding } -3 \textit{ times row (1)} \\ \textit{to row (2)]} \end{array}$$

$$\approx \begin{pmatrix} 1 & -1 & 2 & 6 \\ 0 & 3 & -8 & -19 \\ 0 & 3 & -5 & -13 \end{pmatrix} \quad \begin{array}{l} \textit{[adding } -2 \textit{ times row (1)} \\ \textit{to row (3)]} \end{array}$$

$$\approx \begin{pmatrix} 1 & -1 & 2 & 6 \\ 0 & 3 & -8 & -19 \\ 0 & 0 & 3 & 6 \end{pmatrix} \quad \begin{array}{l} \textit{[adding } -1 \textit{ times row (2)} \\ \textit{to row (3)]} \end{array}$$

From the last row we see that $3x_3 = 6$, and so $x_3 = 2$. From row (2) we obtain

$$3x_2 - 8x_3 = -19$$

and consequently

$$3x_2 = 8x_3 - 19 = 16 - 19 = -3$$

so that $x_2 = -1$. Finally, from row (1)

$$x_1 - x_2 + 2x_3 = 6$$

and hence

$$x_1 = x_2 - 2x_3 + 6 = -1 - 4 + 6 = 1$$

Thus, the solution to the system is $(1, -1, 2)$.

Example 6

Find all (x_1, x_2, x_3) for which

$$2x_1 - 3x_2 + x_3 = -1$$

and

$$3x_1 - \tfrac{9}{2}x_2 + \tfrac{3}{2}x_3 = \tfrac{1}{2}$$

Solution

Let

$$B = \begin{pmatrix} 2 & -3 & 1 & -1 \\ 3 & -\frac{9}{2} & \frac{3}{2} & \frac{1}{2} \end{pmatrix}$$

After a sequence of appropriate transformations, we see that

$$B \approx \begin{pmatrix} 2 & -3 & 1 & -1 \\ 0 & 0 & 0 & 4 \end{pmatrix}$$

Thus, from the last row of B, we see that if the system has a solution, then $0 = 4$, which is, of course, not possible. Thus, the system has *no solution*.

Exercises 12-2

1. Find an echelon matrix equivalent to the given matrix.

(a) $\begin{pmatrix} 1 & 3 \\ 7 & 4 \end{pmatrix}$ (b) $\begin{pmatrix} 3 & 1 & 4 \\ 4 & 0 & 4 \\ 1 & 1 & 2 \end{pmatrix}$

(c) $\begin{pmatrix} 3 & 5 & 6 \\ 7 & 1 & 4 \end{pmatrix}$ (d) $\begin{pmatrix} 1 & 0 & 1 & 0 \\ 0 & 1 & 0 & 1 \\ 0 & 1 & 0 & 1 \\ 1 & 0 & 1 & 0 \end{pmatrix}$

2. Show the following matrices are equivalent.

$$\begin{pmatrix} 1 & 3 & 0 \\ 2 & 1 & 4 \\ 3 & 4 & 4 \end{pmatrix} \quad \begin{pmatrix} 1 & 3 & 0 \\ 0 & 5 & -4 \\ 0 & 0 & 0 \end{pmatrix} \quad \begin{pmatrix} 1 & 3 & 0 \\ 0 & 1 & -\frac{4}{5} \\ 0 & 0 & 0 \end{pmatrix}$$

12-3 RANK AND AUGMENTED MATRICES

The following definition is useful in the investigation of linear systems and matrices.

Definition

A matrix has **rank** r if it can be reduced by elementary transformations to an echelon matrix for which r is the minimum number of nonzero rows.*

---

*It can be shown that the rank of a matrix is unique; that is, it does not depend on the particular sequence of elementary operations used in reduction.

Example 1

Let

$$A = \begin{pmatrix} 1 & 1 \\ 2 & 2 \end{pmatrix}$$

Then A is equivalent to

$$\begin{pmatrix} 1 & 1 \\ 0 & 0 \end{pmatrix}$$

and so the rank of A is 1.

Example 2

Let

$$A = \begin{pmatrix} 1 & 1 & 0 \\ 0 & 2 & 3 \\ 1 & 3 & 3 \end{pmatrix}$$

then, adding -1 times row (1) to row (3), we see that A is equivalent to

$$\begin{pmatrix} 1 & 1 & 0 \\ 0 & 2 & 3 \\ 0 & 2 & 3 \end{pmatrix}$$

and, adding -1 times row (2) to row (3), we see that this matrix is equivalent to

$$\begin{pmatrix} 1 & 1 & 0 \\ 0 & 2 & 3 \\ 0 & 0 & 0 \end{pmatrix}$$

Hence, A is equivalent to an echelon matrix having exactly two nonzero rows, and so, by definition, A has rank 2.

The next result relates the concept of rank of a matrix to the solutions of a linear system. Again, let $a_{11}, a_{12}, \ldots, a_{mn}$ and $c_1, \ldots, c_m$ be real numbers; we wish to find $(x_1, \ldots, x_n)$ for which

$$a_{11}x_1 + a_{12}x_2 + \cdots + a_{1n}x_n = c_1$$
$$a_{21}x_1 + a_{22}x_2 + \cdots + a_{2n}x_n = c_2$$
$$\vdots$$
$$a_{m1}x_1 + a_{m2}x_2 + \cdots + a_{mn}x_n = c_m$$

Let A be the $m \times n$ matrix

$$\begin{pmatrix} a_{11} & a_{12} & \cdots & a_{1n} \\ \cdot & & & \\ \cdot & & & \\ \cdot & & & \\ a_{m1} & a_{m2} & \cdots & a_{mn} \end{pmatrix}$$

which is called the **matrix of coefficients** of the system. Now let A^+ be the matrix

$$\begin{pmatrix} a_{11} & a_{12} & \cdots & a_{1n} & c_1 \\ a_{21} & a_{22} & \cdots & a_{2n} & c_2 \\ \cdot & & & & \\ \cdot & & & & \\ \cdot & & & & \\ a_{m1} & a_{m2} & \cdots & a_{mn} & c_m \end{pmatrix}$$

A^+ is called the **augmented matrix** for the system.

The following theorem is given without proof.

Theorem

1. If the rank of A is the same as the number of columns of A (n in this case), that is, if

   (rank of A) $= n =$ (number of columns of A).

   then the system has exactly one solution.
2. If (rank of A) $< n$, then there is either no solution or there are infinitely many solutions. There is no solution if

   (rank of A^+) $>$ (rank of A)

   whereas there are infinitely many solutions if

   (rank of A^+) $=$ (rank of A)

We shall illustrate this by examples in which A is an echelon matrix, since reduction to this case is always possible.

Example 3

Find (x, y, z, w) for which

$$x + y + z - w = 2$$
$$y + z - 3w = 5$$
$$z + 4w = 0$$
$$w = 4$$

In this case, A is the 4×4 matrix

$$\begin{pmatrix} 1 & 1 & 1 & -1 \\ 0 & 1 & 1 & -3 \\ 0 & 0 & 1 & 4 \\ 0 & 0 & 0 & 1 \end{pmatrix}$$

and since

(rank of A) $= 4 =$ (number of columns of A)

there is exactly one solution, and it is $(-11, 33, -16, 4)$.

Example 4

Solve the following system; that is, find all (x, y, z) for which

$$x + y + z = 1$$
$$0x + y + z = 0$$
$$0x + 0y + 0z = 5$$

Solution

There is *no solution*, since $0 \neq 5$. In this case, the matrices A and A^+ are, respectively,

$$\begin{pmatrix} 1 & 1 & 1 \\ 0 & 1 & 1 \\ 0 & 0 & 0 \end{pmatrix} \quad \text{and} \quad \begin{pmatrix} 1 & 1 & 1 & 1 \\ 0 & 1 & 1 & 0 \\ 0 & 0 & 0 & 5 \end{pmatrix}$$

The rank of A is 2, which is less than the number of columns of A; thus, there is either no solution or infinitely many solutions to the system. Since the rank of A^+ is 3, there is no solution to the system (as was already observed).

Example 5

Find all (x, y, z) for which

$$x + y + z = 1$$
$$0x + y + z = 0$$

Solution

Here the matrices A and A^+ are, respectively,

$$\begin{pmatrix} 1 & 1 & 1 \\ 0 & 1 & 1 \end{pmatrix} \quad \text{and} \quad \begin{pmatrix} 1 & 1 & 1 & 1 \\ 0 & 1 & 1 & 0 \end{pmatrix}$$

First, the rank of A is 2, which is smaller than the number of columns of A. So, as in the previous example, there is either no solution, or there are infinitely many. But, in contrast to the previous example, we now have (rank of A) = (rank of A^+), and so there are infinitely many solutions. In fact, $y = -z$, and we see that

$$x = -y - z + 1 = z - z + 1 = 1$$

Thus $(1, -z, z)$, z any real number, are the solutions to the system.

Exercises 12-3

Find the rank of the matrices in Exercises 1–4.

1. $\begin{pmatrix} 1 & 1 & 1 \\ 3 & 1 & 4 \end{pmatrix}$ 2. $\begin{pmatrix} 1 & 1 & 1 \\ 2 & 2 & 2 \end{pmatrix}$

3. $\begin{pmatrix} 1 & 1 & 0 \\ 0 & 1 & 1 \\ 1 & 0 & 1 \end{pmatrix}$ 4. $\begin{pmatrix} 1 & 2 & 0 & 1 \\ 3 & 0 & 1 & 5 \\ 5 & -2 & 2 & 9 \end{pmatrix}$

5. Apply the theorem of this section to Exercises 1-20 of Section 12-1 to determine how many solutions each system has *without* actually finding the solutions.

6. Repeat Exercise 5 for Exercises 27–32 of Section 12-1.

7. Repeat Exercise 5 for Exercises 35–39 of Section 12-1.

12-4 DETERMINANTS AND CRAMER'S RULE

We shall now consider the special case of linear systems in which the number of equations is the same as the number of unknowns. In the 2×2 case, the problem is to find all ordered paris (x, y) of real numbers for which

$$ax + by = k$$

and

$$cx + dy = l$$

where a, b, c, d, k, and l are real numbers with, say, $a \neq 0$.

As usual, we reduce the system to triangular form by applying elementary operations to the matrix

$$\begin{pmatrix} a & b & k \\ c & d & l \end{pmatrix}$$

Adding $-c/a$ times row (1) to row (2) and then multiplying row (2) by a, we see that the system is equivalent to

$$ax + by = k$$
$$(ad - bc)y = al - ck$$

Thus, *there is exactly one solution if and only if $ad - bc \neq 0$*. Solving, we see that the solution is given by

$$x = \frac{kd - bl}{ad - bc} \quad \text{and} \quad y = \frac{al - kc}{ad - bc}$$

The number $ad - bc$ is called the **determinant** of the 2×2 matrix

$$A = \begin{pmatrix} a & b \\ c & d \end{pmatrix}$$

and is denoted by

$$\begin{vmatrix} a & b \\ c & d \end{vmatrix} \quad \text{or} \quad |A|$$

That is,

$$|A| = \begin{vmatrix} a & b \\ c & d \end{vmatrix} = ad - bc$$

Example 1

$$\begin{vmatrix} 1 & -1 \\ 3 & 5 \end{vmatrix} = 1 \cdot 5 - (-1)3 = 8$$

$$\begin{vmatrix} \frac{1}{2} & 3 \\ 0 & \sqrt{2} \end{vmatrix} = \frac{1}{2}\sqrt{2} - 3 \cdot 0 = \frac{\sqrt{2}}{2}$$

□

The result just obtained for 2×2 linear systems is called **Cramer's Rule.** We restate it for convenience.

Cramer's Rule Let a, b, c, d, k, and l be real numbers, and let

$$A = \begin{pmatrix} a & b \\ c & d \end{pmatrix}$$

Then the system

$$ax + by = k$$
$$cx + dy = l$$

has exactly one solution if and only if $|A| \neq 0$, and that solution is given by

$$x = \frac{\begin{vmatrix} k & b \\ l & d \end{vmatrix}}{|A|} \quad \text{and} \quad y = \frac{\begin{vmatrix} a & k \\ c & l \end{vmatrix}}{|A|}$$

Remark If $k = 0$ and $l = 0$, the system is called *homogeneous*. In this case, if $|A| = 0$ there must be infinitely many solutions. The verification of this is left to the reader.

Example 2

Find all (x, y) for which

$$2x - 3y = 1$$

and

$$-x + y = -3$$

Solution

The determinant of matrix of the system is

$$\begin{vmatrix} 2 & -3 \\ -1 & 1 \end{vmatrix} = 2 \cdot 1 - (-1)(-3) = -1 \neq 0$$

Hence, the solution is given by

$$x = \frac{\begin{vmatrix} 1 & -3 \\ -3 & 1 \end{vmatrix}}{-1} = 8 \quad \text{and} \quad y = \frac{\begin{vmatrix} 2 & 1 \\ -1 & -3 \end{vmatrix}}{-1} = 5$$

Warning Cramer's Rule has very limited application for solving linear systems, since it furnishes a solution only in the "square" case* when $|A| \neq 0$.

---

*The system is called *square* when the matrix of coefficients for the system is $n \times n$.

Example 3

The system

$$2x + 2y = 2$$
$$3x + 3y = 3$$

has solutions (infinitely many, in fact) even though

$$\begin{vmatrix} 2 & 2 \\ 3 & 3 \end{vmatrix} = 0$$

(Give three solutions.)

Example 4

Cramer's Rule does not say anything about the system

$$2x + 3y + 4z = 1$$
$$x + 5y + 7z = 2$$

Explain.

□

The general way to solve linear systems is by reduction to echelon form, using elementary operations. However, for the special case of *square* systems we are led in a natural way to the definition of determinants. Determinants (rather than Cramer's Rule per se) are the objects of interest in the next section.

Exercises 12-4

In Exercises 1–7, compute the determinants.

1. $\begin{vmatrix} 1 & 2 \\ 3 & 5 \end{vmatrix}$ 2. $\begin{vmatrix} 4 & 7 \\ 0 & 1 \end{vmatrix}$

3. $\begin{vmatrix} 3 & 0 \\ 0 & 3 \end{vmatrix}$ 4. $\begin{vmatrix} 1 & 0 \\ 0 & 1 \end{vmatrix}$

5. $\begin{vmatrix} 3 & 0 \\ 0 & 4 \end{vmatrix}$ 6. $\begin{vmatrix} 0 & 1 \\ 1 & 0 \end{vmatrix}$

7. $\begin{vmatrix} 0 & 4 \\ 3 & 0 \end{vmatrix}$

In Exercises 8–11, use Cramer's Rule to find all (x, y) for which

8. $x + y = 1$ 9. $2x + 3y = 5$
   $x - y = 0$ $3x + 4y = 9$

10. $3x + y = 7$ 11. $-2x + 7y = 0$
    $5x - 7y = 10$ $3x - 4y = 1$

12-5 PROPERTIES OF DETERMINANTS. LARGER LINEAR SYSTEMS

We shall first summarize what happens to the determinant of a 2×2 matrix when elementary transformations are performed on the matrix.

Theorem*

1. If the matrix A' is obtained from the matrix A by interchanging two rows (or columns) of A, then $|A'| = -|A|$; for example,

$$\begin{vmatrix} a & b \\ c & d \end{vmatrix} = - \begin{vmatrix} c & d \\ a & b \end{vmatrix}$$

2. If the matrix A' is obtained from the matrix A by multiplying each entry in a row (or column) of A by a number k, then $|A'| = k \cdot |A|$; for example,

$$\begin{vmatrix} a & b \\ kc & kd \end{vmatrix} = k \begin{vmatrix} a & b \\ c & d \end{vmatrix}$$

3. Adding a multiple of some row (or column) of a matrix to another row (or column, respectively) does not change the determinant of the matrix; for example,

$$\begin{vmatrix} a & b \\ c & d \end{vmatrix} = \begin{vmatrix} a & b \\ ka + c & kb + d \end{vmatrix}$$

We have already seen that every $n \times n$ matrix can be reduced to triangular form using elementary operations. Insisting that properties 1, 2, and 3 hold, we now define the determinant of an $n \times n$ matrix A as follows:

Definition

If A is an $n \times n$ matrix then $|A|$ is the number obtained by reducing A to an echelon matrix B using elementary operations (taking into account that after each elementary operation we compensate for its numerical effect), and then multiplying the diagonal elements of B.†

---

*The verification of this theorem is left as an exercise.

†The number obtained in this way is always the same (we shall not prove this), even though the echelon matrix is not; that is, different sequences of elementary operations may lead to different echelon matrices, but $|A|$ will be the same. An alternative, but equivalent, definition of determinants is given in Section 12-7.

Example 1

$$\begin{vmatrix} 2 & 3 \\ 4 & 11 \end{vmatrix} = \begin{vmatrix} 2 & 3 \\ 0 & 5 \end{vmatrix} \quad [\text{to row (2) add } -2 \text{ times row (1)}]$$

$$= 2 \cdot 5 = 10 \quad [\text{the product of} \atop \text{the diagonal elements}]$$

Example 2

$$\begin{vmatrix} 2 & 3 & 7 \\ 2 & 4 & 1 \\ 5 & -3 & 0 \end{vmatrix} = \begin{vmatrix} 2 & 3 & 7 \\ 0 & 1 & -6 \\ 5 & -3 & 0 \end{vmatrix} = \tfrac{1}{2} \begin{vmatrix} 2 & 3 & 7 \\ 0 & 1 & -6 \\ 10 & -6 & 0 \end{vmatrix}$$

$$= \tfrac{1}{2} \begin{vmatrix} 2 & 3 & 7 \\ 0 & 1 & -6 \\ 0 & -21 & -35 \end{vmatrix} = \tfrac{1}{2} \begin{vmatrix} 2 & 3 & 7 \\ 0 & 1 & -6 \\ 0 & 0 & -161 \end{vmatrix}$$

$$= \tfrac{1}{2}(2)(1)(-161) = -161$$

Example 3

$$\begin{vmatrix} 2 & -1 & 5 \\ 0 & 0 & 3 \\ 0 & -3 & 4 \end{vmatrix} = - \begin{vmatrix} 2 & -1 & 5 \\ 0 & -3 & 4 \\ 0 & 0 & 3 \end{vmatrix} \quad [\text{interchanging} \atop \text{rows (2) and (3)}]$$

$$= -(2)(-3)(3) = 18$$

□

We can now generalize Cramer's Rule to the $n \times n$ case.

Cramer's Rule Let $a_{11}, a_{12}, \ldots, a_{nn}$ and $b_1, \ldots, b_n$ be real numbers, and let

$$A = \begin{pmatrix} a_{11} & a_{12} & \cdots & a_{1n} \\ a_{21} & a_{22} & \cdots & a_{2n} \\ \cdot \\ \cdot \\ \cdot \\ a_{n1} & a_{n2} & \cdots & a_{nn} \end{pmatrix}$$

and

$$\overbrace{\qquad\qquad}^{i\text{th column of } A}$$

$$A_i = \begin{pmatrix} a_{11} & \cdots & a_{1,i-1} & b_1 & a_{1,i+1} & \cdots & a_{1n} \\ a_{21} & \cdots & a_{2,i-1} & b_2 & a_{2,i+1} & \cdots & a_{2n} \\ \cdot \\ \cdot \\ \cdot \\ a_{n1} & \cdots & a_{n,i-1} & b_n & a_{n,i+1} & \cdots & a_{nn} \end{pmatrix}$$

Suppose further that $|A| \neq 0$. Then the numbers $x_1, x_2, \ldots, x_n$ for which

$$a_{11}x_1 + a_{12}x_2 + \cdots + a_{1n}x_n = b_1$$
$$a_{21}x_1 + a_{22}x_2 + \cdots + a_{2n}x_n = b_2$$
$$\vdots$$
$$a_{n1}x_1 + a_{n2}x_2 + \cdots + a_{nn}x_n = b_n$$

are

$$x_1 = \frac{|A_1|}{|A|}, \quad x_2 = \frac{|A_2|}{|A|}, \quad \ldots, \quad x_n = \frac{|A_n|}{|A|}$$

and these numbers are the unique solutions to the system. A proof of this is given in Section 12-7. If $|A| = 0$, then there are either no solutions or infinitely many; this follows from the theorem in 12-3.

Notice that Cramer's Rule provides another way to solve a linear system. In practice, it is usually easier to solve such a system by reducing the augmented matrix of the system to an echelon matrix (see Section 12-2). The next example illustrates Cramer's Rule.

Example 4

Find (x, y, z) for which

$$x \qquad + 2z = 1$$
$$y - z = -2$$
$$-x + 2y + z = -1$$

Solution

The matrix of the system, A, is

$$\begin{array}{rrr} 1 & 0 & 2 \\ 0 & 1 & -1 \\ -1 & 2 & 1 \end{array}$$

Using the notation found in the statement of Cramer's Rule, we see that

$$A_1 = \begin{pmatrix} 1 & 0 & 2 \\ -2 & 1 & -1 \\ -1 & 2 & 1 \end{pmatrix}, \quad A_2 = \begin{pmatrix} 1 & 1 & 2 \\ 0 & -2 & -1 \\ -1 & -1 & 1 \end{pmatrix},$$

$$A_3 = \begin{pmatrix} 1 & 0 & 1 \\ 0 & 1 & -2 \\ -1 & 2 & -1 \end{pmatrix}$$

We shall compute $|A|$ and $|A_1|$ explicitly, leaving the details of the computation of $|A_2|$ and $|A_3|$ as an exercise.

$$A = \begin{vmatrix} 1 & 0 & 2 \\ 0 & 1 & -1 \\ -1 & 2 & 1 \end{vmatrix} = \begin{vmatrix} 1 & 0 & 2 \\ 0 & 1 & -1 \\ 0 & 2 & 3 \end{vmatrix} \quad \begin{matrix} [row\ (1)\ added \\ to\ row\ (3)] \end{matrix}$$

$$= \begin{vmatrix} 1 & 0 & 2 \\ 0 & 1 & -1 \\ 0 & 0 & 5 \end{vmatrix} \quad \begin{matrix} [row\ (2)\ multiplied\ by\ -2\ and\ added \\ to\ row\ (3)] \end{matrix}$$

Therefore, $|A| = 1 \cdot 1 \cdot 5 = 5$. Next,

$$A_1 = \begin{vmatrix} 1 & 0 & 2 \\ -2 & 1 & -1 \\ -1 & 2 & 1 \end{vmatrix} = \begin{vmatrix} 1 & 0 & 2 \\ 0 & 1 & 3 \\ 0 & 2 & 3 \end{vmatrix} \quad \begin{matrix} [row\ (1)\ multiplied \\ by\ 2\ and\ added\ to \\ row\ (2);\ row\ (1)\ added \\ to\ row\ (3)] \end{matrix}$$

$$= \begin{vmatrix} 1 & 0 & 2 \\ 0 & 1 & 2 \\ 0 & 0 & -3 \end{vmatrix} \quad \begin{matrix} [row\ (2)\ multiplied\ by\ -2 \\ and\ added\ to\ row\ (3)] \end{matrix}$$

Therefore, $|A_1| = -3$. Similarly, $|A_2| = -6$ and $|A_3| = 4$. Consequently,

$$x = \frac{|A_1|}{|A|} = -\frac{3}{5}$$

$$y = \frac{|A_2|}{A} = -\frac{6}{5}$$

and

$$z = \frac{|A_3|}{|A|} = \frac{4}{5}$$

that is, $(-\frac{3}{5}, -\frac{6}{5}, \frac{4}{5})$ is the solution to the system. The reader might also solve this problem by reducing the augmented matrix to triangular form and then deciding which method seems the easier.

The system

$$a_{11}x_1 + a_{12}x_2 + \cdots + a_{1n}x_n = 0$$
$$a_{21}x_1 + a_{22}x_2 + \cdots + a_{2n}x_n = 0$$
$$\cdot$$
$$\cdot$$
$$\cdot$$
$$a_{n1}x_1 + a_{n2}x_2 + \cdots + a_{nn}x_n = 0$$

always has the solution $(0, 0, \ldots, 0)$. It follows from Cramer's rule that if

$$
\begin{vmatrix}
a_{11} & a_{12} & \cdots & a_{1n} \\
\cdot & & & \cdot \\
\cdot & & & \cdot \\
\cdot & & & \cdot \\
a_{n1} & a_{n2} & \cdots & a_{nn}
\end{vmatrix}
$$

is not 0, then $(0, 0, \ldots, 0)$ is the only solution, and if this determinant is 0, then there are infinitely many solutions.

Exercises 12-5

In Exercises 1–4, compute the determinants.

1. $\begin{vmatrix} 1 & 1 & 0 \\ 0 & 1 & 1 \\ 1 & 0 & 1 \end{vmatrix}$

2. $\begin{vmatrix} 1 & 0 & 1 & 1 \\ 0 & 1 & 1 & 1 \\ 1 & 1 & 0 & 1 \\ 1 & 1 & 1 & 0 \end{vmatrix}$

3. $\begin{vmatrix} 1 & 3 & 5 \\ 2 & 1 & 0 \\ 3 & 4 & 5 \end{vmatrix}$

4. $\begin{vmatrix} 1 & 3 & 0 & 1 \\ 2 & 0 & 0 & 4 \\ 3 & 0 & 1 & 0 \\ 7 & 6 & 1 & 6 \end{vmatrix}$

◇**5.** Let $A = \begin{pmatrix} a & b & c \\ d & e & f \\ g & h & i \end{pmatrix}$. By reducing A to an echelon matrix,

show that

$$
\begin{vmatrix} a & b & c \\ d & e & f \\ g & h & i \end{vmatrix} = a \begin{vmatrix} e & f \\ h & i \end{vmatrix} - b \begin{vmatrix} d & f \\ g & i \end{vmatrix} + c \begin{vmatrix} d & e \\ g & h \end{vmatrix}
$$

6. In Example 4 of this section, compute $|A_2|$ and $|A_3|$.

7. Do Exercises 14 and 15 of Section 12-1 using Cramer's Rule.

12-6 OPERATIONS ON MATRICES

If A is a matrix and a_{ij} is the entry in the ith row and jth column, then, for convenience, we shall sometimes denote A by $[a_{ij}]$ (with the number of rows and columns specified when necessary).

Definition 1

If $A = [a_{ij}]$, then $-A = [-a_{ij}]$; that is, $-A$ is the matrix each of whose entries is the additive inverse of the corresponding entry in A.

Example 1

$$-\begin{pmatrix} 1 & 3 & 7 \\ 2 & -3 & 0 \end{pmatrix} = \begin{pmatrix} -1 & -3 & -7 \\ -2 & 3 & 0 \end{pmatrix}$$

Definition 2

If A is an $m \times n$ matrix all of whose entries are zero, then A is called the **$m \times n$ zero matrix**, and is denoted by **0**.

Example 2

$$\begin{pmatrix} 0 & 0 \\ 0 & 0 \\ 0 & 0 \end{pmatrix} \text{ is the } 3 \times 2 \text{ zero matrix}$$

Definition 3

If c is any number and A is an $m \times n$ matrix, then cA is the $m \times n$ matrix whose entries are c times the corresponding entries of A; that is, if $A = [a_{ij}]$, then $cA = [ca_{ij}]$.

Example 3

$$(-3) \begin{pmatrix} 1 & -2 \\ 0 & -1 \\ \sqrt{2} & 4 \end{pmatrix} = \begin{pmatrix} (-3)(1) & (-3)(-2) \\ (-3)(0) & (-3)(-1) \\ (-3)(\sqrt{2}) & (-3)(4) \end{pmatrix}$$

$$= \begin{pmatrix} -3 & 6 \\ 0 & 3 \\ -3\sqrt{2} & -12 \end{pmatrix}$$

Definition 4

If A and B are matrices of *the same order* (that is, they have the same number of rows and columns), then the **sum** of A and B is the matrix each of whose entries is the sum of the corresponding entries of A and B. That is, if $A = [a_{ij}]$ *and* $B = [b_{ij}]$, then

$$A + B = [a_{ij} + b_{ij}]$$

Example 4

$$
\begin{pmatrix} 1 & 3 & 5 & 0 \\ 3 & -1 & 2 & 4 \\ -2 & 0 & -1 & 2 \end{pmatrix} + \begin{pmatrix} 2 & -1 & 5 & 6 \\ -2 & 1 & 2 & 2 \\ 0 & 0 & 0 & 1 \end{pmatrix}
$$

$$
= \begin{pmatrix} 3 & 2 & 10 & 6 \\ 1 & 0 & 4 & 6 \\ -2 & 0 & -1 & 3 \end{pmatrix}
$$

□

As usual, $A - B$ means $A + (-B)$. Note that $A - A = \mathbf{0}$ for any matrix A. Hence, $-A$ is called the **additive inverse** of A.

We next define matrix multiplication. If A and B are matrices, then the product $A \cdot B$ is defined only if A has the same number of columns as B has rows. *It is convenient to view the rows of A and the columns of B as vectors.* This leads to the following definition:

Definition 5

If A is an $m \times n$ matrix and B is an $n \times p$ matrix, then AB is the $m \times p$ *matrix whose entry in the ith row and jth column is the dot product of the ith row of A and the jth column of B*. That is, if $A = [a_{ik}]$ and $B = [b_{kj}]$, then

$$
AB = [c_{ij}] \qquad \text{where } c_{ij} = \sum_{k=1}^{n} a_{ik}b_{kj}
$$

See the following display.

B^j: the jth
column of B

$c_{ij} = A_i \cdot B^j = \sum_{k=1}^{n} a_{ik}b_{kj}$

A_i: the ith
row of A

$$
\begin{pmatrix} \\ a_{i1}\ a_{i2} \ {-}{-}{-}\ a_{in} \\ \\ \end{pmatrix} \begin{pmatrix} b_{1j} \\ b_{2j} \\ b_{ij} \\ b_{nj} \end{pmatrix} = \begin{pmatrix} \\ A^i \cdot B^j \\ \\ \end{pmatrix}
$$

Example 5

(a) $\begin{pmatrix} 1 & 2 \\ -2 & 0 \\ 2 & 3 \end{pmatrix} \begin{pmatrix} 0 & -3 & 4 & 2 \\ -1 & 2 & \frac{1}{2} & 0 \end{pmatrix}$

$= \begin{pmatrix} (1)(0) + (2)(-1) & (1)(-3) + (2)(2) & (1)(4) + (2)(\frac{1}{2}) & (1)(2) + (2)(0) \\ (-2)(0) + (0)(-1) & (-2)(-3) + (0)(2) & (-2)(4) + (0)(\frac{1}{2}) & (-2)(2) + (0)(0) \\ (2)(0) + (3)(-1) & (2)(-3) + (3)(2) & (2)(4) + (3)(\frac{1}{2}) & (2)(2) + (3)(0) \end{pmatrix}$

$= \begin{pmatrix} -2 & 1 & 5 & 2 \\ 0 & 6 & -8 & -4 \\ -3 & 0 & \frac{19}{2} & 4 \end{pmatrix}$

(b) $\begin{pmatrix} 0 & 1 \\ 0 & 0 \end{pmatrix}\begin{pmatrix} 1 & 0 \\ 0 & 0 \end{pmatrix} = \begin{pmatrix} 0 & 0 \\ 0 & 0 \end{pmatrix}$ but $\begin{pmatrix} 1 & 0 \\ 0 & 0 \end{pmatrix}\begin{pmatrix} 0 & 1 \\ 0 & 0 \end{pmatrix} = \begin{pmatrix} 0 & 1 \\ 0 & 0 \end{pmatrix}$

Hence matrix multiplication is not commutative.

(c) $(-2, 1, 4)\begin{pmatrix} 1 \\ 2 \\ 3 \end{pmatrix} = (-2 + 2 + 12) = (12)$ a 1×1 matrix

(d) $\begin{pmatrix} 1 & 3 \\ 0 & -2 \end{pmatrix}\begin{pmatrix} 1 & \frac{3}{2} \\ 0 & -\frac{1}{2} \end{pmatrix} = \begin{pmatrix} 1 & 0 \\ 0 & 1 \end{pmatrix} = \begin{pmatrix} 1 & \frac{3}{2} \\ 0 & -\frac{1}{2} \end{pmatrix}\begin{pmatrix} 1 & 3 \\ 0 & -2 \end{pmatrix}$

(e) $\begin{pmatrix} 1 & 0 \\ 0 & 1 \end{pmatrix}\begin{pmatrix} a & b \\ c & d \end{pmatrix} = \begin{pmatrix} a & b \\ c & d \end{pmatrix}\begin{pmatrix} 1 & 0 \\ 0 & 1 \end{pmatrix} = \begin{pmatrix} a & b \\ c & d \end{pmatrix}$

for any real numbers a, b, c, and d.

Example 5 illustrates the following facts:

1. The product of an $m \times n$ and an $n \times p$ matrix is an $m \times p$ matrix.
2. Multiplication of matrices is not commutative.
3. $\begin{pmatrix} 1 & 0 \\ 0 & 1 \end{pmatrix}$ is a *multiplicative identity* for 2×2 matrices.
4. $\begin{pmatrix} 1 & 3 \\ 0 & -2 \end{pmatrix}$ and $\begin{pmatrix} 1 & \frac{3}{2} \\ 0 & -\frac{1}{2} \end{pmatrix}$ are *multiplicative inverses* of each other.

□

Note that in general the $n \times n$ matrix whose diagonal entries are 1 and whose other entries are 0 is the *multiplicative identity* for $n \times n$ matrices. It is called the $n \times n$ *identity matrix* and is denoted by I_n, or simply by I.*

*It can be shown that the identity is unique.

Example 6

For any numbers a_{ij} $(i, j = 1, 2, 3)$, we have

$$\begin{pmatrix} 1 & 0 & 0 \\ 0 & 1 & 0 \\ 0 & 0 & 1 \end{pmatrix} \begin{pmatrix} a_{11} & a_{12} & a_{13} \\ a_{21} & a_{22} & a_{23} \\ a_{31} & a_{32} & a_{33} \end{pmatrix} = \begin{pmatrix} a_{11} & a_{12} & a_{13} \\ a_{21} & a_{22} & a_{23} \\ a_{31} & a_{32} & a_{33} \end{pmatrix}$$

that is, for any 3×3 matrix A, $I \cdot A = A$. Similarly, $A \cdot I = A$.
□

In Example 5(d) we encountered two square matrices whose product is I. In general, if A and B are square matrices for which $AB = BA = I$ then A and B are *multiplicative inverses* of each other, and we write*

$$A = B^{-1} \quad \text{and} \quad B = A^{-1}$$

The following example shows how to compute the inverse of a 2×2 matrix if it exists, and gives the condition under which it does exist.

Example 7

Let $A = \begin{pmatrix} a & b \\ c & d \end{pmatrix}$. Then u, v, w, and z are real numbers for which

$$\begin{pmatrix} a & b \\ c & d \end{pmatrix} \begin{pmatrix} u & w \\ v & z \end{pmatrix} = \begin{pmatrix} 1 & 0 \\ 0 & 1 \end{pmatrix} = I$$

if and only if

$$\begin{array}{ll} au + bv = 1 \\ cu + dv = 0 \end{array} \quad \text{and} \quad \begin{array}{ll} aw + bz = 0 \\ cw + dz = 1 \end{array}$$

Now, for both systems of equations, a solution exists (and is unique) if and only if the determinant

$$|A| = \begin{vmatrix} a & b \\ c & d \end{vmatrix} = ad - bc$$

is not zero; and if this is the case, then

$$u = \frac{d}{|A|}, \quad w = \frac{-b}{|A|}, \quad v = \frac{-c}{|A|}, \quad \text{and} \quad z = \frac{a}{|A|}$$

*It can be shown that the inverses are unique.

that is,

$$A^{-1} = \begin{pmatrix} a & b \\ c & d \end{pmatrix}^{-1} = \begin{pmatrix} \dfrac{d}{|A|} & \dfrac{-b}{|A|} \\ \dfrac{-c}{|A|} & \dfrac{a}{|A|} \end{pmatrix}$$

Example 8

Let

$$A = \begin{pmatrix} 1 & 2 \\ -2 & 4 \end{pmatrix}$$

then $|A| = 8 \neq 0$, and

$$A^{-1} = \begin{pmatrix} \frac{1}{2} & -\frac{1}{4} \\ \frac{1}{4} & \frac{1}{8} \end{pmatrix}$$

Check

$$\begin{pmatrix} 1 & 2 \\ -2 & 4 \end{pmatrix}\begin{pmatrix} \frac{1}{2} & -\frac{1}{4} \\ \frac{1}{4} & \frac{1}{8} \end{pmatrix} = \begin{pmatrix} 1 & 0 \\ 0 & 1 \end{pmatrix} = \begin{pmatrix} \frac{1}{2} & -\frac{1}{4} \\ \frac{1}{4} & \frac{1}{8} \end{pmatrix}\begin{pmatrix} 1 & 2 \\ -2 & 4 \end{pmatrix}$$

Example 9

Solve the system
$$x + 2y = 3$$
$$-2x + 4y = 2$$

Solution

(x, y) is a solution if and only if

$$\begin{pmatrix} 1 & 2 \\ -2 & 4 \end{pmatrix}\begin{pmatrix} x \\ y \end{pmatrix} = \begin{pmatrix} 3 \\ 2 \end{pmatrix}$$

Premultiplying by

$$\begin{pmatrix} 1 & 2 \\ -2 & 4 \end{pmatrix}^{-1}$$

we have,* using the result of Example 8,

$$\begin{pmatrix} \frac{1}{2} & -\frac{1}{4} \\ \frac{1}{4} & \frac{1}{8} \end{pmatrix}\begin{pmatrix} 1 & 2 \\ -2 & 4 \end{pmatrix}\begin{pmatrix} x \\ y \end{pmatrix} = \begin{pmatrix} \frac{1}{2} & -\frac{1}{4} \\ \frac{1}{4} & \frac{1}{8} \end{pmatrix}\begin{pmatrix} 3 \\ 2 \end{pmatrix}$$

---

*This computation depends, of course, on the fact that multiplication of matrices, when defined, is associative. The verification of this, and of other algebraic properties, is left for the exercises.

or, equivalently,

$$\begin{pmatrix} 1 & 0 \\ 0 & 1 \end{pmatrix}\begin{pmatrix} x \\ y \end{pmatrix} = \begin{pmatrix} x \\ y \end{pmatrix} = \begin{pmatrix} 1 \\ 1 \end{pmatrix}$$

That is, $x = 1$ and $y = 1$.

□

In the next section inverses of larger square matrices will be computed, leading to a new derivation of Cramer's Rule.

Exercises 12-6

1. Let

$$A = \begin{pmatrix} 1 & -1 & 3 \\ 2 & 4 & 0 \\ 3 & -3 & 1 \\ 2 & 0 & 0 \end{pmatrix} \quad \text{and} \quad B = \begin{pmatrix} 1 & -3 & -3 \\ 2 & 12 & 1 \\ 3 & -9 & 4 \\ 2 & 0 & 0 \end{pmatrix}$$

Find (a) $\frac{1}{2}A$; (b) $A + B$; (c) $A - B$; (d) $3A - B$.

2. Let

$$A = \begin{pmatrix} 1 & -1 & 0 \\ 2 & 0 & -1 \\ 3 & 1 & 1 \end{pmatrix}$$

Compute (a) $A - 5I$; (b) $A - xI$, where x is any number.

3. Compute

$$\begin{pmatrix} 1 & 3 \\ -1 & 4 \end{pmatrix}\begin{pmatrix} 2 & 0 \\ 0 & 1 \end{pmatrix}$$

4. Compute

$$\begin{pmatrix} 0 & 1 \\ 1 & 0 \end{pmatrix}\begin{pmatrix} 0 & 2 \\ 2 & 0 \end{pmatrix}$$

5. Compute

$$\begin{pmatrix} 1 & -1 & 0 & 2 \\ -3 & 0 & -1 & 0 \end{pmatrix}\begin{pmatrix} 1 \\ 0 \\ -1 \\ 2 \end{pmatrix}$$

6. Let

$$A = \begin{pmatrix} 0 & -1 \\ \frac{1}{2} & 3 \\ 7 & 4 \end{pmatrix} \quad \text{and} \quad B = \begin{pmatrix} 5 & -3 & \frac{1}{8} \\ 0 & 2 & 1 \end{pmatrix}$$

Compute AB and BA. (Notice that $AB \neq BA$, and thus matrix multiplication is not commutative.)

7. Let

$$A = \begin{pmatrix} 1 & 0 \\ 1 & 0 \end{pmatrix} \quad \text{and} \quad B = \begin{pmatrix} 0 & 1 \\ 0 & 1 \end{pmatrix}$$

(a) Show that $AB \neq 0$, $BA \neq 0$, and $AB \neq BA$; (b) find two other 2×2 matrices for which $AB \neq 0$, $BA \neq 0$, and $AB \neq BA$.

In Exercises 8–11, find the inverse (if there is one) of the given matrix.

8. $\begin{pmatrix} \frac{1}{2} & 0 \\ 0 & \frac{1}{3} \end{pmatrix}$
    9. $\begin{pmatrix} -7 & 2 \\ 0 & 1 \end{pmatrix}$

10. $\begin{pmatrix} -2 & 1 \\ 4 & -3 \end{pmatrix}$
    11. $\begin{pmatrix} \frac{1}{3} & \frac{1}{2} \\ \frac{2}{5} & \frac{1}{4} \end{pmatrix}$

In Exercises 12 and 13, use the method found in Example 9 to find the numbers x and y for which:

12. $\begin{aligned} -2x + y &= 1 \\ 4x - 12y &= -2 \end{aligned}$
    13. $\begin{aligned} \frac{1}{3}x + \frac{1}{2}y &= \frac{1}{6} \\ \frac{2}{5}x + \frac{1}{4}y &= \frac{1}{5} \end{aligned}$

14. Let

$$A = \begin{pmatrix} 2 & 2 \\ 3 & 1 \end{pmatrix}$$

Find all real numbers x for which the determinant of $A - xI$ is 0 (and hence for which the matrix $A - xI$ has no inverse).

15. Give an example of two nonzero 2×2 matrices A and B *for which* $AB = 0$. (Note that this cannot happen if A and B are real or complex numbers.)

16. Let A and B be 2×2 matrices. Show that $|AB| = |A| \cdot |B|$.

17. Let A be a 2×2 matrix that has an inverse. Show that

$$|A^{-1}| = \frac{1}{|A|}$$

[that is, show that the determinant of the inverse is the (multiplicative) inverse of the determinant].

18. For any matrix A, define

$$A^1 = A,$$

$$A^n = A^{n-1} \cdot A \qquad \text{(where } n \text{ is a positive integer)}$$

If

$$B = \begin{pmatrix} 1 & 1 \\ 1 & 1 \end{pmatrix}$$

find B^2, B^3, and B^4. What is B^n for any positive integer n?

19. Let

$$A = \begin{pmatrix} 0 & 1 \\ 1 & 0 \end{pmatrix}$$

Find A^2, A^3, and A^4 (see Exercise 18). What is A^n (a) when n is a positive even integer? (b) when n is a positive odd integer?

20. Show that matrix addition is commutative and associative.

21. Let A, B, and C be 2×2 matrices. Show that

$$A(BC) = (AB)C$$

and

$$A(B + C) = AB + AC$$

An alternate treatment of complex numbers is to define the complex number $a + bi$ (where a and b are real numbers) to be the 2×2 matrix

$$\begin{pmatrix} a & -b \\ b & a \end{pmatrix}$$

In particular, i is then the matrix

$$\begin{pmatrix} 0 & -1 \\ 1 & 0 \end{pmatrix}$$

In Exercises 22–25, use this definition to prove that

22. $i^2 = -I$

23. $(a + bi)(a' + b'i) = (aa' - bb') + (ab' + a'b)i$
$(a, a', b, b'$ are real numbers)

24. Multiplication of complex numbers is commutative.

25. If $a \neq 0$, $b \neq 0$, then

$$(a + bi)^{-1} = \frac{a - bi}{a^2 + b^2}$$

12-7 DETERMINANTS, INVERSES, AND CRAMER'S RULE

We shall now give another definition of *determinants*, equivalent to the one given in Section 12-4. Although we shall not prove this equivalence, it is easy enough to verify by direct computation for matrices of small order. A few preliminary definitions and observations will be necessary.

Definition

Let $A = [a_{ij}]$ be an $n \times n$ matrix, and denote by M_{ij} the matrix obtained by deleting from A the ith row and the jth column. The determinant $|M_{ij}|$ is called the i, jth **minor** of A, and the number $A_{ij} = (-1)^{i+j}|M_{ij}|$ is called the i, jth **cofactor** of A.

Example 1

Let

$$A = \begin{pmatrix} 1 & 3 & -2 \\ 2 & 0 & 4 \\ 5 & -1 & 3 \end{pmatrix}; \quad \text{then} \quad M_{12} = \begin{pmatrix} 2 & 4 \\ 5 & 3 \end{pmatrix}$$

Now

$$|M_{12}| = 6 - 20 = -14$$

so

$$A_{12} = (-1)^{1+2}(-14) = 14$$

Similarly,

$$A_{31} = (-1)^{3+1} \begin{vmatrix} 3 & -2 \\ 0 & 4 \end{vmatrix} = 12$$

$$A_{22} = (-1)^{2+2} \begin{vmatrix} 1 & -2 \\ 5 & 3 \end{vmatrix} = 13$$

$$A_{23} = (-1)^{2+3} \begin{vmatrix} 1 & 3 \\ 5 & -1 \end{vmatrix} = 16$$

and so on.

Definition

Let

$$
A = \begin{pmatrix}
a_{11} & a_{12} & \cdots & a_{1n} \\
a_{21} & a_{22} & \cdots & a_{2n} \\
\vdots & \vdots & & \vdots \\
a_{n1} & a_{n2} & \cdots & a_{nn}
\end{pmatrix}
$$

Then for any i and j between 1 and n (inclusive) the number

$$
\begin{aligned}
|A| &= a_{i1}A_{i1} + a_{i2}A_{i2} + \cdots + a_{in}A_{in} \\
&= a_{1j}A_{1j} + a_{2j}A_{2j} + \cdots + a_{nj}A_{nj}
\end{aligned}
$$

is the same and is called the **determinant** of A. In other words, the determinant of A is the *dot product* of any row with the vector whose components are the cofactors of that row.*

Example 2

If

$$
A = \begin{pmatrix}
1 & 2 & -1 \\
2 & 3 & 0 \\
-1 & 2 & 3
\end{pmatrix}
$$

then

$$
|A| = 1 \begin{vmatrix} 3 & 0 \\ 2 & 3 \end{vmatrix} - 2 \begin{vmatrix} 2 & 0 \\ -1 & 3 \end{vmatrix} + (-1) \begin{vmatrix} 2 & 3 \\ -1 & 2 \end{vmatrix}
$$
$$
= 9 - 12 - 7 = -10 \quad \text{[using the first row]}
$$

or

$$
|A| = (-2) \begin{vmatrix} 2 & 0 \\ -1 & 3 \end{vmatrix} + 3 \begin{vmatrix} 1 & -1 \\ -1 & 3 \end{vmatrix} - 2 \begin{vmatrix} 1 & -1 \\ 2 & 0 \end{vmatrix}
$$
$$
= -12 + 6 - 4 = -10 \quad \text{[using the second column]}
$$

and so on.

This computation is known as the *expansion of a determinant by cofactors*. Thus, in Example 2, the first expansion was by the cofactors of the first row and the second by the cofactors of the second column. Note that such an expansion reduces the computation to determinants of one lower order.
□

---

*It can be shown that the same result can be obtained if columns are used instead of rows, as is illustrated in Example 2.

The following theorems are stated in their generality, but proved only for the 3 × 3 case. The general proofs are similar.

Theorem 1

If two rows (or columns) of a square matrix are identical, then its determinant is zero.

Proof Note first that for any real numbers a and b

$$\begin{vmatrix} a & a \\ b & b \end{vmatrix} = \begin{vmatrix} a & b \\ a & b \end{vmatrix} = ab - ab = 0$$

If a 3 × 3 matrix has two identical rows (or columns), expand its determinant by the cofactors of the remaining row (or column). Since each of these cofactors is zero, so is the determinant of the matrix. The proof for a general $n \times n$ matrix is left as a miscellaneous exercise.

Example 3

For any real numbers a, b, c, d, e, and f we have

$$\begin{vmatrix} a & d & a \\ b & e & b \\ c & f & c \end{vmatrix} = -d \begin{vmatrix} b & b \\ c & c \end{vmatrix} + e \begin{vmatrix} a & a \\ c & c \end{vmatrix} - f \begin{vmatrix} a & a \\ b & b \end{vmatrix} = 0$$

□

The next theorem leads to an easy computation of inverses.

Theorem 2

The dot product of any row (or column) of a square matrix with the vector of cofactors of a different row (or column) is zero.

The proof of this theorem is illustrated by the following example.

Example 4

Consider the matrix

$$\begin{pmatrix} a & b & c \\ d & e & f \\ g & h & i \end{pmatrix}$$

and compute the dot product of the second row with the cofactors of the third row:

$$(d, e, f) \cdot \left(\begin{vmatrix} b & c \\ e & f \end{vmatrix}, \ -\begin{vmatrix} a & c \\ d & f \end{vmatrix}, \ \begin{vmatrix} a & b \\ d & e \end{vmatrix} \right)$$

$$= d \begin{vmatrix} b & c \\ e & f \end{vmatrix} - e \begin{vmatrix} a & c \\ d & f \end{vmatrix} + f \begin{vmatrix} a & b \\ d & e \end{vmatrix}$$

$$= \begin{vmatrix} a & b & c \\ d & e & f \\ d & e & f \end{vmatrix}$$

This number is zero, because it is the determinant of a matrix with two identical rows.

The general proof is similar.

Remark   We can restate Theorem 2 as follows: *Let $A = [a_{ij}]$ be an $n \times n$ matrix, then*

$$\sum_{j=1}^{n} a_{ij} A_{kj} = 0$$

for $i \neq k$. A similar statement may be made for columns.

Theorem 3

Let $A = [a_{ij}]$ be an $n \times n$ matrix with $|A| \neq 0$. Then A^{-1} is the $n \times n$ matrix whose i, jth entry is $A_{ji}/|A|$, where A_{ji} is the j, ith cofactor of A; that is,[*]

$$A^{-1} = \frac{1}{|A|} \begin{pmatrix} A_{11} & A_{21} & \cdots & A_{n1} \\ A_{12} & A_{22} & \cdots & A_{n2} \\ \cdot & \cdot & & \cdot \\ \cdot & \cdot & & \cdot \\ \cdot & \cdot & & \cdot \\ A_{1n} & A_{2n} & \cdots & A_{nn} \end{pmatrix}$$

[*]Recall that if $A = [a_{ij}]$ and k is a number, then $kA = [ka_{ij}]$.

Proof Considering the 3×3 case for convenience, we have

$$\frac{1}{|A|} \begin{pmatrix} A_{11} & A_{21} & A_{31} \\ A_{12} & A_{22} & A_{32} \\ A_{13} & A_{23} & A_{33} \end{pmatrix} \begin{pmatrix} a_{11} & a_{12} & a_{13} \\ a_{21} & a_{22} & a_{23} \\ a_{31} & a_{32} & a_{33} \end{pmatrix}$$

$$= \frac{1}{|A|} \begin{pmatrix} |A| & 0 & 0 \\ 0 & |A| & 0 \\ 0 & 0 & |A| \end{pmatrix} \begin{matrix} [\text{since } a_{11}A_{11} + a_{21}A_{21} + a_{31}A_{31} \\ = |A|, \quad a_{12}A_{11} + a_{22}A_{21} + a_{32}A_{31} \\ = 0, \text{ and so on, by definition of } |A| \\ \text{and by Theorem 2, respectively}] \end{matrix}$$

$$= \begin{pmatrix} 1 & 0 & 0 \\ 0 & 1 & 0 \\ 0 & 0 & 1 \end{pmatrix}$$

Example 5

Let

$$A = \begin{pmatrix} 1 & 0 & -1 \\ 0 & 2 & -2 \\ -3 & 0 & 4 \end{pmatrix}$$

Expanding, using the first row, we have

$$|A| = \begin{vmatrix} 2 & -2 \\ 0 & 4 \end{vmatrix} - \begin{vmatrix} 0 & 2 \\ -3 & 0 \end{vmatrix} = 8 - 6 = 2$$

$$A_{11} = \begin{vmatrix} 2 & -2 \\ 0 & 4 \end{vmatrix} = 8$$

$$A_{12} = - \begin{vmatrix} 0 & -2 \\ -3 & 4 \end{vmatrix} = 6$$

and so on; finally, we obtain

$$A^{-1} = \frac{1}{2} \begin{pmatrix} 8 & 0 & 2 \\ 6 & 1 & 2 \\ 6 & 0 & 2 \end{pmatrix} = \begin{pmatrix} 4 & 0 & 1 \\ 3 & \frac{1}{2} & 1 \\ 3 & 0 & 1 \end{pmatrix}$$

Check

$$\begin{pmatrix} 1 & 0 & -1 \\ 0 & 2 & -2 \\ -3 & 0 & 4 \end{pmatrix} \begin{pmatrix} 4 & 0 & 1 \\ 3 & \frac{1}{2} & 1 \\ 3 & 0 & 1 \end{pmatrix} = \begin{pmatrix} 1 & 0 & 0 \\ 0 & 1 & 0 \\ 0 & 0 & 1 \end{pmatrix}$$

Theorem 3 leads to Cramer's Rule as follows: If

$$A = \begin{pmatrix} a_1 & b_1 & c_1 \\ a_2 & b_2 & c_2 \\ a_3 & b_3 & c_3 \end{pmatrix}$$

$(|A| \neq 0$ and d_1, d_2, and d_3 are any numbers), then x, y, and z are real numbers for which

$$\begin{pmatrix} a_1 & b_1 & c_1 \\ a_2 & b_2 & c_2 \\ a_3 & b_3 & c_3 \end{pmatrix} \begin{pmatrix} x \\ y \\ z \end{pmatrix} = \begin{pmatrix} d_1 \\ d_2 \\ d_3 \end{pmatrix}$$

if and only if (with A_{11} replaced by A_1, A_{12} by B_1, etc.)

$$\begin{pmatrix} x \\ y \\ z \end{pmatrix} = A^{-1} \begin{pmatrix} d_1 \\ d_2 \\ d_3 \end{pmatrix} = \frac{1}{|A|} \begin{pmatrix} A_1 & A_2 & A_3 \\ B_1 & B_2 & B_3 \\ C_1 & C_2 & C_3 \end{pmatrix} \begin{pmatrix} d_1 \\ d_2 \\ d_3 \end{pmatrix}$$

$$= \frac{1}{|A|} \begin{pmatrix} d_1 A_1 + d_2 A_2 + d_3 A_3 \\ d_1 B_1 + d_2 B_2 + d_3 B_3 \\ d_1 C_1 + d_2 C_2 + d_3 C_3 \end{pmatrix}$$

that is,

$$x = \frac{d_1 A_1 + d_2 A_2 + d_3 A_3}{|A|} = \frac{\begin{vmatrix} d_1 & b_1 & c_1 \\ d_2 & b_2 & c_2 \\ d_3 & b_3 & c_3 \end{vmatrix}}{|A|}$$

$$y = \frac{d_1 B_1 + d_2 B_2 + d_3 B_3}{|A|} = \frac{\begin{vmatrix} a_1 & d_1 & c_1 \\ a_2 & d_2 & c_2 \\ a_3 & d_3 & c_3 \end{vmatrix}}{|A|}$$

and

$$z = \frac{d_1 C_1 + d_2 C_2 + d_3 C_3}{|A|} = \frac{\begin{vmatrix} a_1 & b_1 & d_1 \\ a_2 & b_2 & d_2 \\ a_3 & b_3 & d_3 \end{vmatrix}}{|A|}$$

Exercises 12-7

1. Let

$$A = \begin{pmatrix} 3 & 1 & 0 \\ -1 & 2 & 1 \\ 1 & 1 & 1 \end{pmatrix} \quad \text{and} \quad B = \begin{pmatrix} 1 & 1 & -1 \\ 1 & 1 & 2 \\ 3 & 2 & 1 \end{pmatrix}$$

Verify that

$$|AB| = |BA| = |A| \cdot |B|$$

2. Find all real numbers x for which

$$\begin{vmatrix} -2 & -2x & x+1 \\ x^2 & 1 & x-1 \\ x & 0 & 1 \end{vmatrix} = 0$$

3. Let

$$A = \begin{pmatrix} a_{11} & a_{12} & a_{13} & a_{14} \\ 0 & a_{22} & a_{23} & a_{24} \\ 0 & 0 & a_{33} & a_{34} \\ 0 & 0 & 0 & a_{44} \end{pmatrix}$$

Show by using cofactors that $|A| = a_{11} \cdot a_{22} \cdot a_{33} \cdot a_{44}$.

4. Find all numbers x for which

$$\begin{vmatrix} 2x-1 & 2x & x^2-x+3 & x^4+1 \\ 0 & x^2-4 & 4 & 1 \\ 0 & 0 & x^2+3x & x \\ 0 & 0 & 0 & -3 \end{vmatrix} = 0$$

In Exercises 5–7, find the multiplicative inverse of the given matrix.

5. $\begin{pmatrix} 1 & 1 & 1 \\ 1 & 0 & 1 \\ 0 & 0 & 1 \end{pmatrix}$ **6.** $\begin{pmatrix} 0 & 0 & 0 & 1 \\ 0 & 0 & 2 & 0 \\ 0 & 3 & 0 & 0 \\ 4 & 0 & 0 & 0 \end{pmatrix}$

7. $\begin{pmatrix} a & 0 & 0 & 0 \\ 0 & b & 0 & 0 \\ 0 & 0 & c & 0 \\ 0 & 0 & 0 & d \end{pmatrix}$ where a, b, c, and d are nonzero real numbers

8. Let

$$A = \begin{pmatrix} 4 & 0 & 5 \\ 0 & 1 & -6 \\ 3 & 0 & 4 \end{pmatrix}$$

(a) Find A^{-1}.

(b) Find numbers x, y, and z for which

$$A \begin{pmatrix} x \\ y \\ z \end{pmatrix} = \begin{pmatrix} 1 \\ 0 \\ -1 \end{pmatrix}$$

using the fact that

$$\begin{pmatrix} x \\ y \\ z \end{pmatrix} = A^{-1} \begin{pmatrix} 1 \\ 0 \\ -1 \end{pmatrix}$$

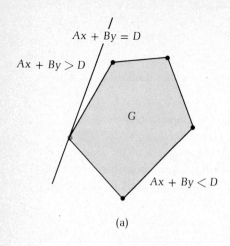

(a)

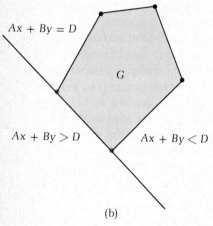

(b)

Figure 12-5

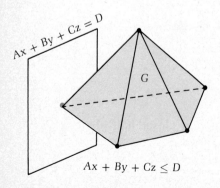

Figure 12-6

The vertices of G are intersections of some pairs of lines taken from the lines given by $x + y = 2$, $x - y = 0$, $x = 0$, and $x + y = 1$. In this example, it is easy to see *which* pairs; in more complicated problems, this is not the case.

The vertex $(1, 1)$ is found by solving the system $\begin{cases} x + y = 2 \\ x - y = 0 \end{cases}$ Similarly the vertex $(\frac{1}{2}, \frac{1}{2})$ is found by solving the system $\begin{cases} x - y = 0 \\ x + y = 1 \end{cases}$. Notice that the system $\begin{cases} x + y = 1 \\ x + y = 2 \end{cases}$ has no solution and the system $\begin{cases} x - y = 0 \\ x = 0 \end{cases}$ has the solution $(0, 0)$ which is not a vertex; thus, $(0, 0)$ is a candidate point that is not a vertex.

Now for any linear function f we need only compute f at the four vertices to find its maximum and minimum values over this region G.

□

We now establish Theorem 1. Suppose that $Ax + By + Cz$ has a maximum in the region G.* Call the maximum value D. Then G is contained in the half-space given by $Ax + By + Cz \leq D$ —that is, every point (x, y, z) in G satisfies the condition $Ax + By + Cz \leq D$ because D is a maximum—and G intersects the plane given by $Ax + By + Cz = D$. Observe that the intersection is either an extreme point (this is the typical case), an edge, or a face, and these correspond to one, two, or at least three maximizing extreme points. See Figure 12-5, which shows the possibilities for a bounded planar G—namely, either one or two maximizing extreme points, while Figure 12-6 shows a typical example in three dimensions.

Exercises 12-8

1. Let G be the region given by $x \geq 0$, $y \geq 0$, and $x + y \leq 1$. Let $f(x, y) = 4x - y$. Determine the maximum and minimum values of f on G.

2. Let G be the region given by $x \geq 0$, $y \geq 0$, $x - y \leq 0$ and $x + y \leq 2$. Let $f(x, y) = -5x + y$. (a) Determine the vertices of G. (b) Find the maximum and minimum of f on G.

3. Let G be the region that is the set of all points (x, y) for which $x + y \leq 2$, $-x - 2y \leq 3$, and $-2x + 3y \leq 6$. Let $f(x, y) = 4x + 8y$. Determine the maximum and minimum of f on G.

---

*We are using the fact that a linear function attains a maximum value on any bounded region which contains its boundary.

Example 3 (continued)

Again, using the fact that f takes on its maximum at the vertices of G, we need only evaluate f at the vertices. Recall that $f(x, y, z) = 3x + 2y + z$, and referring back to Figure 12-3, we have

$$f(200, 0, 300) = 3 \cdot 200 + 2 \cdot 0 + 300 = 900$$
$$f(200, 100, 200) = 3 \cdot 200 + 2 \cdot 100 + 200 = 1000, \text{ etc.}$$

It turns out that $1000 is the maximum possible profit obtained when $x = 200$, $y = 100$ and $z = 200$; that is, the farmer raises 200 turkeys, 100 ducks, and 200 chickens.
□

In more complex problems where the region G cannot be graphed and the vertices easily identified, there is a method that is based on Theorem 1. In three dimensions, the method proceeds as follows: First, get all "candidate" extreme points, that is, intersections of sets of three bounding planes, by solving every set of three equations (in the general case every set of n equations). Next, find all the extreme points (i.e., vertices) by discarding the candidate points that are not in G, that is, the points that do not satisfy all of the other inequalities. Last, calculate $Ax + By + Cz$ for every extreme point and select a point that gives the maximum (or minimum).

The following simple example in the plane illustrates the preceding remarks.

Example 4

Let G be the set of all points (x, y) for which

$$x + y \leq 2$$

and

$$x - y \leq 0$$

and

$$x \geq 0$$

and

$$x + y \geq 1$$

Then G is the intersection of the four closed half-planes given by $x + y \leq 2$, $x - y \leq 0$, $x \geq 0$, and $x + y \geq 1$. See Figure 12-4.

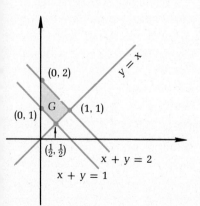

Figure 12-4

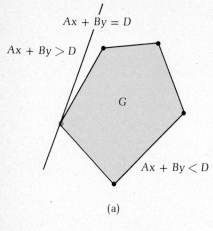

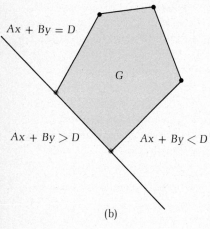

Figure 12-5

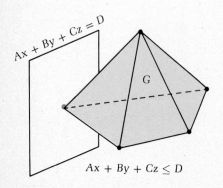

Figure 12-6

The vertices of G are intersections of some pairs of lines taken from the lines given by $x + y = 2$, $x - y = 0$, $x = 0$, and $x + y = 1$. In this example, it is easy to see *which* pairs; in more complicated problems, this is not the case.

The vertex $(1, 1)$ is found by solving the system $\begin{Bmatrix} x + y = 2 \\ x - y = 0 \end{Bmatrix}$ Similarly the vertex $(\frac{1}{2}, \frac{1}{2})$ is found by solving the system $\begin{Bmatrix} x - y = 0 \\ x + y = 1 \end{Bmatrix}$. Notice that the system $\begin{Bmatrix} x + y = 1 \\ x + y = 2 \end{Bmatrix}$ has no solution and the system $\begin{Bmatrix} x - y = 0 \\ x = 0 \end{Bmatrix}$ has the solution $(0, 0)$ which is not a vertex; thus, $(0, 0)$ is a candidate point that is not a vertex.

Now for any linear function f we need only compute f at the four vertices to find its maximum and minimum values over this region G.

□

We now establish Theorem 1. Suppose that $Ax + By + Cz$ has a maximum in the region G.* Call the maximum value D. Then G is contained in the half-space given by $Ax + By + Cz \leq D$ —that is, every point (x, y, z) in G satisfies the condition $Ax + By + Cz \leq D$ because D is a maximum—and G intersects the plane given by $Ax + By + Cz = D$. Observe that the intersection is either an extreme point (this is the typical case), an edge, or a face, and these correspond to one, two, or at least three maximizing extreme points. See Figure 12-5, which shows the possibilities for a bounded planar G—namely, either one or two maximizing extreme points, while Figure 12-6 shows a typical example in three dimensions.

Exercises 12-8

1. Let G be the region given by $x \geq 0$, $y \geq 0$, and $x + y \leq 1$. Let $f(x, y) = 4x - y$. Determine the maximum and minimum values of f on G.

2. Let G be the region given by $x \geq 0$, $y \geq 0$, $x - y \leq 0$ and $x + y \leq 2$. Let $f(x, y) = -5x + y$. (a) Determine the vertices of G. (b) Find the maximum and minimum of f on G.

3. Let G be the region that is the set of all points (x, y) for which $x + y \leq 2$, $-x - 2y \leq 3$, and $-2x + 3y \leq 6$. Let $f(x, y) = 4x + 8y$. Determine the maximum and minimum of f on G.

---

*We are using the fact that a linear function attains a maximum value on any bounded region which contains its boundary.

The farmer's profit

$$= \left\{ \begin{matrix} \text{number} \\ \text{of} \\ \text{turkeys} \end{matrix} \right\} \times \left\{ \begin{matrix} \text{profit} \\ \text{per} \\ \text{turkey} \end{matrix} \right\} + \left\{ \begin{matrix} \text{number} \\ \text{of} \\ \text{ducks} \end{matrix} \right\} \times \left\{ \begin{matrix} \text{profit} \\ \text{per} \\ \text{duck} \end{matrix} \right\}$$

$$+ \left\{ \begin{matrix} \text{number} \\ \text{of} \\ \text{chickens} \end{matrix} \right\} \times \left\{ \begin{matrix} \text{profit} \\ \text{per} \\ \text{chicken} \end{matrix} \right\} = x \cdot 3 + y \cdot 2 + z \cdot 1$$

Call the profit $f(x, y)$. Then

$$f(x, y) = 3x + 2y + z$$

with f defined over the region G. We will find the maximum profit later.

□

In each of the preceding problems, the region G is an example of a *convex polygonal region;* that is, the intersection of closed half-planes and the function f is a *linear function.* In general, f is a **linear function** if $f(x, y, z, \ldots) = Ax + By + Cz + \cdots$.

The following result, which generalizes to any number of quantities, will be proved at the end of this section.

Theorem 1

A linear function over a convex polygonal region G takes on its maximum or minimum at one of the vertices of G.

We use this result to complete the solution of Examples 2 and 3.

Example 2 (continued)

In view of Theorem 1, we need only compute f at the vertices of G to find the maximum possible profit and how to achieve it:

$$f(0, 0) = 300 \cdot 0 + 400 \cdot 0 = 0$$
$$f(0, 5) = 300 \cdot 0 + 400 \cdot 5 = 2000$$
$$f(1, 5) = 300 \cdot 1 + 400 \cdot 5 = 2300$$
$$f(4, 2) = 300 \cdot 4 + 400 \cdot 2 = 2000$$
$$f(4, 0) = 300 \cdot 4 + 400 \cdot 0 = 1200$$

So the maximum possible profit is $2300, achieved by working the gold mine 1 day per week and the silver mine 5 days per week.

found by evaluating f at the vertices, as in Example 1. This will be done later.

Example 3

A farmer raises turkeys, ducks, and chickens. He can raise at most 500 birds with at most 200 turkeys and at most 100 ducks. His profit is $3 per turkey, $2 per duck and $1 per chicken. How many of each bird should he raise to maximize his profit?

Solution

Let x, y, and z be the respective number of turkeys, ducks, and chickens he raises. The restrictions on (x, y, z) are the closed half spaces given by

$$x + y + z \leq 500 \quad \text{(restriction on the total number of birds)}$$

$$x \leq 200 \quad \text{(restriction on the total number of turkeys)}$$

$$y \leq 100 \quad \text{(restriction on the total number of ducks)}$$

$$\left.\begin{array}{l} x \geq 0 \\ y \geq 0 \\ z \geq 0 \end{array}\right\} \quad \text{(the number of birds is not negative)}$$

These restrictions determine the solid shaded region G in Figure 12-3, bounded by the planes given by $x + y + z = 500$, $x = 200$, $y = 100$, $x = 0$, $y = 0$, and $z = 0$.

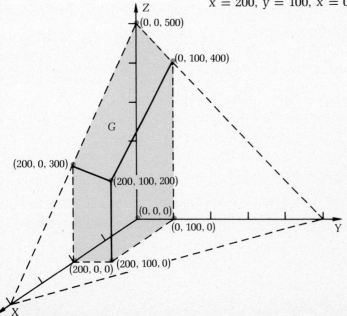

Figure 12-3

need to compute f at the vertices to find its maximum and minimum over G. We obtain

$$f(\tfrac{2}{3}, \tfrac{5}{3}) = 2 \cdot \tfrac{2}{3} + 5 \cdot \tfrac{5}{3} = \tfrac{29}{3}$$
$$f(2, 1) = 2 \cdot 2 + 5 \cdot 1 = 9$$
$$f(0, 1) = 2 \cdot 0 + 5 \cdot 1 = 5$$

So $\tfrac{29}{3}$ is the *maximum* value of f over the region G, and 5 is the *minimum* value of f over the region G.

Example 2

A company owns a gold mine and a silver mine. The gold mine yields a profit of $300 per day, and the silver mine yields a profit of $400 per day. The company has only one mining crew, which will work at most 6 days each week. Also, the gold mine can be worked at most 4 days each week, while the silver mine can be worked at most 5 days each week. How should the company operate to maximize its profit?

Solution

Let x = the number of days the gold mine is operated.
Let y = the number of days the silver mine is operated.

The restrictions on (x, y) are the closed half-planes given by

$x + y \le 6$ (restriction on the total number of days each week)

$x \le 4$ (restriction on the number of days for the gold mine)

$y \le 5$ (restriction on the number of days for the silver mine)

$\left. \begin{matrix} x \ge 0 \\ y \ge 0 \end{matrix} \right\}$ (the number of days each mine is worked is not negative)

This region G in this case is the shaded part of Figure 12-2 and is the intersection of the above closed half-planes.

The company's weekly profit

$$= \left\{ \begin{matrix} \text{the number} \\ \text{of days for} \\ \text{the gold mine} \end{matrix} \right\} \times \left\{ \begin{matrix} \text{profit} \\ \text{per day} \\ \text{for gold} \end{matrix} \right\} + \left\{ \begin{matrix} \text{the number} \\ \text{of days for} \\ \text{silver mine} \end{matrix} \right\} \times \left\{ \begin{matrix} \text{profit} \\ \text{per day} \\ \text{for silver} \end{matrix} \right\}$$

$$= x \cdot 300 + y \cdot 400$$

Call the profit $f(x, y)$ then

$$f(x, y) = 300x + 400y$$

with f defined over the region G. The maximum profit can be

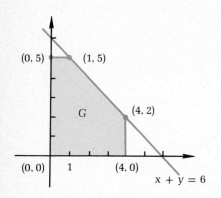

Figure 12-2

12-8 LINEAR PROGRAMMING

Linear programming deals with quantities, say x, y, z, ..., whose values can be freely chosen—subject, however, to a set of conditions of a certain type. The fundamental problem is then to maximize (or minimize) the quantity $Ax + By + Cz + \cdots$, where A, B, C, ... are given numbers.

For the case of two quantities, the restrictions determine a region G in the plane, actually the intersection of closed half-planes.* For the case of three quantities, the restriction determines a region G in space (the intersection of closed half-spaces).

Example 1

Let G be the set of all points (x, y) for which

$$x + 2y \leq 4$$

and

$$-x + y \leq 1$$

and

$$y \geq 1$$

Let $f(x, y) = 2x + 5y$. Find the maximum and minimum values of f over the region G.

Solution

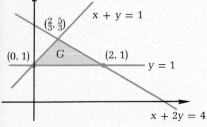

Figure 12-1

G is the shaded region given in Figure 12-1. The vertices are the intersections of the lines given by $x + 2y = 4$, $-x + y = 1$, and $y = 1$, and are found by solving the systems

$$\left\{ \begin{matrix} x + 2y = 4 \\ -x + y = 1 \end{matrix} \right\} \quad \left\{ \begin{matrix} x + 2y = 4 \\ y = 1 \end{matrix} \right\} \quad \left\{ \begin{matrix} -x + y = 1 \\ y = 1 \end{matrix} \right\}$$

The solutions for these systems are, respectively, $(\frac{2}{3}, \frac{5}{3})$, $(2, 1)$, and $(0, 1)$. Thus, these points are the vertices of the triangular region G given in Figure 12-1.

We will show later that the function f has its maximum (or minimum) value over G at one of the vertices. Hence, we only

*A half-plane together with its boundary is called a *closed half-plane*.

2. Find all real numbers x for which

$$\begin{vmatrix} -2 & -2x & x+1 \\ x^2 & 1 & x-1 \\ x & 0 & 1 \end{vmatrix} = 0$$

3. Let

$$A = \begin{pmatrix} a_{11} & a_{12} & a_{13} & a_{14} \\ 0 & a_{22} & a_{23} & a_{24} \\ 0 & 0 & a_{33} & a_{34} \\ 0 & 0 & 0 & a_{44} \end{pmatrix}$$

Show by using cofactors that $|A| = a_{11} \cdot a_{22} \cdot a_{33} \cdot a_{44}$.

4. Find all numbers x for which

$$\begin{vmatrix} 2x-1 & 2x & x^2-x+3 & x^4+1 \\ 0 & x^2-4 & 4 & 1 \\ 0 & 0 & x^2+3x & x \\ 0 & 0 & 0 & -3 \end{vmatrix} = 0$$

In Exercises 5–7, find the multiplicative inverse of the given matrix.

5. $\begin{pmatrix} 1 & 1 & 1 \\ 1 & 0 & 1 \\ 0 & 0 & 1 \end{pmatrix}$

6. $\begin{pmatrix} 0 & 0 & 0 & 1 \\ 0 & 0 & 2 & 0 \\ 0 & 3 & 0 & 0 \\ 4 & 0 & 0 & 0 \end{pmatrix}$

7. $\begin{pmatrix} a & 0 & 0 & 0 \\ 0 & b & 0 & 0 \\ 0 & 0 & c & 0 \\ 0 & 0 & 0 & d \end{pmatrix}$ where a, b, c, and d are nonzero real numbers

8. Let

$$A = \begin{pmatrix} 4 & 0 & 5 \\ 0 & 1 & -6 \\ 3 & 0 & 4 \end{pmatrix}$$

(a) Find A^{-1}.

(b) Find numbers x, y, and z for which

$$A \begin{pmatrix} x \\ y \\ z \end{pmatrix} = \begin{pmatrix} 1 \\ 0 \\ -1 \end{pmatrix}$$

using the fact that

$$\begin{pmatrix} x \\ y \\ z \end{pmatrix} = A^{-1} \begin{pmatrix} 1 \\ 0 \\ -1 \end{pmatrix}$$

9. Let

$$A = \begin{pmatrix} 1 & 3 & 2 \\ -1 & -1 & -1 \\ 2 & 3 & -2 \end{pmatrix}$$

(a) Find A^{-1}.

(b) Find all numbers x, y, and z for which

$$A \begin{pmatrix} x \\ y \\ z \end{pmatrix} = \begin{pmatrix} -9 \\ 0 \\ 18 \end{pmatrix}$$

using the fact that

$$\begin{pmatrix} x \\ y \\ z \end{pmatrix} = A^{-1} \begin{pmatrix} -9 \\ 0 \\ 18 \end{pmatrix}$$

10. Let

$$A = \begin{pmatrix} 1 & 0 & 0 \\ 0 & 1 & 0 \\ 2 & 0 & 0 \end{pmatrix}$$

Show that $|A| = 0$ (and hence, that A has no inverse).

11. Let

$$A = \begin{pmatrix} 4 & -2 & 1 \\ -2 & 1 & 2 \\ 1 & 2 & 4 \end{pmatrix}$$

Find all real numbers x for which the determinant of $A - xI$ is 0 (hence showing that the matrix $A - xI$ has no inverse for these values of x).

12. Let A be a 3 × 3 matrix, and let a be any number. Show that $|a \cdot A| = a^3 |A|$.

◇**13.** Let A be the 5 × 5 matrix

$$\begin{pmatrix} 1 & 1 & 1 & 1 & 1 \\ 1 & 2 & 3 & 4 & 5 \\ 1 & 3 & 6 & 10 & 15 \\ 1 & 4 & 10 & 20 & 35 \\ 1 & 5 & 15 & 35 & 70 \end{pmatrix}$$

Compute $|A|$.

4. Let G be the region given in Exercise 3, and let $g(x, y) = -2x + 3y - 9$. Determine the maximum and minumum of g on G.

5. Let G be the quadrilateral region given by $y \leq (x + 3)/2$, $y \leq -4x + 6$, $y \geq 0$, $x \geq 0$ and let $f(x, y) = \frac{2}{3}x - \frac{1}{3}y$. Determine the minimum and maximum of f on G.

6. A coat manufacturer has 80 square yards of rayon and 120 square yards of wool. A coat requires 1 square yard of rayon and 3 square yards of wool, while a blanket requires 2 square yards of each. How many of each should be made to maximize the manufacturer's income if coats and blankets each sell for $50?

7. A firm owns two factories, say factory A and factory B. Factory A earns the firm $200 every day of operation, whereas factory B earns $300 each day of operation. The company has one staff of employees to run both factories, and this staff works as a unit in one or the other factory. For environmental reasons, factory B cannot be used more than factory A in a week. Furthermore, this staff works only 5 days a week. Because of shortages in supplies, factory A can operate no more than 4 days a week, and factory B no more than 3 days. How many days should the staff operate factory A, and how many days should it operate factory B, in order to earn the maximum profit? What is the maximum profit?

8. Suppose it takes 2 man-hours (m.h.) for an electronics manufacturer to manufacture the parts to a television set and an additional $1\frac{1}{2}$ m.h. to assemble the set. Also suppose it takes 3 m.h. to manufacture the parts to a radio-phonograph console and 1 hour to assemble that. If our electronic manufacturer has at most 120 m.h. a day available for the manufacturing process and 75 m.h. daily for assembly, and makes a $55 profit on each item, how many televisions and how many consoles should he turn out daily to maximize his profit?

9. A man on a meat-and-cottage-cheese diet needs at least 864 calories each day. His daily portion of meat cannot exceed 7 ounces, and he cannot have more than six pints of cottage cheese. An ounce of meat supplies 108 calories and costs 25 cents, and a pint of cottage cheese supplies 72 calories at a cost of 20 cents. What is the least expensive way for him to be on this diet?

Miscellaneous Problems

1. A company manufactures an item that cost $3.50 to produce and market. The fixed costs involved are $82,500. The selling price of each item is $6.00. At what production level will the company break even?

2. Six pounds of tea and eight pounds of coffee cost $8.88. If the price of each goes up 5¢ a pound, then three pounds of coffee and two pounds of tea would cost $3.25. What is the original price per pound of the coffee and tea?

3. An amount totaling $10,000 was invested in two places, one paying 6% and one paying 5%. The total earnings from these investments is $540 (in a year). How much was invested at each rate?

4. A television manufacturer makes three quality sets, A, B, and C, priced, respectively, at $120, $150, and $230. The corresponding costs for producing and marketing these sets are, respectively, $60, $80, and $120. Suppose the manufacturer wishes to make 750 sets. How many of each should he make if the total revenue is to be $119,000, and the total production and marketing costs are $63,000?

5. By finding the inverse of the coefficient matrix, find all x, y, z for which

$$x + y + z = 1$$
$$y + z = 2$$
$$z = 3$$

6. Suppose A is the matrix

$$\begin{pmatrix} 0 & 0 & 1 \\ 0 & 0 & 0 \\ 1 & 0 & 0 \end{pmatrix}$$

Find A^{100}.

7. Let A be the matrix

$$\begin{pmatrix} 1 & 3 & 2 \\ -1 & -1 & -1 \\ 2 & 3 & -2 \end{pmatrix}$$

Find all vectors x for which $Ax = 0$.

8. Show by induction that for any $n \times n$ matrix A, $|A| = 0$ if two rows of A are the same.

9. Find all (x, y) for which

$$\frac{1}{x} + \frac{2}{y} = 1$$

$$\frac{2}{x} - \frac{4}{y} = 0$$

10. Let a and b be nonzero real numbers, where $a^2 \neq b^2$. Find all (x, y, z) for which

$$ax - by = a^2 - b^2$$
$$bx + az = b^2 - a^2$$
$$y + z = 0$$

5

ADDITIONAL TOPICS

Chapter 13 studies the usual operations on sets with some emphasis on counting, including permutations and combinations. Chapter 14 contains a brief introduction to probability theory with an emphasis on the relationship of probability models to real situations.

13 Sets and Counting

13-1 INTRODUCTION:
SETS AND OPERATING ON SETS

A **set** is any *collection* of objects. For example, the letters a, b, and c form the set $\{a, b, c\}$; the first five even natural numbers form the set $\{8, 2, 6, 10, 4\}$; and so on. The objects that compose a set are its *elements*. Given a set S and an element x of S, we shall write

$$x \in S$$

which is read "x is an element of S" or "x belongs to S." Thus $b \in \{a, b, c\}$, $2 \in \{6, 2, 8, 4, 10\}$, and so on. On the other hand,

$$x \notin S$$

means that x is *not* an element of S. For example, $3 \notin \{1, 2, 7\}$. Another common notation is

$$\{x \mid \cdots\}$$

which denotes "the set of all x for which ..."; for example,

$$\{a, b, c\} = \{x \mid x \text{ is one of the first three letters of the alphabet}\},$$

$\{x \mid x \in N \text{ and } x \text{ is a multiple of } 2\}$ is the set of all even natural numbers.

$$\{x \mid x \text{ is a real number and } x^2 = 4\} = \{-2, 2\},$$

and so on.

If every element of a set B is also an element of a set A, then B is a **subset** of A, which is written

$$B \subset A$$

In other words, $B \subset A$ means that *if $x \in B$ then $x \in A$.*

Example 1

(a) $\{a, b\} \subset \{a, b, c\}$
(b) $\{a, b\} \subset \{a, b\}$
(c) $\{a\} \subset \{a, b\}$ and $\{a\} \subset \{a, b, c\}$
(d) $\{x \mid x \text{ is a student of mathematics}\} \subset \{x \mid x \text{ is a student}\}$
(e) $\{x \mid x \in N \text{ and } x \leq 23\} \subset \{x \mid x \text{ is a real number and } x < 27\}$

□

As a matter of convenience, we define the **empty set** (or **null set**), *denoted by $\varnothing$, as the set with no elements,* and agree that $\varnothing$ *is a subset of every set;* that is,

$$\varnothing \subset A \text{ for every set } A$$

For example,

$$\{x \mid x \text{ is a square triangle}\}$$
$$= \{x \mid x \in N \text{ and } 1 < x < 2\} = \varnothing$$

Given sets A and B, their *union* and *intersection* are defined as follows:

The **union** of A and B, denoted by $A \cup B$, is the set of all elements each of which belongs to *either A or B* (or both); and the **intersection** of A and B, denoted by $A \cap B$, is the set of all elements each of which belongs to *both* A and B; that is,

$$A \cup B = \{x \mid x \in A \text{ or } x \in B\}$$

and

$$A \cap B = \{x \mid x \in A \text{ and } x \in B\}$$

Many definitions and facts concerning sets may be conveniently illustrated by means of Venn diagrams; some examples are given in Figures 13-1 and 13-2.

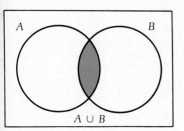

$A \cup B$

Figure 13-1

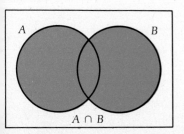

$A \cap B$

Figure 13-2

Note that $A \cup B = B \cup A$ and $A \cap B = B \cap A$; that is, the operations $\cup$ and $\cap$ are *commutative*.

Example 2

Let $A = \{1, 2, 4\}$, $B = \{1, 2, 5\}$, and $C = \{3, 5\}$; then $A \cup B = \{1, 2, 4, 5\}$, $A \cap B = \{1, 2\}$, $A \cup C = \{1, 2, 3, 4, 5\}$, and $A \cap C = \varnothing$.

□

Two sets that have no elements in common (that is, whose intersection is $\varnothing$, as in the case of A and C of Example 2) are said to be **disjoint**.

The operations $\cup$ and $\cap$ may be extended to any number of sets. Thus, given sets $A_1, A_2, \ldots, A_n$, we have

$$A_1 \cup A_2 \cup \cdots \cup A_n = \bigcup_{k=1}^{n} A_k$$

$$= \{x \mid x \in A_1 \text{ or } x \in A_2 \cdots \text{ or } x \in A_n\}$$

and

$$A_1 \cap A_2 \cap \cdots \cap A_n = \bigcap_{k=1}^{n} A_k$$

$$= \{x \mid x \in A_1 \text{ and } x \in A_2 \cdots \text{ and } x \in A_n\}$$

Note at this point that $\cup$ and $\cap$ are *associative*.

Example 3

Given the sets A, B, and C of Example 2, we have

$$A \cup B \cup C = \{1, 2, 3, 4, 5\} \quad \text{and} \quad A \cap B \cap C = \varnothing$$

Furthermore,

$$(A \cup B) \cup C = A \cup (B \cup C)$$

and

$$(A \cap B) \cap C = A \cap (B \cap C)$$

□

An interesting difference between the algebra of numbers and the algebra of sets lies in the fact that $\cup$ and $\cap$ are *distributive with respect to each other*; that is, for any sets A, B, and C:

$$A \cup (B \cap C) = (A \cup B) \cap (A \cup C)$$

and

$$A \cap (B \cup C) = (A \cap B) \cup (A \cap C)$$

The reader is asked to verify this fact by drawing the appropriate Venn diagrams.

Example 4

Using once more the sets of Example 2, we have

$$A \cup (B \cap C) = \{1, 2, 4\} \cup \{5\} = \{1, 2, 4, 5\}$$

and also

$$(A \cup B) \cap (A \cup C) = \{1, 2, 4, 5\} \cap \{1, 2, 3, 4, 5\}$$
$$= \{1, 2, 4, 5\}$$

Furthermore,

$$A \cap (B \cup C) = \{1, 2, 4\} \cap \{1, 2, 3, 5\} = \{1, 2\}$$

and also

$$(A \cap B) \cup (A \cap C) = \{1, 2\} \cup \varnothing = \{1, 2\}$$

□

Certain logical considerations dictate that all the discussed sets considered in a given context be looked upon as subsets of some fixed superset, usually called the **universal set** and denoted by **U**. For example, the union of all the sets under discussion may serve as **U**.

Now, given **U** and $A \subset U$, the **complement** of A (relative to **U**) is the set of all those elements of **U** that are *not* in A, and is denoted by A'; that is

$$A' = \{x \mid x \in U \text{ and } x \notin A\}$$

See Figure 13-3.

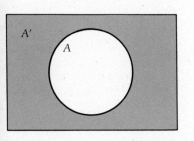

Figure 13-3

Example 5

Let **U** be the set **N** of natural numbers, **O** the set of odd natural numbers, and **E** the set of even natural numbers; then $E' = O$, $O' = E$, $O \cup E = N(=U)$, and $O \cap E = \varnothing$.

□

Note that for *any* set A we have $A \cup A' = U$ and $A \cap A' = \varnothing$. Note further that $(A')' = A$.

The following interesting relationships between $\cup$, $\cap$, and $'$, known as the *De Morgan Laws,* can be proved, but the reader is asked to verify them by means of Venn diagrams:

1. The complement of a union is the intersection of complements.
2. The complement of an intersection is the union of complements.

That is, for any sets A and B,

$$(A \cup B)' = A' \cap B' \quad \text{and} \quad (A \cap B)' = A' \cup B'$$

In general, given subsets $A_1, A_2, \ldots, A_n$ of a universal set U, we have

$$\left(\bigcup_{k=1}^{n} A_k\right)' = \bigcap_{k=1}^{n} A_k' \quad \text{and} \quad \left(\bigcap_{k=1}^{n} A_k\right)' = \bigcup_{k=1}^{n} A_k'$$

Example 6

Let $U = \{1, 2, 3, 4, 5, 6, 7, 8, 9, 10\}$, $A = \{1, 2, 3, 4, 5\}$, and $B = \{4, 5, 6, 7\}$. Then

$$(A \cup B)' = \{1, 2, 3, 4, 5, 6, 7\}' = \{8, 9, 10\}$$

and

$$A' \cap B' = \{6, 7, 8, 9, 10\} \cap \{1, 2, 3, 8, 9, 10\} = \{8, 9, 10\}$$

Furthermore,

$$(A \cap B)' = \{4, 5\}' = \{1, 2, 3, 6, 7, 8, 9, 10\}$$

and

$$A' \cup B' = \{6, 7, 8, 9, 10\} \cup \{1, 2, 3, 8, 9, 10\}$$
$$= \{1, 2, 3, 6, 7, 8, 9, 10\}$$

□

Venn diagrams are particularly useful in determining numerical relationships among several sets, as shown in the following example.

Example 7

Suppose a group of students enroll for courses in mathematics, physics, or philosophy, as follows:

    24 take mathematics
    15 take philosophy
    16 take physics
    12 take math and physics
     3 take philosophy and physics
    10 take philosophy and mathematics
     2 take mathematics, philosophy, and physics

How many students are in the group? How many students take philosophy, but *not* mathematics? How many students take only mathematics or physics?

Solution

The Venn diagram in Figure 13-4 illustrates the given information. We must fill in the number of students taking all three subjects, then the number of students taking the various com-

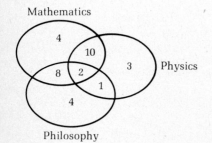

Mathematics

Physics

Philosophy

Figure 13-4

binations of two subjects (for example, since 10 students take philosophy *and* mathematics, and 2 of those take all three subjects, it follows that 8 students take philosophy and mathematics but do *not* take physics), and so on.

We see from this Venn diagram that there are 32 students in the group, 5 of whom take philosophy but not mathematics, and 17 of whom take *only* mathematics or physics.

□

If A is a finite set (that is, a set with a finite number of elements), then *the number of elements in A is denoted by* $n(A)$. Thus, if in the foregoing example we denote by A, B, and C, respectively, the sets of students taking mathematics, philosophy, and physics, then the problem may be stated as follows: $n(A) = 24$, $n(B) = 15$, $n(C) = 16$, $n(A \cap C) = 12$, $n(B \cap C) = 3$, $n(A \cap B) = 10$, and $n(A \cap B \cap C) = 2$. Furthermore, the solution shows that $n(A \cup B \cup C) = 32$. Note that in this case $N(A \cup B \cup C) \neq N(A) + N(B) + N(C)$.

A methodical way of computing the number of elements in a union of *finite* sets is based on the direct observation that *if A and B are finite sets that have no elements in common* (that is, $A \cap B = \varnothing$), then $n(A \cup B) = n(A) + n(B)$ and, more generally, **if $A_1, A_2, \ldots, A_n$ are pairwise disjoint finite sets** (that is, $A_i \cap A_j = \varnothing$ if $i \neq j$), **then**

$$n\left(\overset{n}{\underset{k=1}{\cup}} A_i \right) = n(A_1 \cup A_2 \cup \cdots \cup A_n) = \sum_{i=1}^{n} n(A_i)$$

In the case where A and B are not necessarily disjoint, we have the following theorem.

Theorem

If A and B are finite sets, then

$$n(A \cup B) = n(A) + n(B) - n(A \cap B)$$

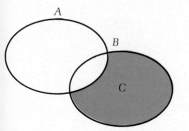

Figure 13-5

Proof   Refer to Figure 13-5 and note that the set C, obtained from B by deleting from it all elements in $A \cap B$, satisfies

$$n(C) = n(B) - n(A \cap B)$$

and that $A \cup B = A \cup C$, with $A \cap C = \varnothing$. Hence, we have

$$n(A \cup B) = n(A \cup C) = n(A) + n(C)$$
$$= n(A) + n(B) - n(A \cap B)$$

Example 8

Let $A = \{a, b, c\}$ and $B = \{b, c, d, e\}$. Then

$$n(A) = 3, \quad n(B) = 4, \quad n(A \cap B) = n(\{b, c\}) = 2$$

and hence

$$n(A \cup B) = 3 + 4 - 2 = 5$$

Check

$$n(A \cup B) = n(\{a, b, c, d, e\}) = 5$$

Example 9

If A, B, and C are finite sets, then

$$
\begin{aligned}
n(A \cup B \cup C) &= n[(A \cup B) \cup C] \\
&= n(A \cup B) + n(C) - n[(A \cup B) \cap C] \\
&= n(A) + n(B) - n(A \cap B) + n(C) \\
&\qquad\qquad\qquad - n[(A \cap C) \cup (B \cap C)] \\
&= n(A) + n(B) + n(C) - n(A \cap B) \\
&\qquad - n(A \cap C) - n(B \cap C) + n(A \cap B \cap C)
\end{aligned}
$$

If, in particular, A, B, and C are the sets of Example 7, then

$$n(A \cup B \cup C) = 24 + 15 + 16 - 12 - 3 - 10 + 2 = 32$$

A generalization to larger numbers of finite sets is left for the exercises.

□

Frequently, it is the subsets, or subsets of a given size, rather than the elements of a set that must be counted. Systematic procedures for such counting are developed in Sections 13-2 and 13-3.

Example 10

Find the number of subsets of each size that a four-element set has.

Solution

The set $\{1, 2, 3, 4\}$ has

1 subset with no elements:	$\varnothing$
4 one-element subsets:	$\{1\}, \{2\}, \{3\}, \{4\}$
6 two-element subsets:	$\{1, 2\}, \{1, 3\}, \{1, 4\},$ $\{2, 3\}, \{2, 4\}, \{3, 4\}$
4 three-element subsets:	$\{1, 2, 3\}, \{1, 2, 4\},$ $\{1, 3, 4\}, \{2, 3, 4\}$
1 four-element subset:	$\{1, 2, 3, 4\}$

Since $1 + 4 + 6 + 4 + 1 = 16$,* a four-element set has $16 = 2^4$ subsets.

Example 11

The set $\{1, 2\}$ has $4 = 2^2$ subsets: $\{1, 2\}, \{1\}, \{2\}, \varnothing$; the set $\{1\}$ has $2 = 2^1$ subsets: $\{1\}, \varnothing$; the empty set $\varnothing$ has $1 = 2^0$ subset: $\varnothing$.

Example 12

Noting that a two-element set has $4 = 2^2$ subsets, we may infer that a three-element set has $2^2 \cdot 2 = 2^3$ subsets, since adding one element to a set simply *doubles* the number of subsets. For example, since $\{1, 2\}$ has the subsets $\{1, 2\}, \{1\}, \{2\}$, and $\varnothing$, adding 3 to $\{1, 2\}$ creates the additional subsets $\{1, 2, 3\}$, $\{1, 3\}, \{2, 3\}$, and $\{3\}$—one new subset for each old one.

It follows, then, that a four-element set has $2^3 \cdot 2 = 2^4$ subsets, a five-element set has $2^4 \cdot 2 = 2^5$ subsets, and, in general, *an n-element set has 2^n subsets.* We shall arrive at this fact somewhat differently in the next section.

Exercises 13-1

In Exercises 1–9, let $U = \{a, b, c, d\}$, $A = \{a, b, c\}$, $B = \{c, d\}$ and $C = \{a, d\}$. Find each of the given sets.

1. $A \cup B$ **2.** $A \cap B$

3. A', B', C' **4.** $A \cap (B \cup C)$

5. $(A \cap B) \cup (A \cap C)$ **6.** $A' \cup B'$

---

*As we shall see later, it is not mere coincidence that the numbers 1, 4, 6, 4, 1 are the coefficients of the polynomial function

$$x \rightarrow (1 + x)^4 = 1 + 4x + 6x^2 + 4x^3 + x^4$$

and that 1, 2, 1 are the coefficients of

$$x \rightarrow (1 + x)^2 = 1 + 2x + x^2$$

and so on.

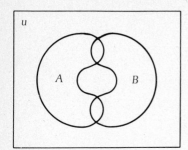

Exercises 10, 11, 12

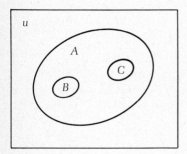

Exercises 13, 14

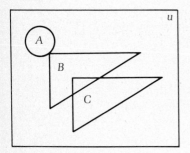

Exercises 15, 16, 17

7. $(A \cap B)'$

8. $(A \cup B)'$

9. $A' \cap B'$

In Exercises 10–17, shade the required set.

10. $A \cap B$

11. $A' \cap B$

12. $(A \cap B)'$

13. $(A \cap (B \cup C))'$

14. $B' \cap C'$

15. $A \cup (B \cap C)$

16. $(B \cup C)' \cap A$

17. $(B \cap C') \cup (C \cap B')$

18. In the figure for Exercise 18, U is the set of all points inside the square, S is the set of all points inside the rectangle, and T is the set of all points inside the circle. Write an expression for each of the shaded regions.

In Exercises 19–23, let the universal set U be the interval $(-5, 3]$, and let $A = (-5, -1)$, $B = (-\frac{1}{2}, 2]$, $C = [0, 3]$, $D = (-1, 3]$, and $E = (-5, 0)$. Find the given sets.

19. $A \cap B$

20. $E \cup C$

21. $E \cap D$

22. $C \cap D'$

23. $B' \cap C'$

24. List all subsets of the set $\{1, 2, 3\}$.

25. List all the subsets of $\{1, 2, 3, 4, 5\}$.

26. How many two-element subsets does a set with six elements have? How many three-element subsets does it have?

In Exercises 27–31, let N denote the set of natural numbers, let R denote the set of real numbers, and let C denote the set of complex numbers. List the elements of each of the given sets.

27. $\{x: x \in N \text{ and } 1 \le x \le 7\}$

28. $\{x: x \in R \text{ and } (x^2 - 2)(x^2 - x - 6) = 0\}$

29. $\{(x, y): x \in R, y \in R, x + y = 1 \text{ and } 2x - 3y = 4\}$

30. $\{t: t \in C \text{ and } (t^2 + t + 1)(t^2 + t - 1) = 0\}$

31. $\{x: 0 \le x \le 2\pi \text{ and } 2\sin^2 x + \sin x = 0\}$

In Exercises 32–35, let U be the real line and let

$A = \{x: x \text{ in } R \text{ and } x > 4\}$

$B = \{x: x \text{ in } R \text{ and } 2 - x > 0\}$

$C = \{x: x \text{ is an even integer and } 9 - 2x < 0\}$

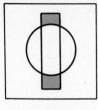

(a)

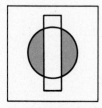

(b)

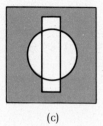

(c)

Exercise 18

Find the given sets

32. $A \cap B$

33. $A \cap C$

34. $A \cup B$

35. $A' \cap C$

36. Assume every language major takes at least one course in Russian, French, or German. Also suppose that these majors are distributed as follows:

    31 take Russian
    25 take French
    20 take German
    10 take Russian and French
    8 take Russian and German
    4 take French and German
    2 take Russian, French, and German

How many language majors are there? How many take only one course? How many take exactly two courses?

37. Let U, A, and B be finite sets, and suppose $A \subset B$. Show that $n(A \cup B') = n(A) + n(B')$.

13-2 THE FUNDAMENTAL PRINCIPLE OF COUNTING

Given sets A and B, the Cartesian product of A and B, denoted by $A \times B$, is the set of all ordered pairs (a, b) with $a \in A$ and $b \in B$; that is,

$$A \times B = \{(a, b) \mid a \in A \text{ and } b \in B\}$$

More generally, the Cartesian product of the sets $A_1, A_2, \ldots, A_n$, denoted by $A_1 \times A_2 \times \cdots \times A_n$, is the set of all ordered n-tuples $(a_1, a_2, \ldots, a_n)$ with $a_1 \in A_1$, $a_2 \in A_2$, $\ldots$, $a_n \in A_n$; that is,

$$A_1 \times A_2 \times \cdots \times A_n$$
$$= \{(a_1, a_2, \ldots, a_n) \mid a_1 \in A_1, a_2 \in A_2, \ldots, a_n \in A_n\}$$

Example 1

(a) Let $A = \{a, b\}$ and $B = \{1, 2, 3\}$. Then

$$A \times B = \{(a, 1), (a, 2), (a, 3), (b, 1), (b, 2), (b, 3)\}$$

whereas

$$B \times A = \{(1, a), (2, a), (1, b), (2, b), (3, a), (3, b)\}$$

Notice that $A \times B \neq B \times A$*; in fact, no element in $A \times B$ is in $B \times A$. Notice, however, that the *number* of elements in $A \times B$ and $B \times A$ is the same.

(b) Let $A = \{1, 2\}$. Then $A \times A = \{(1, 1), (1, 2), (2, 1), (2, 2)\}$.

(c) Let $\boldsymbol{R}$ be the set of real numbers. Then the plane can be defined as $R \times R$, since

$$\boldsymbol{R} \times \boldsymbol{R} = \{(x, y): x \in \boldsymbol{R} \text{ and } y \in \boldsymbol{R}\}$$

that is, $\boldsymbol{R} \times \boldsymbol{R}$ is the set of all ordered pairs of real numbers.

Example 2

(a) Let $A = \{1, 2\}$, $B = \{a, b\}$, and $C = \{x, y, z\}$. Then

$$\begin{aligned} A \times B \times C = \{&(1, a, x), (1, a, y), (1, a, z), \\ &(1, b, x), (1, b, y), (1, b, z), (2, a, x), (2, a, y), \\ &(2, a, z), (2, b, x), (2, b, y), (2, b, z)\} \end{aligned}$$

(b) $R \times R \times R$, being the set of all ordered triples of real numbers, may be identified with the three-dimensional space.
□

Note that in Example 1(a) we have

$$n(A \cdot B) = 6 = 2 \cdot 3 = n(A) \cdot n(B)$$

and in Example 2(a) we have

$$n(A \cdot B \cdot C) = 12 = 2 \cdot 2 \cdot 3 = n(A) \cdot n(B) \cdot n(C)$$

It is true, in general, that if $A_1, A_2, \ldots, A_n$ are finite sets, then

$$n(A_1 \cdot A_2 \cdot \ldots \cdot A_n) = n(A_1) \cdot n(A_2) \cdot \ldots \cdot n(A_n)$$

That is, the number of elements in a Cartesian product is the product of the numbers of elements in the component sets.

Example 3

If there are five roads from city X to city Y and two roads from city Y to city Z, how many paths are there from city X to city Z passing through city Y?

*Thus, the operation $\times$ is not commutative.

Solution

Let A, the set of roads from X to Y, be $\{a, b, c, d, e\}$, and let B, the set of roads from Y to Z, be $\{f, g\}$. Then the paths from X to Z via Y are the ordered pairs $(a, f), (b, f), (c, f), (d, f), (e, f), (a, g)$, and so on; that is, the set of paths from X to Z is $A \times B$. Hence, $n(A \times B)$, the number of elements in $A \times B$, is precisely the number of paths from X to Z, and so

$$n(A \times B) = n(A) \cdot n(B) = 5 \cdot 2 = 10$$

Thus, there are ten ways to go from X to Z.
□

The fact given in Example 3 is sometimes phrased as follows:

If one thing can be done in k_1 different ways, and if, for each of these ways, a second thing can be done in k_2 different ways, and if, for each of the first two, a third thing can be done in k_3 different ways, and so on, then the number of ways the entire sequence of n things can be done is the product of the number of ways each can be done, and hence is $k_1 \cdot k_2 \cdot k_3 \cdot \ldots \cdot k_n$.

This is often called the **fundamental principle of counting.**

Example 4

Find the number of ways in which five books can be placed on a shelf next to one another.

Solution

The first place can be filled in *five* ways, and once it is filled, the next place can be filled in *four* ways, and so on. Hence, the answer is $5 \cdot 4 \cdot 3 \cdot 2 \cdot 1 = 120$.
□

The product of the first n natural numbers, for a given n, occurs so frequently in counting problems that there is a special name and notation for it: **n factorial,** written **$n!$.** Thus,

$$n! = n \cdot (n - 1) \cdot (n - 2) \cdot \ldots \cdot 3 \cdot 2 \cdot 1$$

(In Example 4 the answer is 5!.)

Example 5

Given three slots on a bookshelf and *seven* books, find the number of ways in which the slots can be filled.

Solution

The first of the three slots can be filled in seven ways, and so on. The answer is $7 \cdot 6 \cdot 5 = 210$. Note that

$$7 \cdot 6 \cdot 5 = \frac{7 \cdot 6 \cdot 5 \cdot 4 \cdot 3 \cdot 2}{4 \cdot 3 \cdot 2 \cdot 1} = \frac{7!}{4!}$$

In general, if $r \leq n$, we have

$$n \cdot (n - 1) \cdot \cdots \cdot (n - r + 1) = \frac{n!}{(n - r)!}$$

Example 6

In how many ways can four persons be seated in a row of nine seats?

Solution

$$9 \cdot 8 \cdot 7 \cdot 6 = \frac{9!}{5!}$$

Example 7

Four seats, labeled t, d, h, and j, are reserved, respectively, for Tom, Dick, Harry, and John. Each of them may or may not come to a given performance. How many possible seating situations are there?

Solution

For each seat there are precisely two possibilities—occupied or unoccupied. Since there are four seats, the total number of possibilities is $2 \cdot 2 \cdot 2 \cdot 2 = 2^4 = 16$.
□

Note that Example 7 is equivalent to the problem of finding the number of subsets of a four-element set; for given such a set, say $\{x_1, x_2, x_3, x_4\}$, all of its subsets will be constructed when all possible "seating arrangements" for x_1, x_2, x_3, and x_4 have been exhausted. Generalizing this argument, we see again that an n-element set has precisely 2^n subsets.

Exercises 13-2

1. Suppose five flags, each of a different color, are arranged in a row for a signal. How many different ways may the flags be arranged?

2. How many ways can three balls be placed in seven boxes if each box contains at most one ball?

3. On a ten-question, true-false test, a student guesses on every question. What are his chances of answering all the questions correctly?

4. A fair coin is tossed five times. What are the chances that all heads show up?

5. A menu lists two soups, three meat dishes, and five desserts. How many meals are there consisting of soup, meat, and dessert (in that order)?

◇**6.** How many signals can be given using, for each signal, all the following flags arranged in a row: two red, three blue, and five yellow?

7. A coin is tossed four times. One possible outcome for this experiment is *HHTH* meaning heads on the first toss, heads on the second toss, tails on the third toss and heads on the fourth toss. How many outcomes are there altogether for this experiment? List them.

8. A die is thrown three times. One possible outcome for this experiment is $(1, 3, 5)$ meaning 1 shows up on the first throw, 3 on the second throw, and 5 on the third throw. How many outcomes are there altogether for this experiment?

9. In how many ways can five people be assigned to five rooms if only one person is assigned to each room?

10. An automobile manufacturer produces four different models A, B, C, and D. Models A and B can come in any of four body styles—sedan, hardtop, convertible, and station wagon, while models C and D come only as sedans or hardtops. Each car comes in any of five colors. How many different cars are available altogether?

13-3 PERMUTATIONS AND SUBSETS

A *permutation* of a finite nonempty set is an ordered arrangement of the elements of the set and is therefore an ordered n-tuple in which each element of the set occurs exactly once. For example, if $A = \{a, b, c\}$ then there are six permutations of A, namely, (a, b, c), (a, c, b), (b, a, c), (b, c, a), (c, a, b), and (c, b, a). Notice that we can construct these six permutations (3-tuples) by filling slots. Thus the first slot may be filled in three ways, then the second slot in two ways, and then the last slot one way. Hence, by the fundamental principle of counting, there are $3! = 6$ ways to fill the three slots.

The above example is generalized in the following theorem.

Theorem 1

The number of permutations of a set of n objects is $n!$.

Proof   The proof imitates the foregoing example.

Example 1

In how many ways can five people be assigned to five rooms if exactly one person is assigned to each room?

Solution

Think of the rooms as "slots" in a 5-tuple; then each assignment may be regarded as a 5-tuple. Thus, $(1, 2, 3, 4, 5)$, $(2, 1, 3, 4, 5)$, $(4, 5, 3, 1, 2)$, and so on, are the assignments, and there are $5! = 120$ altogether.

□

The following example leads to a generalization of Theorem 1

Example 2

Let $A = \{1, 2, 3, 4, 5\}$. How many 3-tuples with no entries the same can be constructed using elements of A?

Solution

The first slot can be filled in five ways, the second slot in four ways, and the third slot in three ways; hence, there are $5 \cdot 4 \cdot 3 = 60$ 3-tuples.

For convenience we define

$$5_3 = 5 \cdot 4 \cdot 3, \qquad 8_2 = 8 \cdot 7, \qquad 9_4 = 9 \cdot 8 \cdot 7 \cdot 6$$

and, in general,

$$n_r = \overbrace{n(n-1)(n-2) \cdots (n-r+1)}^{r \text{ factors}} = \frac{n!}{(n-r)!}$$

(See Section 13-2.)

□

The above example generalizes to the following result.

Theorem 2

The number of r-tuples with no entries the same which can be constructed from a set of n elements is n_r.

Another way of stating this is: *The number of permutations of n things taken r at a time is n_r.*

Example 3

A president, a vice-president, and a secretary are chosen from a group of ten people. In how many ways can this be done?

Solution

Each choice can be regarded as a 3-tuple (president, vice-president, secretary). Hence, there are

$$10_3 = \frac{10!}{7!} = 10 \cdot 9 \cdot 8 = 720 \text{ ways}$$

◻

We can now obtain a convenient formula for the number of r-element subsets of a set. In this connection the following notation is widely used:

$\binom{n}{r}$ denotes the number of **r-element subsets of a set of n elements**

Example 4

We have already seen that

$\binom{3}{2}$ = the number of two-element subsets of a set of three elements = 3

$\binom{5}{3}$ = the number of three-element subsets of a set of five elements = 10

$\binom{4}{0}$ = the number of empty subsets of a set of four elements = 1

Theorem 3

The number of r-element subsets of a set of n elements is

$$\frac{n!}{r!(n-r)!}$$

That is,

$$\binom{n}{r} = \frac{n!}{r!(n-r)!}$$

Proof Let A be a set of n elements. Each permutation of r elements of A is a permutation of some r-element subset of A, and each r-element subset of A has $r!$ permutations. Hence, since $n!/(n-r)!$ is the **total** number of permutations of n things taken r at a time, we have

$$\binom{n}{r} \cdot r! = \frac{n!}{(n-r)!}$$

Therefore,

$$\binom{n}{r} = \frac{n!}{r!(n-r)!}$$

Example 5

The number of three-element subsets of a set of seven elements is

$$\binom{7}{3} = \frac{7!}{3!4!} = \frac{7 \cdot 6 \cdot 5 \cdot 4 \cdot 3 \cdot 2 \cdot 1}{3 \cdot 2 \cdot 1 \cdot 4 \cdot 3 \cdot 2 \cdot 1} = 35$$

Example 6

How many five-card poker hands are there?

Solution

Each hand is a five-element subset of a set of 52 cards; hence, there are $\binom{52}{5} = \dfrac{52!}{5!47!}$ poker hands.

Example 7

How many five-card poker hands are there which contain three kings and two aces?

Solution

To construct such a hand, first choose three kings from the four kings in the deck. This can be done in $\binom{4}{3}$ ways.

Then choose two aces. This can be done in $\binom{4}{2}$ ways. Hence, by the fundamental principle of counting, both can be done in $\binom{4}{3}\binom{4}{2}$ ways. Therefore, there are $\binom{4}{3}\binom{4}{2} = 4 \cdot 6 = 24$ such hands.

Example 8

How many five-card poker hands are there which are full houses? (A full house has three cards of equal rank and the other two cards of equal rank.)

Solution

To construct such a hand, first choose a rank for the three cards. This can be done in 13 ways since there are 13 ranks. Then choose the three cards of that rank. This can be done in $\binom{4}{3}$ ways. Then choose a rank for the two cards. This can be done in 12 ways. Then choose the two cards of that rank. This can be done in $\binom{4}{2}$ ways. Hence, by the fundamental principal of counting, the total number of full houses is

$$13 \binom{4}{3} \cdot 12 \binom{4}{2} = 13 \cdot 4 \cdot 12 \cdot 6 = 3744$$

Exercises 13-3

1. How many lines are determined by ten points, no three of which lie on a line?

2. Consider ten points, no four of which are in a plane. How many planes are determined by these points? (Three points determine a plane.)

3. In a league of ten teams, each team plays each other team four times. How many games are played?

4. A coin is tossed ten times. In how many ways can this be done so that
(a) all heads show up?
(b) exactly three heads show up?
(c) exactly five heads show up?
(d) at least one head shows up?

5. How many five-card poker hands contain three red and two black cards?

6. How many committees of two men and three women can be selected from a group consisting of ten men and fifteen women?

7. A bridge hand consists of thirteen cards. How many bridge hands contain:
(a) four spades, three hearts, three diamonds, and three clubs?
(b) at least three cards of each suit?

8. A professor tells three jokes each year he teaches a certain course, and does not tell the same three jokes in two different years. How many jokes must he have to do this for 35 years?

◇**9.** If four men are assigned randomly to five rooms, what is the most likely distribution? (The possible distributions are: all men in one room, three in one room and one in some other room, two in one room and two in some other room, etc.)

◇**10.** Five three-person teams are formed from a group consisting of ten women and five men. Each team has two women and one man. How many ways can this be done?

13-4 THE BINOMIAL THEOREM

We shall digress briefly from our path to probability theory to prove the following important theorem.

Theorem 1

For any real number x and any natural number n, we have

$$(1 + x)^n = \sum_{k=0}^{n} \binom{n}{k} x^k$$

Proof   Since $(1 + x)^n = (1 + x)(1 + x) \cdots (1 + x)$ taken n times, $(1 + x)^n$ is the sum of integer multiples of the numbers 1, x, x^2, $\ldots$, x^n. Furthermore, the coefficient of, say, x^k (for $1 \leq k \leq n$) is simply the number of times that x^k occurs in the resulting sum (obtained after multiplying out the entire product); and this, in turn, is the number of ways in which k of the factors $(1 + x)$ can be chosen from a total of n factors. Hence, this number is $\binom{n}{k}$.

Corollary

Binomial Theorem   For any real numbers x and y, and any natural number n, we have

$$(x + y)^n = \sum_{k=0}^{n} \binom{n}{k} x^{n-k} y^k$$

Proof   If $x = 0$, the result is obvious. If $x \neq 0$, we have

$$(x + y)^n = x^n \left(1 + \frac{y}{x}\right)^n = x^n \sum_{k=1}^{n} \binom{n}{k} \frac{y^k}{x^k} = \sum_{k=1}^{n} \binom{n}{k} x^{n-k} y^k$$

Example 1

If a and b are real numbers, then

$$(2a - b)^4 = \sum_{k=0}^{4} \binom{4}{k} (2a)^{4-k}(-b)^k$$

$$= \binom{4}{0}(2a)^4 + \binom{4}{1}(2a)^3(-b) + \binom{4}{2}(2a)^2(-b)^2$$

$$+ \binom{4}{3}(2a)(-b)^3 + \binom{4}{4}(-b)^4$$

Since $\binom{n}{r} = \binom{n}{n-r}$, where n and r are nonnegative integers with $r \le n$, we see that

$$\binom{4}{0} = \binom{4}{4} = 1, \quad \binom{4}{1} = \binom{4}{3} = \frac{4!}{3!} = 4,$$

and

$$\binom{4}{2} = \frac{4!}{2!2!} = 6$$

Hence,

$$(2a - b)^4 = 16a^4 - 32a^3b + 24a^2b^2 - 8ab^3 + b^4$$

$\square$

Note that if $x = y = 1$, then $(x + y)^n = 2^n$, and thus, from the Binomial Theorem,

$$\sum_{k=0}^{n} \binom{n}{k} = 2^n$$

On the other hand, $\binom{n}{0}$ is precisely the number of 0-element subsets (i.e., $\varnothing$) in an n-element set, $\binom{n}{1}$ is precisely the number of one-element subsets in an n-element set, and so on. Thus,

an n-element set has exactly 2^n subsets

Example 2

The number of subsets of the set $\{1, 2, 3, 4, 5, 6, 7, 8\}$ is 2^8.

Exercises 13-4

In Exercises 1–3, let a be a real number and let f be the polynomial function given by $f(x) = (2a + 3x)^{100}$. Find the coefficients of the following terms of f:

1. x^7 **2.** x^{10}

3. x^{51}

In Exercises 4–7, let x and y be positive real numbers. Expand, using the Binomial Theorem:

4. $(x + 3y)^7$

5. $\left(x - \dfrac{1}{y^2}\right)^8$ **6.** $\left(\sqrt{x} - \dfrac{4}{\sqrt{y}}\right)^6$

7. $(\sin x + \cos x)^5$

8. Calculate $\left(1 + \dfrac{1}{n}\right)^n$ for $n = 1, 2, 3, 4$.

9. Let a, b, and c be numbers. Find $(a + b + c)^3$.

10. Approximate

(a) $(.98)^6$ (*Hint:* Consider $\left(1 - \dfrac{2}{100}\right)^6$ and use the first three terms of the Binomial expansion.)

(b) $(1.03)^{10}$

14 Probability

14-1 INTRODUCTION. EVENTS

Questions of probability arise in situations where there is uncertainty. Typically, these situations are concerned with a chance process (that is, a process whose result is uncertain, or equivalently, a process with more than one possible result), and they ask for the probabilities of various conditions on the result. For example, the process might be shuffling an ordinary deck of 52 cards and dealing out a hand of five cards. We might ask for the probability that the cards dealt are all of the same suit, or the probability that all four aces are in the hand, etc.

In constructing a mathematical model for such a process, we first introduce a set S of **possible outcomes.** These must be mutually exclusive and exhaustive so that exactly one of them occurs. Associated with any condition on the outcome of the process is the subset of S, or the **event,** consisting of all possible outcomes which satisfy the condition. The original condition is equivalent to the condition that the outcome belong to the associated subset. This is an important conceptual simplification, for it assures that no matter how complex the condition, it reduces to a subset of S.

Example 1

Let the chance process be the roll of a die (fair or not), and suppose we are interested in the probability that the upface is odd, or the probability that it is an ace, or that it is ≥ 3. Naturally, we take as the set of possible outcomes $S =$ (1, 2, 3, 4, 5, 6), where the possible outcomes are the possible numbers on the upface. The condition that the upface be odd determines the subset, or event, $\{1, 3, 5\}$. The upface is odd if and only if the outcome is in $\{1, 3, 5\}$ or, equivalently, if and only if the event $\{1, 3, 5\}$ occurs. Similarly, the upface being an ace corresponds to the singleton $\{1\}$ and the upface being ≥ 3 corresponds to $\{3, 4, 5, 6\}$.

Implicit in our description of S are certain assumptions that usually go unstated: that the die will not land on an edge, for instance, or that the die will not shatter. Such assumptions inevitably arise in fitting an ideal mathematical model to a real physical situation.

Note also that if we are interested only in whether the upface is an ace or not, we can, if we choose, take as the set of possible outcomes {ace, non-ace}, which is simpler and yet still adequate for the question at hand. Whether we make this choice or not is usually a matter of mathematical convenience.

Example 2

A coin, fair or not, is tossed twice. We take for the set of possible outcomes $S = \{HH, HT, TH, TT\}$. The possible HT, for instance, is heads on the first toss and tails on the second. Observe that exactly one head appears if and only if the outcome is in $\{HT, TH\}$. If we were interested only in the number of heads that appear, we could take as our possible outcomes 0, 1, or 2 heads. It is usually not advisable to do so, however, because the probability situation can be better understood in the context of the larger S.

Example 3

A coin is tossed until a head appears. We take $S = \{H, TH, TTH, \ldots\}$ or, equivalently, $S = \{1, 2, \ldots\} = N$, the possible outcome n being that the first head appears on the nth toss.

14-2 PROBABILITY FUNCTIONS

Intuitively, probability is a measure of likelihood relative to certainty. For instance, probabilities 1, .7, .4, and 0 correspond to 100% certainty, 70% certainty, 40% certainty, and impossibility, respectively. Now, let S be the set of possible outcomes in a model of some chance process, and let A be any subset of S (or event). Associated with A is the probability that the outcome of the process is one of the possibilities in A or, equivalently, that A occurs. This number is denoted by $P(A)$. Observe that P is a function with domain the class of all subsets of S. The set S together with the **probability function** P constitutes the mathematical model of the chance process.

We now seek the general properties that P should possess in order to be consonant with our intuitive notion of probability. We have

$$0 \leq P(A) \leq 1 \text{ for every subset } A$$

since $P(A)$ is a probability. Since S is exhaustive, some possibility in S is certain to occur; that is, S is certain to occur. Hence, $\boldsymbol{P(S) = 1}$. Now $P(\varnothing)$ is the probability that the outcome is in $\varnothing$, that is, that there is no outcome, which is impossible. Hence, $\boldsymbol{P(\varnothing) = 0}$.

Let A and B be **disjoint** subsets of S, and let $C = A \cup B$. The occurrence of C is tantamount to the occurrence of A, or the occurrence of B and these occurrences are exclusive possibilities. Intuition dictates that $P(C) = P(A) + P(B)$. This property is the **additivity** of P. For example, suppose that two men and a woman are candidates in a political election. Suppose further that man A has a 40% chance of being elected and man B has a 20% chance. Then the probability that a man is elected is 60%. In particular, since A and its complement A' are disjoint and $S = A \cup A'$, we have

$$P(A) + P(A') = P(S) = 1 \quad \text{or} \quad P(A') = 1 - P(A)$$

The additivity of P generalizes to any number n of disjoint subsets $A_1, A_2, \ldots, A_n$. If $A = A_1 \cup A_2 \cup \cdots \cup A_n$, then

$$P(A) = P(A_1) + P(A_2) + \cdots + P(A_n)$$

This can be argued directly, as for the case $n = 2$, or derived from that case by repeated application.

Now consider the probabilities of the individual possible outcomes of S (that is, strictly speaking, the probabilities of the singleton subsets). Since any subset A is the disjoint union of the singletons contained in it, it follows from the above that

$P(A)$ = sum of the probabilities of the possible outcomes in A

and, in particular,

$1 = P(S)$ = sum of the probabilities of all the possible outcomes

The structure of P is now transparent: total probability 1 is distributed among the possible outcomes, and the probability of any subset is the subtotal contained in it. We shall assume that these facts hold even when there is an infinite number of possible outcomes. In this case, certain sums will be infinite series.*

The following list summarizes the basic properties of a probability function P.

1. For every subset A (of S), $0 \le P(A) \le 1$.
2. $P(S) = 1$ and $P(\varnothing) = 0$.
3. **Additivity of P:** If A and B are disjoint subsets of S, then $P(A \cup B) = P(A) + P(B)$.

3'. If $A_1, A_2, \ldots, A_n$ are disjoint subsets of S, then

$$P(A_1 \cup A_2 \cup \cdots \cup A_n)$$
$$= P(A_1) + P(A_2) + \cdots + P(A_n)$$

that is,

$$P\left(\bigcup_{i=1}^{n} A_i\right) = \sum_{i=1}^{n} P(A_i)$$

4. For any subset A of S, $P(A) + P(A') = 1$.

The distribution of the total probability 1 among the possible outcomes depends, of course, on the special features of the chance process under study. Some of the considerations that enter are illustrated in the following examples.

*See Chapter 4 for an introduction to infinite sums.

Example 1

A fair die is rolled. As before, $S = \{1, 2, 3, 4, 5, 6\}$. Because of the physical symmetry of the die, it is reasonable to assume probabilistic symmetry; that is, any face is as likely to come up as any other. Since the probabilities of the six possible outcomes are all the same and their total is 1, each must be $\frac{1}{6}$. Therefore,

$P(\text{upface is an ace}) = P(1) = \frac{1}{6}$

$P(\text{upface is odd}) = P(1, 3, 5) = \frac{1}{6} + \frac{1}{6} + \frac{1}{6} = \frac{3}{6}$

$P(\text{upface} \geq 3) = P(3, 4, 5, 6) = \frac{1}{6} + \frac{1}{6} + \frac{1}{6} + \frac{1}{6} = \frac{4}{6}$

$\square$

More generally, suppose S consists of N equally likely possible outcomes. Then each outcome has probability $1/N$ (since the total probability is 1), and if A has n possible outcomes, then

$$P(A) = \underbrace{\frac{1}{N} + \frac{1}{N} + \cdots + \frac{1}{N}}_{n \text{ times}} = n\frac{1}{N} = \frac{n}{N}$$

and the calculations of probabilities are reduced to counting problems.

Example 2

A hand of five cards is dealt at random from an ordinary deck of 52 cards. (The phrase "at random" implies that the process is completely symmetric with respect to the cards, no card in any way being treated differently from any other card.) Taking S to be all possible (unordered) sets of five cards, the possible outcomes are equally likely, and hence the probability of any given event A is

$$P(A) = \frac{\text{number of possible outcomes in } A}{\binom{52}{5}}$$

For instance,

$$P(\text{all 4 aces dealt}) = \frac{48}{\binom{52}{5}}$$

since there are 48 possible hands with all four aces, one corresponding to each of the possibilities for the non-ace. Also

$$P(\text{4 of a kind dealt}) = \frac{(13)(48)}{\binom{52}{5}}$$

$$P(\text{no spade dealt}) = \frac{\binom{39}{5}}{\binom{52}{5}}$$

$$P(\text{at least 1 spade dealt}) = 1 - \frac{\binom{39}{5}}{\binom{52}{5}}$$

$$P(\text{5 spades dealt}) = \frac{\binom{13}{5}}{\binom{52}{5}}$$

$$P(\text{5 of some suit dealt}) = \frac{4\binom{13}{5}}{\binom{52}{5}}$$

The justifications of these results are left as exercises for the reader.

Example 3

A fair coin is tossed twice. What is $P(\text{exactly one head is tossed})$?

Solution

As before, we take $S = \{HH, HT, TH, TT\}$. Since the faces of the coin are symmetric, and since S "preserves the symmetry," the possible outcomes of S are equally likely, and hence

$$P(\text{exactly one head tossed}) = P(HT, TH) = \tfrac{2}{4} = \tfrac{1}{2}$$

Had we classified the possible results by the number of heads that appear (and thus taken $S = \{0, 1, 2\}$, in which the symmetry is not maintained), we might not have been able to exploit the symmetry of the coin and see that the probabilities of these possible outcomes are $\tfrac{1}{4}, \tfrac{1}{2}, \tfrac{1}{4}$, respectively. Having found the probabilities, we might then use the smaller S if so desired.

Example 4

A person is selected at random from a population of N people. We naturally take as our possible outcomes the N people; therefore, S is the population itself, and all persons are equally likely to be the person selected. Corresponding to the condition that the person selected possess a certain attribute (related, for example, to age, sex, height, weight, income, hair color, and so forth) is the subset A of S consisting of those people in the population who possess the attribute, and we have

$$P(\text{person selected has attribute}) = P(A)$$
$$= \frac{\text{number people in } A}{N}$$

Example 5

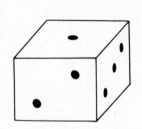

Figure 14-1

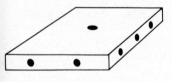

Figure 14-2

A die with one dimension shortened is rolled. See Figure 14-1. Suppose the 1 and 6 faces are the two larger square faces. Again $S = \{1, 2, 3, 4, 5, 6\}$, but obviously, the possible outcomes are not equally likely. (For an extreme case, see Figure 14-2.) The die, however, is symmetric with respect to the 1 and 6 faces and also with respect to the 2, 3, 4, and 5 faces. Consequently, we assume $P(1) = P(6) = a$, say, and $P(2) = P(3) = P(4) = P(5) = b$, say; that is, the probabilities of 1, 2, 3, 4, 5, 6 are a, b, b, b, b, a, respectively, where $2a + 4b = 1$. There is really only one undetermined number here since either a or b can be obtained from the other. We now have, for instance

$$P(\text{upface is odd}) = P(1, 3, 5) = a + 2b = \tfrac{1}{2}(2a + 4b) = \tfrac{1}{2}$$

and

$$P(\text{upface} \geq 3) = P(3, 4, 5, 6) = 3b + a$$
$$= 3\frac{1 - 2a}{4} + a = \frac{3 - 2a}{4}$$

We cannot determine a theoretically, even if we were given the amount by which the die was shortened. Some conclusions are possible, however. For instance, the square faces 1 and 6 are more likely than the rectangular faces 2, 3, 4, and 5; that is, $a > b$. To see this, fix attention on the down face that is rolled. It is more likely to be square (which is equivalent to the upface being square), since the square faces not only have greater area but correspond to more stable positions of the die. (See Figure 14-2.) From $a > b$, it follows that

$$2a + 4a > 2a + 4b = 1$$

that is, $6a > 1$, or $a > \frac{1}{6}$ and, similarly, $b > \frac{1}{6}$. In other words, the shortening increases the probabilities of the 1 and 6 faces, a fact which is quite obvious. Also, the greater the departure from a cube, the greater the probability a, until, in the extreme case where the height is almost 0, we would have $2a$ close to 1, or a close to $\frac{1}{2}$ (see Figure 14-2).

Thus, in the absence of any further information, our model involves an unknown number. But our model may still be very useful. The presence of such undetermined numbers or *parameters* is, in fact, typical in practical application, not only of probability theory but also of many other branches of mathematics.

There are other possibilities with regard to the determination of a. Suppose that we have had vast experience with this die (or identical dice) and have observed that in many thousands of rolls a square face has come up 40% of the time. It would seem reasonable then to take $2a = .4$, that is, $a = .2$, and hence $b = \dfrac{1 - 2a}{4} = .15$. There is also the intermediate possibility that we have some, but not very much, experience with the shortened die; for example, we might have observed that a square face comes up 40% of the time, but out of a total of only 50 tosses. We would expect, then, that $2a$ was fairly close to .4, but it would not be a firm conclusion. The field of *statistical inference* deals with precisely such situations in which there is only partial information.

Example 6

A fair die is rolled twice. We take

$$
S = \left\{
\begin{array}{l}
(1, 1), (1, 2), \ldots, (1, 6) \\
(2, 1), \quad \ldots, \ldots, (2, 6) \\
\quad \cdot \\
\quad \cdot \\
\quad \cdot \\
(6, 1), \quad \ldots, \ldots, (6, 6)
\end{array}
\right\}
$$

where the first and second components refer to the first and second rolls, respectively. We would use the same S for the chance process of one roll of two dice. The two components would refer to the two dice, with one die being regarded as "first" in some arbitrary but fixed way. On the basis of the physical symmetry, we take the possible outcomes to be equally likely. Let X be the sum of the two upfaces rolled. The possibilities for X are 2, 3, 4, ..., 11, 12. We have, for instance,

X = 6   if and only if
$$\{(1, 5), (2, 4), (3, 3), (4, 2), (5, 1)\} \text{ occurs}$$

and

X = 7   if and only if
$$\{(1, 6), (2, 5), (3, 4), (4, 3), (5, 2), (6, 1)\} \text{ occurs}$$

Hence,

$$P(X = 6) = \tfrac{5}{36}, \qquad P(X = 7) = \tfrac{6}{36} = \tfrac{1}{6}$$
$$P(X \neq 7) = 1 - \tfrac{1}{6} = \tfrac{5}{6}$$
$$P(X = 6 \text{ or } 7) = P(X = 6) + P(X = 7) = \tfrac{5}{36} + \tfrac{6}{36} = \tfrac{11}{36}$$

Example 7

A fair die is rolled five times. Naturally, we take the outcome of the process to be the ordered set of five upfaces. By the fundamental principle of counting, the number of possible outcomes is $6 \cdot 6 \cdot 6 \cdot 6 \cdot 6 = 6^5$, and, for the usual reasons, we attach equal probability to all of them. What is the probability that all five upfaces are the same? What is the probability that they are all different?

Solution

We count the possible outcomes with all upfaces the same. There are six choices for the first upface, and for each such choice there is only one choice for the other upfaces. Hence,

$$P(\text{all upfaces are same}) = \frac{6}{6^5} = \frac{1}{6^4}$$

To count the possible outcomes with all upfaces different, observe that there are six choices for the first upface and, for each of these, five choices for the second upface, and so on. Therefore,

$$P(\text{all upfaces are different}) = \frac{6 \cdot 5 \cdot 4 \cdot 3 \cdot 2}{6^5}$$

Let X be the number of upfaces that are aces. Then, letting $\overline{\text{ace}}$ denote non-ace,

$$P(X = 0) = P(\overline{\text{ace}} \; \overline{\text{ace}} \; \overline{\text{ace}} \; \overline{\text{ace}} \; \overline{\text{ace}}) = \frac{5^5}{6^5} = \left(\frac{5}{6}\right)^5$$

and, hence,

$$P(\text{at least one upface is an ace}) = P(X > 0) = 1 - (\tfrac{5}{6})^5$$

Observe that the event $X = 1$ breaks down into five *cases* (that is, 5 disjoint subsets):

$$
\begin{array}{ccccc}
\text{ace} & \overline{\text{ace}} & \overline{\text{ace}} & \overline{\text{ace}} & \overline{\text{ace}} \\
\overline{\text{ace}} & \text{ace} & \overline{\text{ace}} & \overline{\text{ace}} & \overline{\text{ace}} \\
\overline{\text{ace}} & \overline{\text{ace}} & \text{ace} & \overline{\text{ace}} & \overline{\text{ace}} \\
\overline{\text{ace}} & \overline{\text{ace}} & \overline{\text{ace}} & \text{ace} & \overline{\text{ace}} \\
\overline{\text{ace}} & \overline{\text{ace}} & \overline{\text{ace}} & \overline{\text{ace}} & \text{ace}
\end{array}
$$

All five cases have the same probability, so

$$P(X = 1) = 5P(\text{ace } \overline{\text{ace}}\ \overline{\text{ace}}\ \overline{\text{ace}}\ \overline{\text{ace}})$$

$$= 5 \cdot \frac{1 \cdot 5 \cdot 5 \cdot 5 \cdot 5}{6^5} = 5\left(\frac{1}{6}\right)\left(\frac{5}{6}\right)^4$$

The event $X = 2$ breaks down into $\binom{5}{2}$ equally likely cases, one for each choice of two positions for "ace" out of the five positions. Hence,

$$P(X = 2) = \binom{5}{2} P(\text{ace ace } \overline{\text{ace}}\ \overline{\text{ace}}\ \overline{\text{ace}})$$

$$= \binom{5}{2} \frac{1 \cdot 1 \cdot 5 \cdot 5 \cdot 5}{6^5}$$

$$= \binom{5}{2}\left(\frac{1}{6}\right)^2\left(\frac{5}{6}\right)^3$$

We find, in general,

$$P(X = k) = \binom{5}{k}\left(\frac{1}{6}\right)^k\left(\frac{5}{6}\right)^{n-k} \qquad k = 0, 1, \ldots, 5$$

The reader should check this for $k = 0, 5$.

Example 8

What is the probability that in a group of n people there are at least two with the same birthday?

Solution

We shall make some simplifying assumptions that are approximately valid and should lead to a reasonably accurate answer. First of all, we imagine that the people are selected at random from the population at large as far as birthdays are concerned. If we are considering the group of people present at a party for Virgos and Leos at the local astrology club, then this assumption is patently ridiculous. It is justified, however, if the birthdays of the people in the group have no bearing, direct or indirect, on their being in the group. We assume further that the birthdays of people as a whole are evenly distributed over the 365 days of the year. (We are ignoring February 29 birthdays.) This is only approximately true since seasonal factors do enter in the incidence of human births. Thinking of the people as numbered from 1 to n, the outcome of our chance process is the ordered sequence of the n birthdays. There are $(365)^n$ possible outcomes, and under our assumptions they are equally likely. The event that there is a common birthday in the group is the complement of the event that all the birthdays are different. Thus, the probability we seek is

$$1 - P(\text{all } n \text{ birthdays are different})$$

In complete analogy with the second problem in Example 7 (n corresponds to 5, 365 to 6), we have

$$P(\text{all the birthdays are different}) = \frac{(365)_n}{(365)^n}$$

$$= \frac{(365)(364) \cdots (365 - n + 1)}{365^n}$$

Hence,

$$P(\text{there is a common birthday}) = 1 - \frac{(365)_n}{(365)^n}$$

For $n = 10$, the probability is .12, it is approximately $\frac{1}{2}$ for $n = 23$, and for $n = 30$, it is .71.

Exercises 14-2

In Exercises 1–3, assume that a fair die is rolled once. Compute:

1. $P(\text{the upface} \geq 2)$ 2. $P(\text{the upface} \neq 6)$

3. $P[2 \leq (\text{the upface}) < 5]$

In Exercises 4–7, a hand of 13 cards is dealt at random from an ordinary deck of 52 cards. Compute:

4. P(all spades are dealt)

5. P(four aces are dealt)

6. P(all black cards are dealt)

7. P(at least one spade is dealt)

In Exercises 8–12, assume that a fair coin is tossed three times. What is the probability that:

8. exactly one head is tossed?

9. exactly two heads are tossed?

10. three heads are tossed?

11. at least one head is tossed?

12. at least two heads are tossed?

13. A die is loaded so that 1 and 6 are each twice as likely to show up as any of the other numbers. What then is the probability that a number greater than 2 shows up if the die is rolled once?

In Exercises 14–17, assume that two fair dice are rolled. Let X be the sum of the upfaces. Compute:

14. $P(X = 3)$ **15.** $P(X \neq 3)$

16. $P(X > 3)$ **17.** $P(2X + 1 = 5)$

In Exercises 18–28, assume that a fair die is rolled ten times. Let X be the number of upfaces that are greater than 4. Compute:

18. $P(X = 0)$ **19.** $P(X = 1)$

20. $P(X = 2)$ **21.** $P(X = 3)$

22. $P(X = 7)$ **23.** $P(X = 10)$

24. $P(X \neq 10)$ **25.** $P(X > 0)$

26. $P(1 \leq X \leq 3)$ **27.** $P(X \geq 0)$

28. $P(3X - 1 < 8)$

29. What is the probability that at least two people in a group of n people were born in the same month? (Make the simplifying assumptions of Example 8.)

14-3 DEPENDENT AND INDEPENDENT TRIALS

It should not be inferred from our choice of examples so far that the main application of probability theory is to games of chance, although the origins of the theory do lie there. This would be far from the case. Probability theory and its offshoot, statistical inference, have today an enormous and rapidly expanding range of applications in science and industry. Rather, we have chosen these examples for their value in illustrating the basic ideas in simple settings, the more important applications usually being too complicated. The study of the somewhat more involved chance processes that we will examine now will provide access to more practical applications.

We consider chance processes built up successively from component chance processes, called **trials.** The outcome of the first trial determines what the second trial will be, and the ordered pair of outcomes of the first two trials determines what the third trial will be (if there is one), and so on. The (overall) outcome of the chance process is the **ordered set** of outcomes of the trials.

Example 1

A form of Russian roulette is played with three revolvers, each having one bullet, but varying numbers of chambers, namely, 4, 5, and 6. A gun is selected at random, the cylinder is spun, and the trigger is pulled. This is a chance process consisting of two trials. The first trial is the selection of the gun, and since this selection is made at random, the three possible outcomes, 4, 5, 6 (with the obvious meaning) are equally likely. Accordingly, as the outcome of the first trial is 4, 5, or 6, the second trial is a chance process with 4, 5, or 6 equally likely possible outcomes (the actual outcome being the chamber that lines up with the barrel). We are assuming here that the weight of the bullet in each gun has a negligible effect on the symmetry of the cylinder with respect to the chamber. Also, let us denote the chambers of gun 4 (that is, the gun with four chambers) by 1, 2, 3, 4 and similarly for guns 5 and 6, in each case letting 1 denote the occupied chamber. The set of possible outcomes of the overall process is the set of ordered pairs

$$S = \left\{ \begin{array}{l} (4, 1), (4, 2), (4, 3), (4, 4) \\ (5, 1), \quad \ldots \quad , (5, 4), (5, 5) \\ (6, 1), \quad\quad \ldots \quad\quad , (6, 6) \end{array} \right\}$$

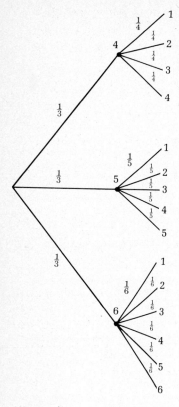

Figure 14-3

Now, how do we put the probabilities associated with the trials together to get the probabilities of the possible outcomes of S? Take $(4, 2)$, for instance. The probability that gun 4 is selected is $\frac{1}{3}$; furthermore, **given** that gun 4 is selected, the probability that chamber 2 is selected is $\frac{1}{4}$. It is intuitively evident that the probability of the outcome $(4, 2)$ is $\frac{1}{3} \cdot \frac{1}{4} = \frac{1}{12}$. The probabilities of all the other possible outcomes are similarly calculated.

The construction of S and its probability distribution can be represented in a tree diagram (see Figure 14-3). We can think of the total probability 1 being distributed as follows: $\frac{1}{3}$ each to the disjoint subsets

$$A_4 = \{(4, 1), \ldots, (4, 4)\}, A_5, A_6$$

corresponding to the selection of gun 4, 5, or 6, respectively; and then within each of A_4, A_5, and A_6 the probability $\frac{1}{3}$ is distributed according to the probabilities of the possible outcomes of the corresponding second trial. We have

$$1 = \tfrac{1}{3} + \tfrac{1}{3} + \tfrac{1}{3} = \tfrac{1}{3}(\tfrac{1}{4} + \tfrac{1}{4} + \tfrac{1}{4} + \tfrac{1}{4}) + \tfrac{1}{3}(\tfrac{1}{5} + \tfrac{1}{5} + \tfrac{1}{5} + \tfrac{1}{5} + \tfrac{1}{5})$$
$$+ \tfrac{1}{3}(\tfrac{1}{6} + \tfrac{1}{6} + \tfrac{1}{6} + \tfrac{1}{6} + \tfrac{1}{6} + \tfrac{1}{6})$$
$$= \tfrac{1}{12} + \tfrac{1}{12} + \tfrac{1}{12} + \tfrac{1}{12} + \tfrac{1}{15} + \tfrac{1}{15} + \tfrac{1}{15} + \tfrac{1}{15} + \tfrac{1}{15}$$
$$+ \tfrac{1}{18} + \tfrac{1}{18} + \tfrac{1}{18} + \tfrac{1}{18} + \tfrac{1}{18} + \tfrac{1}{18}$$

which equals the sum of probabilities of all possible outcomes. Now,

$$P(\text{gun fires}) = P[(4, 1), (5, 1), (6, 1)]$$
$$= \frac{1}{3} \cdot \frac{1}{4} + \frac{1}{3} \cdot \frac{1}{5} + \frac{1}{3} \cdot \frac{1}{6} = \frac{1}{3}\left(\frac{15 + 12 + 10}{60}\right)$$
$$= \frac{37}{180}$$

□

The techniques of the preceding example apply in general. Let O be a possible outcome of the first trial, having probability p, say, in that trial, and let O' be a possible outcome of that second trial, which is determined by the occurrence of O, and probability p' in that trial. Then the probability that (O, O') occurs is pp'. If there is a third trial, O'', then the probability that (O, O', O'') occurs is $pp'p''$, where p'' is the probability that O'' occurs in that third trial, which is determined by the occurrence of (O, O') in the first two trials. The same principle extends to any number of trials.

A very important special case arises when *the second trial is the same chance process regardless of the outcome of the first trial.* The trials are then said to be **independent.** The definition extends to any number of trials, n. The essential point is that the kth trial is the same chance process regardless of the outcomes of the preceding $k - 1$ trials, for $k = 2, 3, \ldots, n$. Of course, in the case of independent trials, their time order is irrelevant. They can take place simultaneously, or even in reverse order for that matter. For example, the roll of a die followed by the toss of a coin is the same, as far as our model is concerned, as the simultaneous roll of a die and toss of a coin.

We now explore some further examples of dependent trials.

Example 2

It is known that fraternal twins are about twice as likely to occur as identical twins. What, approximately, is the probability that a pair of twins will be of the same sex?

Solution

The probabilities are $\frac{1}{3}$ and $\frac{2}{3}$ that the twins are identical and fraternal, respectively. If the twins are identical, they are certain to be of the same sex. If they are fraternal, the probability is about $\frac{1}{2}$, since both boys or both girls are 2 of the 4 approximately* equally likely possibilities. (See Figure 14-4.) It follows that

$$P(\text{twins are of same sex}) \approx \tfrac{1}{3} \cdot 1 + \tfrac{2}{3} \cdot \tfrac{1}{2} = \tfrac{2}{3}$$

*Male births are slightly (yet definitely) more likely than female births—about 51% vs. 49%.

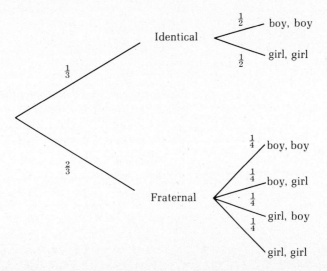

Figure 14-4

Example 3

For the sake of illustration, imagine that every person has either 0, 1, or 2 children, with fixed probabilities p_0, p_1, p_2, respectively, where $p_0 + p_1 + p_2 = 1$. Let X be the number of grandchildren of a given person. Find $P(X = h)$, for $h = 0, 1, 2, 3, 4$.

Solution

We take

$$S = \begin{cases} (0, 0) \\ (1, 0), (1, 1), (1, 2) \\ (2, 0), (2, 1), (2, 2), (2, 3), (2, 4) \end{cases}$$

where the first component of each possible outcome is the number of children of the given person and the second component is the number of their children, that is, the number of grandchildren. We immediately have

$$P(0, 0) = p_0 \cdot 1$$
$$P(1, 0) = p_1 p_0$$
$$P(1, 1) = p_1 p_1$$
$$P(1, 2) = p_1 p_2$$

To get the probabilities of the other possible outcomes (in terms of p_0, p_1, p_2, that is), we need the probabilities p_i', say, of i grandchildren if there are two children. See Figure 14-5.

We now consider the subsidiary process that arises when the given person has two children. The possibilities for the numbers of offspring of two persons are:

$$(0, 0) \quad (0, 1) \quad (0, 2)$$
$$(1, 0) \quad (1, 1) \quad (1, 2)$$
$$(2, 0) \quad (2, 1) \quad (2, 2)$$

[(1, 2), for example, now means one person has 1 offspring and the other has 2] and their probabilities are

$$p_0 p_0 \quad p_0 p_1 \quad p_0 p_2$$
$$p_1 p_0 \quad p_1 p_1 \quad p_1 p_2$$
$$p_2 p_0 \quad p_2 p_1 \quad p_2 p_2$$

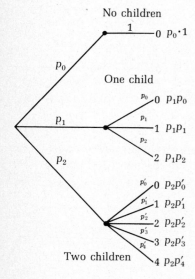

Figure 14-5

Therefore,

$$p'_0 = p_0 p_0 = p_0^2$$
$$p'_1 = p_0 p_1 + p_1 p_0 = 2 p_0 p_1$$
$$p'_2 = p_0 p_2 + p_1 p_1 + p_2 p_0 = 2 p_0 p_2 + p_1^2$$
$$p'_3 = p_1 p_2 + p_2 p_1 = 2 p_1 p_2$$
$$p'_4 = p_2 p_2 = p_2^2$$

Now the event that $X = 0$ is $\{(0, 0), (1, 0), (2, 0)\}$, so

$$P(X = 0) = p_0 + p_1 p_0 + p_2 p_0^2$$

Similarly,

$$P(X = 1) = p_1 p_1 + p_2(2 p_0 p_1) = p_1^2 + 2 p_0 p_1 p_2$$
$$P(X = 2) = p_1 p_2 + p_2(2 p_0 p_2 + p_1^2) = p_1 p_2 + 2 p_0 p_2^2 + p_1^2 p_2$$
$$P(X = 3) = p_2(2 p_1 p_2) + 2 p_1 p_2^2$$
$$P(X = 4) = p_2 p_2^2 = p_2^3$$

Example 4

A fair die is rolled until an ace appears. Find

P(ace appears on an even numbered roll)

Solution

We take $S = \{a, \bar{a}a, \bar{a}\bar{a}a, \ldots\}$ where a, $\bar{a}$ denote ace, nonace, respectively. On a single roll, the probability that the upface is an ace is $\frac{1}{6}$, and the probability that it is not is $1 - \frac{1}{6} = \frac{5}{6}$. We have

$$P(a) = \tfrac{1}{6}, \qquad P(\bar{a}a) = \tfrac{5}{6} \cdot \tfrac{1}{6}, \qquad P(\bar{a}\bar{a}a) = \tfrac{5}{6} \cdot \tfrac{5}{6} \cdot \tfrac{1}{6},$$

$$P(\underbrace{\bar{a} \ldots \bar{a}a}_{n-1 \text{ times}}) = (\tfrac{5}{6})^{n-1} \cdot \tfrac{1}{6}, \qquad n \in N$$

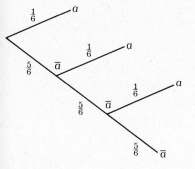

We can make a tree diagram here also (see Figure 14-6). The sum of the probabilities, that is, the sum of the infinite series $\frac{1}{6} + \frac{5}{6} \cdot \frac{1}{6} + (\frac{5}{6})^2 \cdot \frac{1}{6} + \cdots$ gives the probability that an ace will

Figure 14-6

eventually appear. In fact, the sum, being a geometric series with first term $\frac{1}{6}$ and common ratio $\frac{5}{6}$, is

$$\frac{1}{6} \cdot \frac{1}{1 - \frac{5}{6}} = 1$$

(see Section 4-4). Were this not so, S would not be exhaustive, and we would have to include as a possible outcome $\bar{a}\bar{a}\bar{a} \ldots$, that is, ace never appearing. Now,

$$P(\text{ace appears on an even roll})$$
$$= P(\bar{a}a, \bar{a}\bar{a}\bar{a}a, \ldots)$$
$$= \tfrac{5}{6} \cdot \tfrac{1}{6} + (\tfrac{5}{6})^3 \cdot \tfrac{1}{6} + (\tfrac{5}{6})^5 \cdot \tfrac{1}{6} + \cdots$$
$$= \tfrac{5}{6} \cdot \tfrac{1}{6}[1 + (\tfrac{5}{6})^2 + (\tfrac{5}{6})^4 + \cdots]$$
$$= \frac{5}{6} \cdot \frac{1}{6} \frac{1}{1 - (\tfrac{5}{6})^2} = \frac{5}{36 - 25} = \frac{5}{11}$$

The following examples deal with independent trials.

Example 5

The shortened die of Example 5 in Section 14-2 and a fair die are rolled. Let $X =$ sum of upfaces rolled; then

$$P(X = 6) = P[(1, 5), (2, 4), (3, 3), (4, 2), (5, 1)]$$
$$= a\tfrac{1}{6} + b\tfrac{1}{6} + b\tfrac{1}{6} + b\tfrac{1}{6} + b\tfrac{1}{6} = \tfrac{1}{6}(a + 4b)$$
$$= \tfrac{1}{6}(2a + 4b - a) = \tfrac{1}{6}(1 - a)$$

and

$$P(X = 7) = a\tfrac{1}{6} + b\tfrac{1}{6} + b\tfrac{1}{6} + b\tfrac{1}{6} + b\tfrac{1}{6} + a\tfrac{1}{6}$$
$$= \tfrac{1}{6}(2a + 4b) = \tfrac{1}{6}$$

Example 6

In the sections of electrical circuits shown in Figure 14-7, $\bigcirc$ and $\square$ represent components which are subject to chance failure. Whether a component fails or not is assumed to depend only on its internal constitution. Consequently, the associated trials, one for each component, are independent of each other. Let p and p', respectively, be the probabilities that $\bigcirc$ and $\square$ fail during the desired period of operation of the circuit. In each case below, we calculate the probability that the section works, that is, does not break the circuit.

(a) $P(\text{section works}) = P(\text{both } \bigcirc\text{'s work}) = [P(\bigcirc \text{ works})]^2$
$$= [1 - P(\bigcirc \text{ fails})]^2 = (1 - p)^2$$

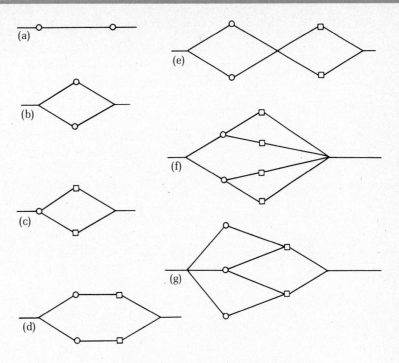

Figure 14-7

(b) P(section works) $= P$(not both $\bigcirc$'s fail)
$$= 1 - P(\text{both } \bigcirc\text{'s fail}) = 1 - p^2$$

(c) P(section works) $= P(\bigcirc$ works and not both $\square$'s fail)
$$= P(\bigcirc \text{ works})P(\text{not both } \square\text{'s fail})$$
$$= (1 - p)(1 - p'^2)$$

(d) P(section works) $= P$(not both-$\bigcirc$—$\square$-'s fail)
$$= 1 - [P(\text{-}\bigcirc\text{—}\square\text{- fails})]^2$$
$$= 1 - [1 - P(\bigcirc, \square \text{ both work})]^2$$
$$= 1 - [1 - P(\bigcirc \text{ works}) P(\square \text{ works})]^2$$
$$= 1 - [1 - (1 - p)(1 - p')]^2$$

(e) P(section works) $= P$(left half works)P(right half works)
$$= (1 - p^2)(1 - p'^2)$$

(f) P(section works) $= 1 - P$(upper and lower half both fail)
$$= 1 - [P(\text{upper half fails})]^2$$
$$= 1 - [1 - (1 - p)(1 - p'^2)]^2 \text{ by (c)}$$

(g) P(section works) $= P$(works and middle $\bigcirc$ works)
$$\qquad\qquad + P(\text{works and middle } \bigcirc \text{ fails})$$
$$= (1 - p)P(\text{not both } \square\text{'s fail})$$
$$\qquad\qquad + pP[(\text{d}) \text{ works}]$$
$$= (1 - p)(1 - p'^2) + p1$$
$$\qquad\qquad -p[1 - (1 - p)(1 - p')]^2$$

Exercises 14-3

1. What is the probability that a pair of fraternal twins will both be girls? (*Hint:* See Example 2.)

2. A fair coin is tossed ten times. What is the probability that heads show up exactly five times?

3. A single die is thrown five times. What is the probability that the up face is a six exactly three times?

4. An urn contains five red balls and five black balls. A ball is drawn and *not* replaced. This is done six times. What is the probability that exactly three red balls are drawn?

5. Two teams, call them A and B, play a series until one team has won three games. What is the probability that A wins the series, given that in each game the probability is two-thirds that A will win? (*Hint:* P[A wins] = P[A wins in exactly three games] + P[A wins in exactly four games] + P[A wins in exactly five games].)

6. Two evenly matched players play a set of tennis. What is the probability that the final score is 7–5? (*Note:* To win a set, one player must win at least six games *and* win by at least two games.)

7. (Gambler's ruin) A gambler with an initial capital of $2 repeatedly bets $1 that a tossed coin will show up heads until he goes broke or has $3. What is the probability that he goes broke?

8. A fair coin is tossed until heads appears. Compute:
(a) P(the coin is tossed exactly 5 times)
(b) P(the coin is tossed an even number of times)
(c) P(the coin is tossed < 5 times)
(d) P(the coin is tossed > 5 times)

9. A, B, and C toss a coin until one of them tosses heads. This person is the winner. What is the probability that C wins?

◇**10.** A missing person is in one of two regions, A or B. The probability that he is in region A is $\frac{4}{5}$. Ten helicopters are available independently to search the regions. If the person is in the region that the helicopter searches, the probability that the helicopter will spot him is $\frac{1}{2}$. Each helicopter can search in only one region. How many helicopters should be assigned to region A?

Answers to Odd-Numbered Exercises

Answers 1-2

1. -27

3. -10

5. $-\frac{1}{55}$

7. $\frac{1}{6}$

9. $-\frac{714}{95}$

11. $\frac{87}{290}$ (i.e., $\frac{3}{10}$)

13. $\frac{2680}{153}$

15. $3xy - 4x$

17. $x^2 + 2x + 1$

19. $5x^2 + 5$

21. $2x$

23. $x = -\frac{4}{3}$

25. $x = 2$

27. $x = 1$ or π or -3

29. $x = \frac{1}{15}$ or 15

31. $x = -\frac{1}{10}$

33. $-\frac{1}{5}$

35. $x = \frac{14}{13}$

37. 1

41. $\dfrac{4a - b}{7a - 7b + 1}$

43. $\dfrac{axy + axz - xz}{axy + axz - xz + ay}$

45. True

47. False

49. True

Answers 1-3

1. False

3. True

5. True

7. True

9. True

11. False or c positive

13. $x \geq 3$

15. $-5x - 10 \geq 3x + 4$
$$-8x \geq 14$$
$$-x \geq \tfrac{14}{8} = \tfrac{7}{4}$$
$$x \leq -\tfrac{7}{4}$$

17. $x > \tfrac{3}{2}$

19. $\quad 4 < 3x + 9 \quad$ or $\quad 3x + 9 < 5$
$$-5 < 3x \qquad\qquad 3x < -4$$
$$-\tfrac{5}{3} < x \qquad\qquad x < -\tfrac{4}{3}$$
Thus $\quad -\tfrac{5}{3} < x < -\tfrac{4}{3}$

21. $3 < x < 5$

23. $\tfrac{2}{5} < x < \tfrac{4}{5}$

25. $-3 \leq x \leq 3$

27. $2x - 4 = 0 \quad$ or $\quad y - 3 = 0$
$$x = 2 \quad \text{and} \quad y = 3$$

29. $x > 4 \quad$ or $\quad x < -3$

31. $\dfrac{-1 - 3x}{x - 2} \geq 0$

$x - 2 > 0 \;$ and $\; -1 - 3x \geq 0$
$$x > 2, \qquad -1 \geq 3x$$
$$-\tfrac{1}{3} > x$$
$x - 2 < 0 \;$ and $\; -1 - 3x \leq 0$
$$x < 2, \qquad -1 \leq 3x$$
$$-\tfrac{1}{3} \leq x$$
$$-\tfrac{1}{3} \leq x < 2$$

33. $x \leq -3 \quad 2 \leq x \leq 3$

35. True

37. False

39. True

Answers 1-4

1. 3^6

3. 2^6

5. 6^0 (i.e., 1)

7. 3^5

9. 6^2

11. 3^{-1}, or, $\tfrac{1}{3}$

13. $\dfrac{3^2 + 2^2}{2^3 \times 3^2}$

15. $x^{-3}y^{15}$

17. $\dfrac{1}{x^6 y^6}$

19. $\dfrac{3y}{2x}$

21. $\dfrac{y}{4z^4}$

23. $\dfrac{x}{y^n}$

25. x^{2+n}

27. $x^{n-5}y^{3-n}$

29. 2^2

31. $xy(x^{-1} + y^{-1}) = xy\left(\dfrac{1}{x} + \dfrac{1}{y}\right) = y + x, \ x \neq 0, \ y \neq 0$

33. $\dfrac{(x^{-1} + y^{-1})(x + y)^{-1}}{(xy)^{-1}} = \dfrac{\left(\dfrac{1}{x} + \dfrac{1}{y}\right)\left(\dfrac{1}{x+y}\right)}{\dfrac{1}{xy}}$

$$\begin{cases} x \neq 0, \ y \neq 0, \ x + y \neq 0, \\ xy \neq 0 \end{cases}$$

$$= xy\left(\dfrac{1}{x} + \dfrac{1}{y}\right)\left(\dfrac{1}{x+y}\right)$$

$$= (y + x)\left(\dfrac{1}{x+y}\right)$$

$$= 1$$

35. True

37. True

39. False

Answers 1-5

1. 12

3. 3

5. -2

7. $\dfrac{-4}{5}$

9. $2\sqrt{6}$

11. $5\sqrt{3}$

13. $-12\sqrt[3]{4}$

15. 1

17. $(18^3)\sqrt{3}$

19. $(3^3)\sqrt{5}$

21. $3^5 \cdot 2^4 \sqrt[5]{3^3 \cdot 2^2}$

23. 4

25. $2\sqrt{2}$

27. $9\sqrt{6}$

29. $4 - 8\sqrt{2}$

31. $20\sqrt{3} + 9\sqrt{2} - 15\sqrt{15}$

33. $\frac{5}{2}\sqrt{3} - 2$

35. $x^2 y \sqrt{xy}$

37. $-xyz^{-2}\sqrt[7]{x^2(-y)^4}$

39. $\dfrac{x^2}{zy}\sqrt[16]{\dfrac{1}{zy^4}}$

41. $(x - y)^2$

43. $(-3x)(yz)^{-1}\sqrt[7]{(-3x)^2(yz)^{-1}}$

45. $(x - 2x^2)\sqrt{x}$

47. $(x^2 + 3x)\sqrt[3]{x^2 y}$

49. 9, -5

51. $x \geq -2$ or $x \leq -6$

53. $-4 \leq x \leq 3$

55. True

57. True

59. False

61. $\frac{7}{2}, -\frac{3}{2}$

Answers 1-6

1. 25

3. 2

5. -1024

7. $\sqrt[5]{\frac{1}{2}}$

9. $2\sqrt[3]{9}x^6y^4z^4$

11. $\frac{1}{4}x^3\sqrt[3]{y}$

13. $2\sqrt[3]{x}\,y^2$

15. 4

17. $\dfrac{\sqrt[7]{x^6}}{2\sqrt[5]{y}}$

19. x^m

21. True

23. False

Answers 1-7

1. (Details for Example 2, special product 3)

$$(x - y)(x^2 + xy + y^3) = x(x^2 + xy + y^3) - y(x^2 + xy + y^2)$$
$$= (x^3 + x^2y + xy^3) - (yx^2 + xy^2 + y^3)$$
$$= x^3 + x^2y + xy^3 - yx^2 - xy^2 - y^3$$
$$= x^3 - y^3$$

(Details for special product 7)

$$(x + y)^2 = (x + y)(x + y)$$
$$= (x + y)x + (x + y)y$$
$$= (x^2 + xy) + (xy + y^2)$$
$$= x^2 + xy + xy + y^2$$
$$= x^2 + 2xy + y^2$$

3. $x^6 - y^6$

5. 0 if $x \geq 0$; $4x^2$ if $x < 0$

7. $x^3 + 27$

9. $x^4 - 81$

11. $27x^3 - 243x^2 + 729x - 729$

13. $2(2x - 3)(x - 2)$

15. $(3x + 2)(2x - 3)$

17. $(x + \sqrt{7})(x - \sqrt{7})$

19. $(2x^2 - 3)(x - 2)$

21. $(x^2y + xy^2)(x^2y - xy^2) = x^2y^2(x + y)(x - y)$

23. $(x^2 + 3y^2)(xy - 1)$

25. True

27. True

Answers 1-8

3. $\dfrac{4y - 10x^2}{5x^5y^2}$

5. $\dfrac{xy - y^3 + y^4 + x^2}{x^2y^2}$

7. $\dfrac{-2x^2 + 8x - 3}{x(-6x^2 + 5x - 1)}$

9. $\dfrac{x^2 + 11x + 11}{2x^3 + 3x^2 - 2x - 3}$

11. $\dfrac{-16x^4 - 4x^3 - 4x^2 - 17x - 1}{16x^4 - 8x^2 + 1}$

13. $\dfrac{x + 9}{x + 5}$

15. $\dfrac{x + 2}{x}$

17. $\dfrac{-1}{x^2 - 5x + 7}$

19. $\dfrac{1 - b}{a}$

Answers 1-9

1. $x = 9, -\frac{3}{2}$ **3.** $x = -3, -4$

5. $x = 0, 2$ **7.** $x = 0, 1$

9. $x = \dfrac{-2}{3}$ **11.** $x = 3, -2, 0$

13. $x = \pm 3, \pm 2$ **15.** $x = 3, \pm 2$

17. $x = -3$ **19.** $x = \dfrac{-10}{3}$

21. $x = -3$ **23.** $x = \pm \sqrt{\frac{11}{2}}$

25. $x = 4, -6, 3$

27. $-\sqrt{7} \le x \le -1,$ or, $1 \le x \le \sqrt{7}$

Answers 1-10

1. $x = 2 + \sqrt{6}, x = 2 - \sqrt{6}$

3. No solution

5. $x = \dfrac{3 + \sqrt{37}}{4}, x = \dfrac{3 - \sqrt{37}}{4}$

7. $x = -3, x = \frac{5}{3}$ **9.** $x = -\frac{5}{2}$

11. $x = \frac{1}{6}$ **13.** No solution

15. $x = \dfrac{-1 + \sqrt{89}}{4}, x = \dfrac{-1 - \sqrt{89}}{4}$

17. No solution   **19.** $x = \dfrac{\sqrt{6}}{3}$

21. $x = -5$ **23.** $\pm \dfrac{2\sqrt{3}}{3}, \pm \sqrt{\frac{2}{5}}$

25. $a = 1, b = -14, c = 45$

Answers 2-2

1. $\sqrt{2}$ **3.** $\sqrt{20}$

5. $\sqrt{2}|a - b|$ **7.** $\sqrt{2}(a - 1)^2$

9. -1 and 2

11. Use the converse of the Pythagorean Theorem.

13. $(2 + 2\sqrt{3}, 4 - \sqrt{3})$, or $(2 - 2\sqrt{3}, 4 + \sqrt{3})$

15. $6x - 4y = 1$

Answers 2-3

1.

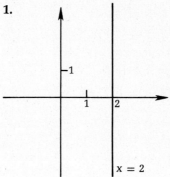

3.

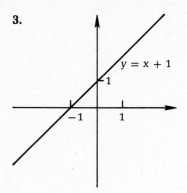

5.

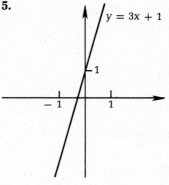

7.

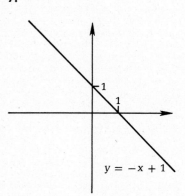

9.

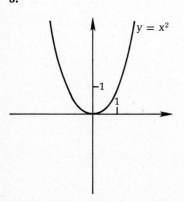

11.

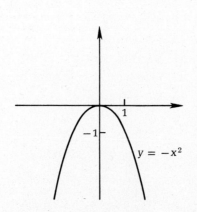

13.

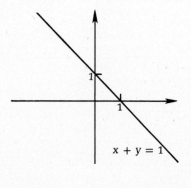

15. 1 circle: center $(1, -3)$, radius 2

17. Interior and boundary of the circle whose center is $(-2, 3)$ and whose radius is $\sqrt{2}$

19. Circle, center at $(-3, 2)$, radius $3\sqrt{2}$

21. A point: $(\frac{3}{2}, 1)$

23. Circle: center $(\frac{1}{2}, -\frac{1}{4})$, radius $\frac{3}{4}$

25. Circle: center $(\frac{3}{10}, -\frac{1}{10})$, radius $\sqrt{\frac{87}{500}}$

27. $x^2 + y^2 - 2y = 24$ **29.** $x^2 - 2x + y^2 = 15$

31.

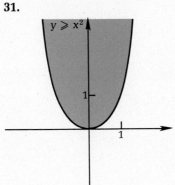

33.

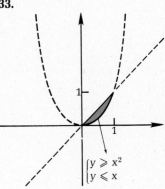

35.

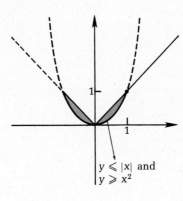

37.

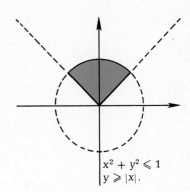

39.

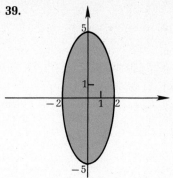

41.

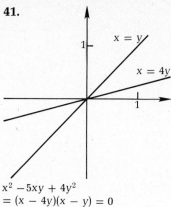

$$x^2 - 5xy + 4y^2$$
$$= (x - 4y)(x - y) = 0$$

43.

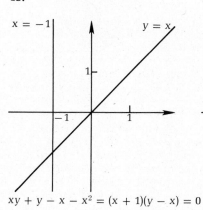

$$xy + y - x - x^2 = (x + 1)(y - x) = 0$$

45.

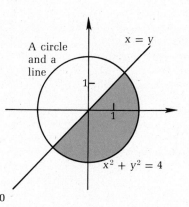

A circle
and a
line

$$x^2 + y^2 = 4$$

47.

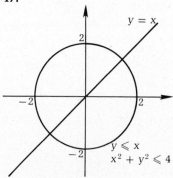

$$y \leqslant x$$
$$x^2 + y^2 \leqslant 4$$

49.

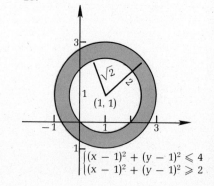

$$\begin{cases} (x - 1)^2 + (y - 1)^2 \leqslant 4 \\ (x - 1)^2 + (y - 1)^2 \geqslant 2 \end{cases}$$

51. $\dfrac{x^2}{a^2 - c^2} + \dfrac{y^2}{a^2} = 1$

Answers 2-4

1. 0

3. 2

5. $-6 + 8i$

7. $5i$

9. $(-5, 3)$

11. $(\sqrt{6} - \sqrt{4}, 2)$

13. -10

15. $-1 - i2\sqrt{2}$

17. $-3 + 4i$

19. $7i$

21. 13

23. $\frac{1}{3}i$

25. $\frac{1}{5} + \frac{2}{5}i$

27. $-\frac{3}{34} + \frac{5}{34}i$

29. i

31. 5

33. -1

35. $-1 \pm \sqrt{3}i$

37. $\pm \pi i$

39. $\frac{-1 \pm \sqrt{3}i}{2}$

41. $\frac{1 \pm \sqrt{11}i}{6}$

43. $2, -1 \pm \sqrt{3}i$

45. It is a real number

47. *Hint:* Consider x + iy = x − iy.

53. 2

55. 1, 2

57. 1, -3

59. 4

61. 2

Answers 3–1

1. 0

3. 0

5. 210

7. $\sqrt{2}$

9. $786 + 556\sqrt{2}$

11. $7750\sqrt{2} + 6328\sqrt{3}$

13. 1

15. 7

17. 73

19. $\frac{193}{256}$

21. 1

23. 0

25. $\left(\frac{13}{16}\right)^3 - \frac{13}{16} = -\frac{1131}{4096}$

27. $\frac{10421}{9840} \approx 1.06$

29. ± 4

31. $-\frac{4}{3}$

33. $-\frac{1}{3}$ and 1

35. All reals

37. $[3, 5]$

39. $(-\infty, 2]$

41. 3

43. *(a)* $\begin{cases} (1) \ f: f(a) = 1, f(b) = 1, f(c) = 1 \\ (2) \ f: f(a) = 2, f(b) = 2, f(c) = 2 \end{cases}$

(b) $\begin{cases} (1) \ f: f(a) = 1, f(b) = 1, f(c) = 2 \\ (2) \ f: f(a) = 1, f(b) = 2, f(c) = 2 \\ (3) \ f: f(a) = 2, f(b) = 1, f(c) = 1 \\ (4) \ f: f(b) = 2, f(b) = 2, f(c) = 1 \\ (5) \ f: f(a) = 2, f(b) = 1, f(c) = 2 \\ (6) \ f: f(a) = 1, f(b) = 2, f(c) = 1 \end{cases}$

Answers 3-2

1.

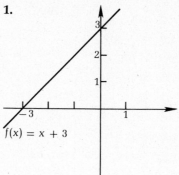

$f(x) = x + 3$

3.

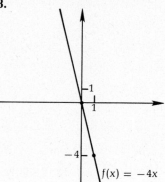

$f(x) = -4x$

5.

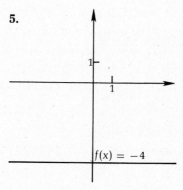

$f(x) = -4$

7.

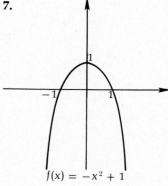

$f(x) = -x^2 + 1$

9.

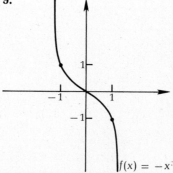

$f(x) = -x^3$

11.

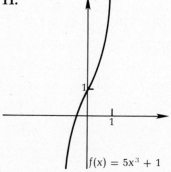

$f(x) = 5x^3 + 1$

13.

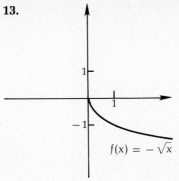

$$f(x) = -\sqrt{x}$$

15.

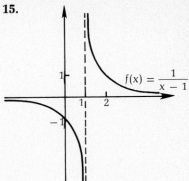

$$f(x) = \frac{1}{x-1}$$

17. $x \neq 1, x \neq 2$ **19.** $x > -\frac{4}{3}$ **21.** $x > 2$

23. $D: x \geq -\frac{3}{2}, R: y > 0$

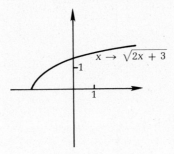

$$x \to \sqrt{2x + 3}$$

25. $D:$ all real numbers, $R: y \geq 7$

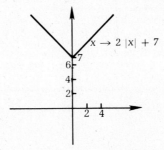

$$x \to 2 |x| + 7$$

27. $D: |x| > \sqrt{\frac{5}{2}}, \ R: y > 0$

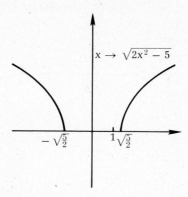

29.

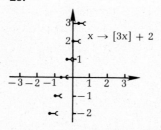

31.

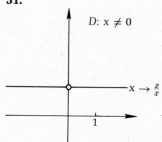

33.

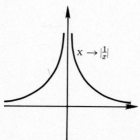

35.

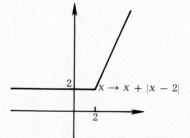

37. Yes, $x \to -1$

39. No

41.

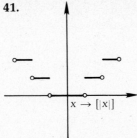

$x \to [|x|]$

43.

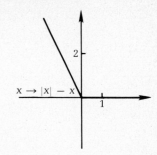

$x \to |x| - x$

45. Note that $\dfrac{x^2}{|x|} = x$ if $x > 0$, $= -x$ if $x < 0$, and is undefined at 0. Hence, graph of $x \to \dfrac{x^2}{x}$ is graph of $x \to |x|$ with $(0,0)$ removed.

47. $f(x) = g(x)$ (i.e., $|x| = [x]$) when $x = 0$ or a positive integer. To see when $f(g(x)) = g(f(x))$, compare graphs obtained in problems 37 and 38: $f(g(x)) = g(f(x))$ for $x \geq 0$, or x a negative integer.

Answers 3-3

1. $x \to x^3 + 1$

3. $x \to -3x^3$

5. $x \to 3x^3 + 1$

7.

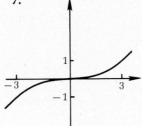

9.

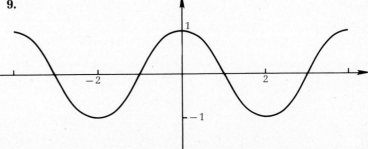

11.

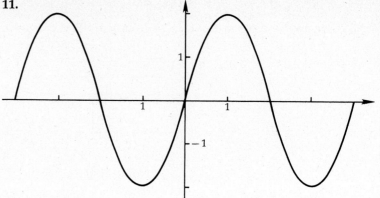

13.

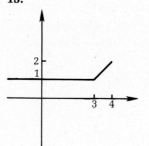

15.

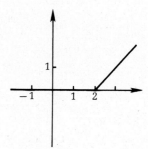

17.

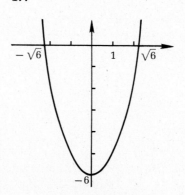

at $-\sqrt{6}$ and $\sqrt{6}$

19.

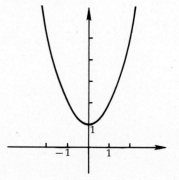

21.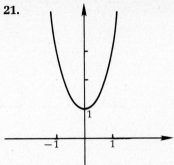

23.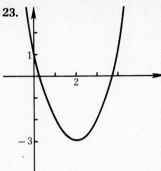

25.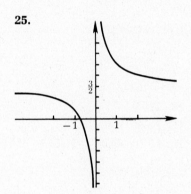

27.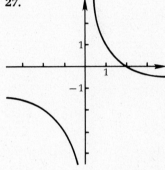

29.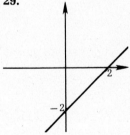

31. $\dfrac{-1}{2}$

33. No solution   **35.** 0

37. $x \to x^2 - 6x + 7$

39. $g(x) = 2x^2 + 7x + 1, \ \dfrac{-7 \pm \sqrt{41}}{4}$

Answers 4-1

1. 3, 5, 9, 17, 33

3. $\frac{5}{6}, \frac{13}{36}, \frac{35}{216}, \frac{97}{1296}, \frac{275}{7776}$

5. $-\frac{1}{2}, \frac{2}{3}, -\frac{3}{4}, \frac{4}{5}, -\frac{5}{6}$

7. $a_n = 7 - 2(n - 1) = 9 - 2n$

9. $a_n = 5^{n-1} \cdot 3 = \frac{3}{5} \cdot 5^n$　　　**11.** $a_n = 1$

13. $a_n = 3n - 2$　　　**15.** $a_n = 2^n$

17. $a_n = 1/n^2$

19. $a_2 < ra_1 = r$;
$a_3 < ra_2 < r^2 a_1 = r^2$, etc.

21. $\left(\frac{21}{20}\right)^{10} \cdot 1000$

Answers 4-2

1. $\frac{25}{12}$　　　**3.** $3 + \sqrt{2} + \sqrt{3}$

5. 0　　　**7.** 1

9. -103　　　**11.** 44, -97

13. 129　　　**15.** True

17. False

19. *Hint:* Write out some of the terms.

Answers 4-3

1. 77　　　**3.** 297

5. 650　　　**7.** 20,300

9. 800　　　**11.** 67

13. -38　　　**15.** 208

17. 39　　　**19.** $\frac{25}{2}$

21. $\frac{25}{2}$　　　**23.** $2n + 2$

25. $(n + 3)\sqrt{2}$　　　**29.** 1295

31. 5149　　　**39.** 1

41. 10　　　**43.** $\dfrac{n(2n^2 + 9n + 13)}{6}$

Answers 4-4

1. $3^6 - 1$

3. -364

5. $\frac{1093}{2187}$

7. $\frac{1}{4}(1 + (\frac{1}{3})^{11})$

9. 3^{13}

11. $-\frac{1}{1000}$

13. 2^{19}

15. $\frac{1}{3}$ and $\frac{1}{3}i$

17. $27, 81, 243$ $(a_n = 3^{n-1})$

19. $3, -3, 3$ $(a_n = 3(-1)^n)$

21. $a_n = 4^{n-2}$

23. $a_n = 6(-2)^{n-2}$

25. $a_6 = \frac{2}{5}$

27. $a_{21} = 16\sqrt{2}$

29. $2^{14} - 2$

31. $\frac{243}{16} - \frac{4}{3}$

33. $\frac{1}{150}\left(1 + \frac{1}{5^5}\right)$

37. $a_1 = \frac{36}{121}$

39. r

41. $16,105$

Answers 4-5

1. Converges to 2

3. Converges to 0

5. Converges to 2

7. Diverges

9. Converges to 1

11. 6

13. No limit

15. 2

17. $\frac{5}{6}$

19. 9

21. 48

23. No limit

25. No limit

27. $\frac{22}{9}$

29. $\frac{34}{9}$

31. $\frac{31}{99}$

33. $\frac{141}{990}$

35. $6\frac{415}{999}$

37. $\dfrac{3}{3 - 2x}$

39. $\frac{7}{15}$

41. 32

43. $\frac{1}{2}$

Answers 4-6

1. True

3. False

5. Lower bound, 1; no upper bound

7. Lower bound, 0; upper bound, $\frac{1}{2}$

9. If $a = 0$, then 0 is an upper and lower bound.
   If $a = 1$, then 1 is a lower and upper bound.
   If $0 < a < 1$, then 0 is a lower bound and a is an upper bound.

15. The limit is 4

Answers 5-1

3. $f(x) = x^6 + x^3$
$g(x) = -x^6$

5. deg $(f + g) = 3$, $(f + g)(x) = x^3$
deg $(2f - 3g) = 3$, $(2f - 3g)(x) = 2x^3 - 10x^2 + 5$
deg $(\frac{1}{2}f \cdot g) = 5$, $(\frac{1}{2}f \cdot g)(x) = x^5 - 2x^4 - \frac{1}{2}x^3 + 2x^2 - \frac{1}{2}$

7. $g(x) = 12x^2 + 32x + 26$

9. degree $= 6$; in general it is mn

Answers 5-2

1. $y = \dfrac{5}{2}x$

3. $y = -x + 3$

5. $y = -\dfrac{110}{56}x + \dfrac{13}{56}$

7. $y = \dfrac{b}{a}x$

9.

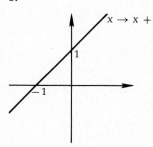

11.

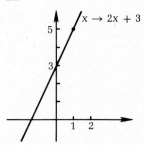

13.

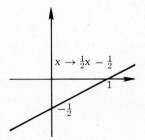

15.

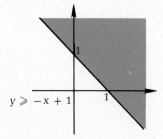

17.

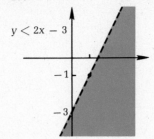

$y < 2x - 3$

19.

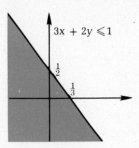

$3x + 2y \leqslant 1$

21.

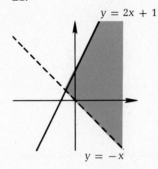

$y = 2x + 1$

$y = -x$

23.

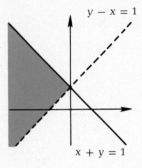

$y - x = 1$

$x + y = 1$

25. $(-7/5, 0)$

27. $f(x) = 5x - 3$

29. $y = \dfrac{-3}{5}x + \dfrac{1}{50}$

31. neither

33. perpendicular

35. $a = \dfrac{-3}{2}$

43. -2

45. $\frac{1}{6}$

47. $g(x) = \dfrac{x + 7}{3}$

Answers 5-3

1.

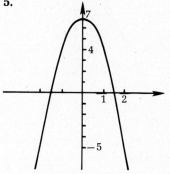

3.

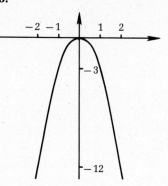

5.

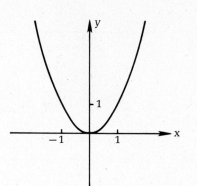

7.

$$f(x) = x^2 + 3x + 2$$

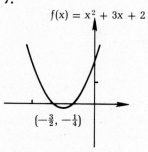

$\left(-\frac{3}{2}, -\frac{1}{4}\right)$

9.

$$f(x) = 2x^2 + 4x - 3$$

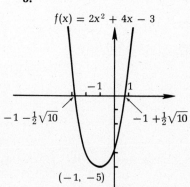

$-1 - \frac{1}{2}\sqrt{10}$ $-1 + \frac{1}{2}\sqrt{10}$

$(-1, -5)$

11.

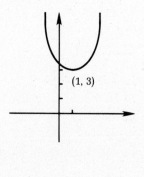

$(1, 3)$

13.

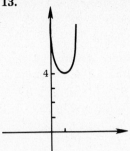

15.

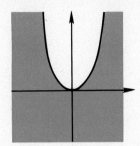

17.

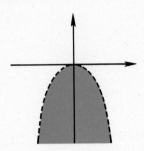

19.

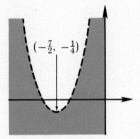

21.

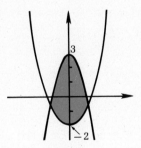

23. $x > 2$ or $x < -3$ **25.** $-\pi < x < \pi$

27. $-3 \leq x \leq 2$ **29.** $-1 < x < 0$

31. All reals   **33.** $x > 3$ or $x < -1$

35. Entirely above: $-4 < k < 4$

Touches exactly once: $k = \pm 4$

37. (a) $k < -1$ or $k > 0$

(b) $-1 < k < 0$

39. $r_1 + r_2 = \dfrac{-b + \sqrt{b^2 - 4ac}}{2a} + \dfrac{-b - \sqrt{b^2 - 4ac}}{2a} = -\dfrac{b}{a}$

41. From Exercise 39, note that $\dfrac{1}{2}(r_1 + r_2) = -\dfrac{b}{2a}$

43. $\frac{72}{25}$ hours past midnight   **45.** $(\frac{9}{5}, \frac{18}{5})$

47. Maximum area: put all the wire into the semicircle.

Minimum area: side of triangle $= \dfrac{2\sqrt{3}\,l}{(\pi + 6\sqrt{3})}$

49. $\dfrac{2}{\sqrt{5}}$

Answers 5-4

1. $r = 0$

$g(x) = x^2 - 101x + 1$

3. $r = -20$

$g(x) = x^3 + x^2 - 9x - 19$

5. $r = 34$

$g(x) = x^4 + 2x^3 + 4x^2 + 8x + 16$

7. $r = -25$

$g(x) = -x^5 - 2x^4 - 2x^3 - 4x^2 - 7x - 14$

9. $2x^2 - 5x + 12 - \dfrac{42}{x + 2}$

11. $x^3 + 3x^2 + 11x + 30 + \dfrac{95}{x - 3}$

13. $2x^3 + 10x^2 + 50x + 249 + \dfrac{1251}{x - 5}$

15. $x^4 + x^3 + x^2 + x + 1$

17. $g(x) = x + 3$ **19.** $g(x) = x^2 + x + 3$

21. $g(x) = 3x^2 + x + 5$; the zero is 2

23. $-\frac{3}{2}, 5, -3$ **29.** $k = -\frac{3}{2}$

31. $f(-2) = -63, f(-1) = -5, f(0) = -3, f(+1) = -3, f(2) = 1$

Answers 5-5

1. -3, mult. 6; 3, mult. 2; $-\pi$, mult. 4
3. $-\frac{2}{3}$, mult. 4; $-\frac{1}{30}$, mult. 7
5. $2, -2$
7. $\dfrac{1}{\sqrt{3}}, -\dfrac{1}{\sqrt{3}}$
9. None
11. 1, 1, 1
13. 0, 0, 2, 2, -2, -2
15. -1, 1, 2
17. -3, 3, 3
19. $-\pi$
21. -2, -2
23. 2, 2, -2, -2
25. 3, $\sqrt{3}$, $-\sqrt{3}$
27. 1, 1, -2, -2
29. 2, -2, $-\frac{3}{2}$, $\frac{1}{3}$
31. $f(x) = 2(x - 1)(x - 2)(x + 2)$
33. $f(x) = 2(x + 1)^5(x - 3)^2$
35. -1 mult. 1; 2 mult. 1;
    0 mult. 1; 1 mult. 2
    Zero lies between -2 and 4

Answers 5-6

1. One way to show $3i$ is a zero of f is to use synthetic division. The other zeros are $-3i$ and $\frac{7}{3}$.
3. $1 + 2i$, $1 - 2i$, $1 - i$
5. $x^4 - 2x^3 + 3x^2 - 2x + 2$
7. $2i$ is also a root, but $\overline{2i} = -2i$ is not. There is no contradiction, since some of the coefficients of f are complex nonreal numbers.
9. Four times

Answers 5-7

3. $-\dfrac{7}{2}, \pm\sqrt{\dfrac{3}{2}}$
5. -4
7. 1
9. $-\frac{1}{3}$, $1 \pm \sqrt{5}$
11. None
13. Upper bound 2, lower bound -3
15. Upper bound 2, lower bound 0

Answers 5-8

1.

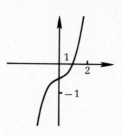

3.

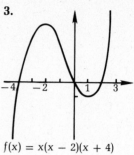

$f(x) = x(x - 2)(x + 4)$

5.

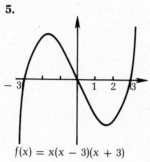

$f(x) = x(x - 3)(x + 3)$

7.

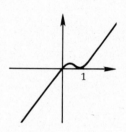

9.

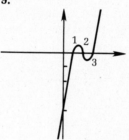

11.

$f(x) = (2x + 1)^2(x - 3)$

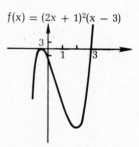

13.

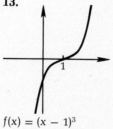

$f(x) = (x - 1)^3$

15.

$f(x) = x^4$

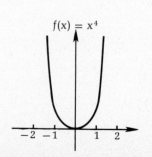

17.

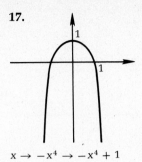

$$x \rightarrow -x^4 \rightarrow -x^4 + 1$$

19.

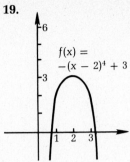

$$f(x) = -(x - 2)^4 + 3$$

21. All real numbers

23. $x \geq 1$ or $-4 \leq x \leq 0$

25. $x \geq 2$ or $-3 \leq x \leq 1$

27. $|x| \geq 1$

29. None

31. $0 < x < 2$ or $x < -4$

33. $x < 1$

35. $1 < x < 2$ or $-3 < x < -2$

Answers 6-1

1.

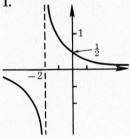

3.

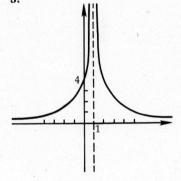

5.

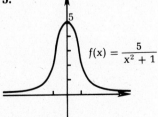

$$f(x) = \frac{5}{x^2 + 1}$$

7.

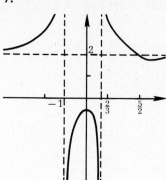

9.

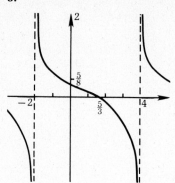

11.

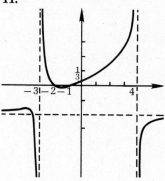

13.

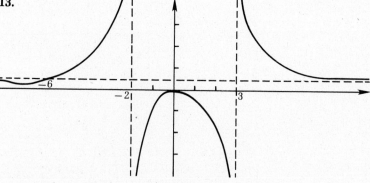

15.

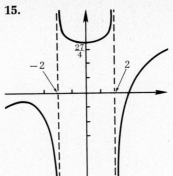

17.

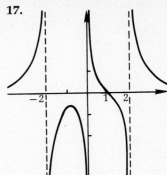

19.

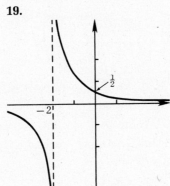

21.

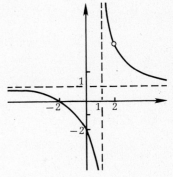

23.

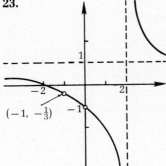

$$f(x) = x(x + 1)(x + 2)/x(x + 1)(x - 2)$$

25.

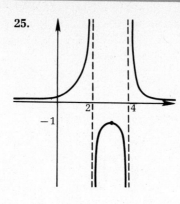

27.

$$f(x) = \frac{(x + 1)(x - 1)(x - 3)}{(x + 1)(x - 1)(x - 3)}$$

29. $\dfrac{-a \pm a\sqrt{13}}{6}$

31. $-3, 0$

33. $\dfrac{-1 \pm \sqrt{46}}{3}$

35. 11

37. $0, 3$

39. 0 (note that -1 and $-\frac{3}{2}$ are *not* in the domain of f)

41. $3a$ (Note that $-a$ and $-3a$ are not in the domain of $f \cdot g$.)

43. $x > -\frac{1}{2}$

45. $x < -1$ or $x > 0$

47. $\frac{1}{2} < x < 1$

49. $-2 < x < -1$ or $x > 2$

51. $-4 < x < 2, x \neq 0$

53. $-\frac{5}{2} < x < -2$

Answers 6-2

1. $h(x) = 2, r(x) = -2x^2 + 3x + 2$

3. $h(x) = x^3 + x - 2, r(x) = 0$

5. $h(x) = x^2 + x + \frac{5}{6}, r(x) = \frac{7}{12}$

7. $h(x) = x^2 - 1, r(x) = -6x^2 + 3x + 6$

9. $h(x) = x^2 + 7x + 34, r(x) = 138$

11. $h(x) = 2x^3 + 9x + 3, r(x) = 6x + 11$

13. $h(x) = 2x^3 + x^2 + 4, r(x) = 2x - 1$

15. $h(x) = x + 2, r(x) = -x - 6$

17. $h(x) = x^2 + \sqrt{7}x + (7 + \sqrt{7}), r(x) = 7 + 10\sqrt{7}$

19. $h(x) = x^3 - ix^2 - x + i, r(x) = 10$

21. $h(x) = x^2 - 3ax + a^2, r(x) = 0$

23. $h(x) = x^3 - ax^2 + 2a^3, r(x) = 2a^4$

Answers 7-1

1. $x \to \dfrac{x}{5}$

3. $x \to \dfrac{x+1}{3}$

5. $x \to \sqrt[3]{x+1}$

7. $x \to \dfrac{1}{\sqrt[3]{x}}$

9. True

11. True

13. $f^{-1}(x) = \dfrac{1 + \sqrt{4x+25}}{2}$ $x \geq \dfrac{-25}{4}$

$= x + \dfrac{27}{4}$ $x \leq \dfrac{-25}{4}$

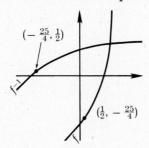

Answers 7-2

1.,3.

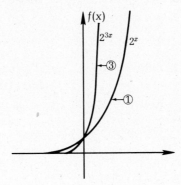

5.,7.

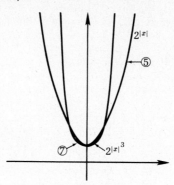

9.,11.

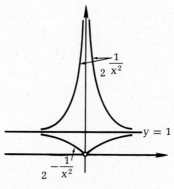

13. $\frac{1}{4}$

15. -1

17. None

19. One

21.

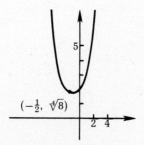

Answers 7-3

1.,3.,5.

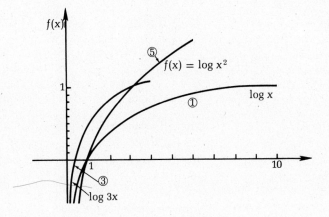

7.,9.

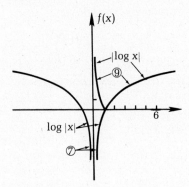

11.

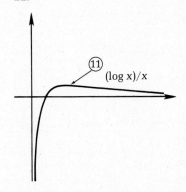

13. 4

15. -1

17. -2

19. 1.1761

21. 0.3495

23. 1.3980

25. -0.3010

27. $\dfrac{1}{\sqrt{10} - 1}$

29. $x = 2$

31. 1.303

33. .6303

35. 1.77

37. 8.52×10^{10}

39. $\left(-\dfrac{4}{5}, -\dfrac{6 \log 2}{5 \log 5}\right)$

41. (a) 4.64×10^{10}
    (b) 0.226 hr.

43. 7.2%

45. $\log 1000 = 3$

47. 773

49. 7

53. 1.0238 and 237.57

Answers 8-1

1. $\dfrac{2\pi}{300} \approx .021$ rad.

3. $\dfrac{7\pi}{6} \approx 3.67$ rad.

5. $\dfrac{\pi^2}{90} \approx .11$ rad.

7. $\dfrac{5\pi^2}{9} \approx 5.48$ rad.

9. $68.75° \left[\text{i.e.,} \left(\dfrac{216}{\pi}\right)°\right]$

11. $114.6° \left[\text{i.e.,} \left(\dfrac{360}{\pi}\right)°\right]$

13. $\left(\dfrac{18,000}{\pi}\right)° \sim 5,732.48°$

15. $20,636.94°$

17. $-2520°$

19. 25.7 rad.

21. $131.85°$

23. $7.5°$

25. 2.26

27. $\dfrac{36}{\pi}$, or $11.46°$

29. $\dfrac{360}{\pi}$ ft. (or approx. 114.6 ft.)

31. $\dfrac{100}{3\pi} \approx 10.61$ turns

Answers 8-2

1. $f(0) = 1, f\left(\dfrac{\pi}{4}\right) = 0, f\left(\dfrac{\pi}{6}\right) = \dfrac{1}{2}, f\left(\dfrac{\pi}{8}\right) = \dfrac{\sqrt{2}}{2},$

$f\left(\dfrac{\pi}{3}\right) = -\dfrac{1}{2}, f\left(\pi + \dfrac{\pi}{8}\right) = \dfrac{\sqrt{2}}{2}, f\left(-\dfrac{\pi}{2}\right) = -1,$

$f\left(2\pi + \dfrac{\pi}{3}\right) = -\dfrac{1}{2}, f(10\pi) = 1, f(-10\pi) = 1$

3. $\frac{\pi}{4} + n\pi$ (n any integer)

5. $\frac{\sqrt{3}}{2}$, $-\frac{1}{2}$

7. $\frac{\sqrt{3}}{2}$, $-\frac{1}{2}$

9. $\frac{\sqrt{2}}{2}$

11. $-\frac{1}{2}$

13. True

15. True

17. 0.8415

19. -0.9582

21. -0.9856

23. 0.286

25. $\pi/6$

27. 0.05

29. 0.88

31. $\frac{\pi}{3} + 2n\pi$, $\frac{2\pi}{3} + 2m\pi$ (m, n integers)

33. $\frac{7\pi}{6} + 2n\pi$, $\frac{11\pi}{6} + 2m\pi$ (m, n integers)

35. $\frac{\pi}{12} + \frac{2n\pi}{3}$, $\frac{\pi}{4} + \frac{2m\pi}{3}$ (m, n integers)

37. True

39. False

41. $\frac{\pi}{4} + m\pi$, m an integer

43. All real numbers

45. $\frac{\pi}{4} + n\pi$ (n an integer)

Answers 8–3

1. $\frac{5\sqrt{3}}{2}$

3. 60°

5. ≈ 17.01

7. ≈ 38

9. ≈ 3.1

Answers 8-4

1.

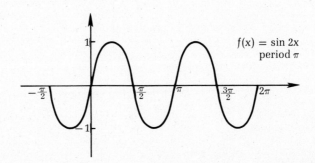

$f(x) = \sin 2x$
period π

3.

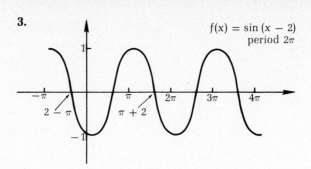

$f(x) = \sin(x - 2)$
period 2π

5.

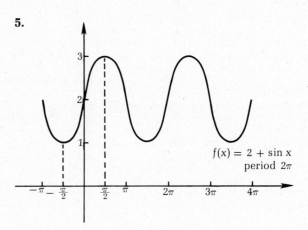

$f(x) = 2 + \sin x$
period 2π

7.

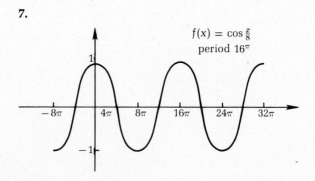

$f(x) = \cos \frac{x}{8}$
period 16π

9.

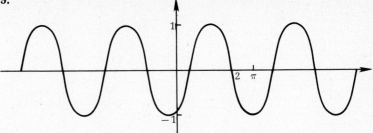

11.

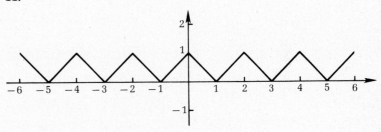

13. -11

15.

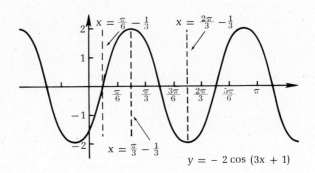

$y = -2\cos(3x + 1)$

17.

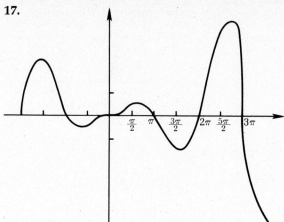

19.

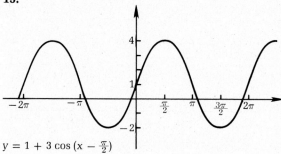

$y = 1 + 3 \cos \left(x - \frac{\pi}{2}\right)$

21.

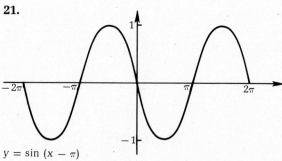

$y = \sin (x - \pi)$

23.

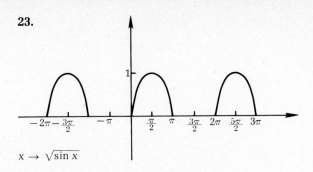

$x \to \sqrt{\sin x}$

25. 3, at about 2.7, 7.3, and 8.3

27. 63   (31 for $x > 0$, 1 at $x = 0$, and 31 for $x < 0$)

Answers 8-5

1. $\dfrac{\sqrt{3} - 1}{2\sqrt{2}}$

3. $\dfrac{-(1 + \sqrt{3})}{2\sqrt{2}}$

5. $-\dfrac{\sqrt{2} + \sqrt{6}}{4}$

7. $x = \dfrac{2n + 1}{6}\pi$ (n is an integer)

9. True

11. True

13. $x = \dfrac{\pi}{6}$, or $\dfrac{5\pi}{6}$, any y in $[0, 2\pi]$

19. $\cos x = \dfrac{2 \pm \sqrt{14}}{6}$

21. $0, 2\pi, \dfrac{4\pi}{3}$

Answers 8-6

3. Expand $(\sin^2 x)^2 = \left(\dfrac{1 - \cos 2x}{2}\right)^2$

5. $\dfrac{\pi}{3}, \dfrac{2\pi}{3}, \dfrac{4\pi}{3}, \dfrac{5}{3}\pi$

7. $\dfrac{\pi}{4}$

9. $x = \dfrac{\pi}{10} + \dfrac{2n\pi}{5}$, or $x = -\dfrac{\pi}{2} - 2n\pi$, where n is an integer

11. Show that the problem is equivalent to finding all numbers x for which $\cos 3x + \dfrac{1 - \sin^2 2x}{\cos^2 2x} = 0$, and thus cos $3x + 1 = 0$. The solutions are $\dfrac{\pi}{3}$, π, and $\dfrac{5\pi}{3}$.

Answers 8-7

1. $\dfrac{1}{\sqrt{3}}, \dfrac{2}{\sqrt{3}}, \sqrt{3}, 2$

3. $-\dfrac{1}{\sqrt{3}}, -\dfrac{2}{\sqrt{3}}, -\sqrt{3}, 2$

5. $-1, -\sqrt{2}, -1, \sqrt{2}$

7. $-\sqrt{3}, 2, -\dfrac{1}{\sqrt{3}}, -\dfrac{2}{\sqrt{3}}$

9. Undefined, undefined, 0, 1

11. Undefined, undefined, 0, -1

13. $2 + \sqrt{3}$ 15. $-2 - \sqrt{3}$

17. 5.36, 2.22   19. 1.87, 4.41

21. 3.05, 6.19   23. -0.79, 0.98

33. Approx. .58   35. $\dfrac{p - q}{p + q}$

37. (a) 27°   39. 25°
    (b) 45°
    (c) 63°

41. 260 yd.   43. $\sim$33 miles

45. 9.03 ft.   47. True

49. False   51. False

53. True

Answers 9-1

	$\sin^{-1} x$	$\cos^{-1} x$
1.	$\dfrac{\pi}{6}$	$\dfrac{\pi}{3}$
3.	$\dfrac{\pi}{4}$	$\dfrac{\pi}{4}$
5.	$\dfrac{\pi}{3}$	$\dfrac{\pi}{6}$
7.	$\dfrac{\pi}{2}$	0
9.	0	$\dfrac{\pi}{2}$

	$\tan^{-1} x$	$\cot^{-1} x$
11.	$\dfrac{\pi}{3}$	$\dfrac{\pi}{6}$
13.	$\dfrac{\pi}{6}$	$\dfrac{\pi}{3}$
15.	$\dfrac{\pi}{4}$	$\dfrac{\pi}{4}$

17. $\dfrac{\sqrt{3}}{2}$

19. $\dfrac{1}{\sqrt{3}}$

21. $\sqrt{1 - (.47)^2} \approx .88$

23. Same as 20; $\sqrt{1 - (.73)^2}$

25. $\dfrac{\pi}{2} - \dfrac{1}{3}$

27. 3

29. $\dfrac{2\sqrt{6}}{5}$

31. $\dfrac{7}{8}$

33. $\dfrac{1 + 6\sqrt{10}}{20}$

35. $\dfrac{2x}{1 - x^2}$

37. 1

45. $\dfrac{1}{3}$

47. $-1, 0, 1$

49. $\sqrt{3} - 1$

51. $\dfrac{49}{65}$

61. $x = \dfrac{1 + \cos y}{-1 + \cos y}$

63. False

65. False

67. False

69. False

71. (a) 2.3; (b) 0.9

Answers 9-2

1. 75.5″

3. 140″, 21.8°

5. 8.83

7. 254.1

9. $a = 4.05$, $b = 3.62$

11. $a = 4.48$, $b = 1.57$

13. $a = 38.6$, $c = 58.0$, $\alpha = 41.6°$

15. $\alpha = 68°55'$, $b = 201.6$, $a = 194.0$

17. 100°57′, 24°9′

19. $\alpha = 97°55'$

21. 11.65 miles

Answers 9-3

1. $(\sqrt{3}, 1)$

3. $(-4, 0)$

5. $\left(-\dfrac{5\sqrt{2}}{2}, \dfrac{5\sqrt{2}}{2}\right)$

7. $(2, \frac{4}{3}\pi)_p$

9. $(0, t)_p$, t any number

11. Circle of rad. 3, center at 0

13. Line at $0°$

15. 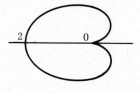   **17.**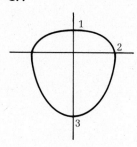

19. $x^2 + 4x + y^2 = 0$ **21.** $4x^2 + 3y^2 - 12y = 36$

23. $r = 2a \tan t \sin t$

25. $r^2 \sin^2 t = 4ar \cos t + 4a^2$

27. $r^2 = 2a^2 \cos 2t$

Answers 9-4

1. $|z| = \pi$, $\arg(z) = 0$

3. $|z| = 5$, $\arg(z) = 306°52'$

5. $|z| = 2\pi$, $\arg(z) = \dfrac{\pi}{2}$ **7.** $\sqrt{3} + i\sqrt{3}$

9. $-2i$ **11.** $4\mathbf{u}\left(\dfrac{11\pi}{6}\right)$

13. $\dfrac{1}{2}\mathbf{u}\left(\dfrac{5\pi}{6}\right)$ **15.** 4 [i.e., $4\mathbf{u}(0)$]

17. $2\mathbf{u}\left(\dfrac{\pi}{4}\right)$, $2\mathbf{u}\left(\dfrac{3\pi}{4}\right)$, $2\mathbf{u}\left(\dfrac{5\pi}{4}\right)$, $2\mathbf{u}\left(\dfrac{7\pi}{4}\right)$

19. $2\mathbf{u}(165°)$, $2\mathbf{u}(345°)$

21. $z = (3 - i)/2$, $(-3 - i)/2$

23. (a) $\arg(z_1) - \arg(z_2) = \dfrac{(2n + 1)\pi}{2}$, $n = 0, \pm 1, \pm 2, \ldots$

  (b) $\arg(z_1) - \arg(z_2) = n\pi$, $n = 0, \pm 1, \pm 2, \ldots$

Answers 10-1

1.

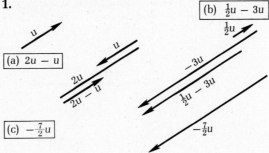

3.

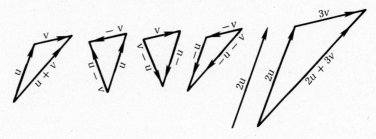

5.

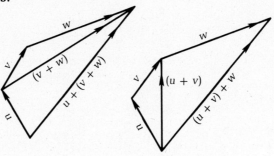

7. 6.57, 12.4°   **9.** 8.1, 17.7°

11. 5.93, 13.4°   **13.** 3.86, 21.5°

15.

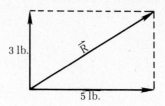

17. 31°

19. $40\sqrt{3}$ pounds, 30°

21.

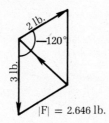

$|F| = 2.646$ lb.

Answers 10-2

1. $\langle -18, -10 \rangle$

3. $\langle -4, -1 \rangle$

5. $\sqrt{2}$

7. $\sqrt{317}$

9. $\sqrt{113}$

11. 25

13. $a = 0$, $b = -5$

15. $a = 2$, $b = -6$

17. $a = 2$, $b = 1$

19. $a = \frac{1}{2}$ or 1, $b = \frac{5}{3}$ or -1

21. $a = -1$, $b = 2$

23. 90°

25. 22.5°

27. 82.9°

29. $\left\langle \dfrac{1}{\sqrt{101}}, \dfrac{10}{\sqrt{101}} \right\rangle$, $\left\langle -\dfrac{1}{\sqrt{101}}, -\dfrac{10}{\sqrt{101}} \right\rangle$

31. $\sqrt{34}$, 4.4°

33. $\sqrt{34}$, 57.6°

35. $\left\langle \dfrac{12}{5}, \dfrac{6}{5} \right\rangle$

39. 7

41. 0

Answers 11-2

1.

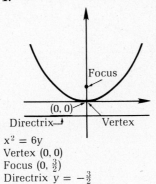

$x^2 = 6y$
Vertex $(0, 0)$
Focus $(0, \frac{3}{2})$
Directrix $y = -\frac{3}{2}$

3.

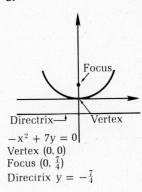

$-x^2 + 7y = 0$
Vertex $(0, 0)$
Focus $(0, \frac{7}{4})$
Directrix $y = -\frac{7}{4}$

5.

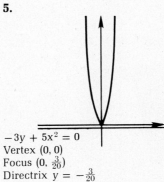

$-3y + 5x^2 = 0$
Vertex $(0, 0)$
Focus $(0, \frac{3}{20})$
Directrix $y = -\frac{3}{20}$

7.

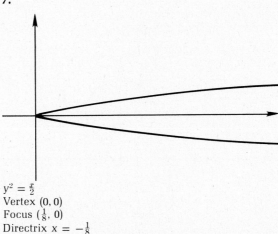

$y^2 = \frac{x}{2}$
Vertex $(0, 0)$
Focus $(\frac{1}{8}, 0)$
Directrix $x = -\frac{1}{8}$

9.

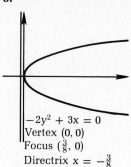

$-2y^2 + 3x = 0$
Vertex $(0, 0)$
Focus $(\frac{3}{8}, 0)$
Directrix $x = -\frac{3}{8}$

11. $y^2 = 4\sqrt{2}x$

13. $2x^2 = y$

15.

$3y + 2x^2 \geqslant 0$

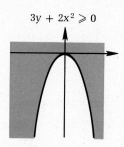

17.

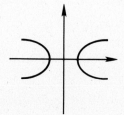

$2x - 5y^2 < 0$

19. $5x^2 = 16y$

21. $(1, \pm\sqrt{3})$

Answers 11-3

1.

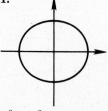

$\dfrac{x^2}{4} + \dfrac{y^2}{3} = 1$
Vertices $(\pm 2, 0), (0, \pm\sqrt{3})$
Foci $(\pm 1, 0)$

3.

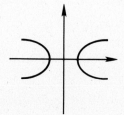

$4x^2 - 9y^2 = 36$
Vertices $(\pm 3, 0)$
Foci $(\pm\sqrt{13}, 0)$

5.

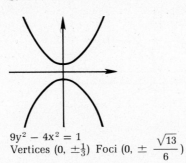

$9y^2 - 4x^2 = 1$
Vertices $(0, \pm\frac{1}{3})$ Foci $(0, \pm \dfrac{\sqrt{13}}{6})$

7. Upper half of ellipse whose equation is $\dfrac{x^2}{4} + \dfrac{z^2}{16} = 1$

(vertices: $(0, \pm 4)$, $(\pm 2, 0)$)

9. Lower half of hyperbola whose equation is $\dfrac{y^2}{16} - \dfrac{x^2}{4} = 1$

11. $(3, 2)$ and $(4, 3/2)$ **13.** None

15. $\dfrac{x^2}{36} + \dfrac{y^2}{20} = 1$

Answers 11-4

1.

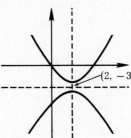

$4(y + 3)^2 - 9(x - 2)^2 = 1$
Hyperbola with vertices
$(2, -3 \pm \frac{1}{2})$

3.

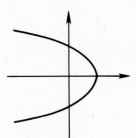

$y^2 + 4x = 8$
Parabola with vertex $(2, 0)$

5.

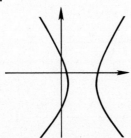

$25x^2 - 4y^2 - 150\,x - 16y$
$+ 109 = 0$
Hyperbola with
vertices $(1, -2)$, $(5, -2)$

7.

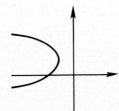

$3y^2 = 6y - 5x - 13$
Parabola with
vertex $(-2, 1)$

9.

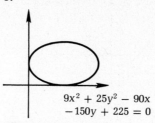

$$9x^2 + 25y^2 - 90x$$
$$- 150y + 225 = 0$$

Ellipse with
vertices
$(0, 3)$ $(10, 3)$ $(5, 0)$ $(5, 6)$

11.

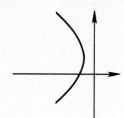

$2y^2 + 5x - 3y + 4 = 0$
Vertex $\left(-\frac{23}{40}, \frac{3}{4}\right)$

13. $x^2 + 2x + 9y + 10 = 0$

15.

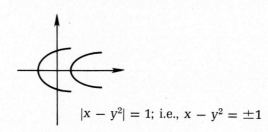

$|x - y^2| = 1$; i.e., $x - y^2 = \pm 1$

Answers 11-5

1. $30°$: $\left(\dfrac{\sqrt{3} - 3}{2}, \dfrac{1 + 3\sqrt{3}}{2}\right)$

    $45°$: $(-\sqrt{2}, 2\sqrt{2})$

    $60°$: $\left(\dfrac{1 - 3\sqrt{3}}{2}, \dfrac{3 + \sqrt{3}}{2}\right)$

3. $30°$: $\left(\dfrac{3 + \sqrt{3}}{2}, \dfrac{1 - 3\sqrt{3}}{2}\right)$

    $45°$: $(2\sqrt{2}, -\sqrt{2})$

    $60°$: $\left(\dfrac{1 + 3\sqrt{3}}{2}, \dfrac{\sqrt{3} - 3}{2}\right)$

5. $\left[\dfrac{\sqrt{2}(-3 - \sqrt{3})}{4}, \dfrac{\sqrt{2}(1 - 3\sqrt{3})}{4}\right]$

7. $\left(\dfrac{3 - \sqrt{3}}{2\sqrt{2}}, \dfrac{3\sqrt{3} + 1}{2\sqrt{2}}\right)$

9. $x(-3\sqrt{3} - 2) + y(3 - 2\sqrt{3}) = 2$

13.

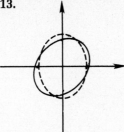

$5x^2 - 6xy + 5y^2 - 32 = 0$
Ellipse

15.

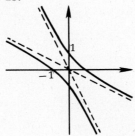

The hyberbola $x^2\left(\dfrac{3\sqrt{3}}{4} + 1\right) - y^2\left(\dfrac{3\sqrt{3}}{4} - 1\right) = 1$ rotated through $45°$.

Answers 11-6

1.

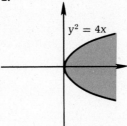

$y^2 = 4x$

3.

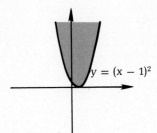

$y = (x - 1)^2$

5.

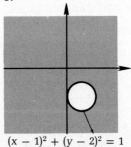

$(x - 1)^2 + (y - 2)^2 = 1$

7.

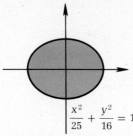

$$\frac{x^2}{25} + \frac{y^2}{16} = 1$$

region doesn't include
the ellipse itself

9.

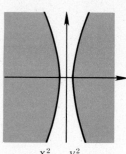

$$\frac{x^2}{4} - \frac{y^2}{49} = 1$$

11.

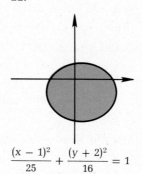

$$\frac{(x-1)^2}{25} + \frac{(y+2)^2}{16} = 1$$

13.

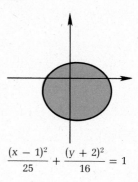

$$\frac{(x-1)^2}{25} + \frac{(y+2)^2}{16} = 1$$

Answers 11-7

1. $x^2 + y^2 + z^2 = 81$
3. $(x-1)^2 + y^2 + (z+3)^2 = 16$
5. $2x - 8y + 8z = 19$

7.

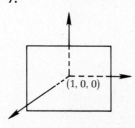

(1, 0, 0)

Fig. S-137

9.

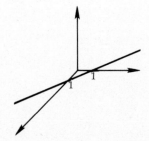

11.

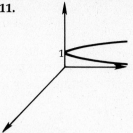

The parabola
$y = x^2$ in the
plane z^2.

13. A sphere of radius 1 centered at $(3, 2, 1)$

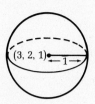

Answers 12-1

1. $(t + 10, -t - 4, t)$, t any real number

3. $(t, \frac{3}{2}t - 2, t + 1)$, t any real number

5. $(3, t - 1, t)$, t any real number

7. $\left(1 + \dfrac{5t}{2}, 2 - \dfrac{t}{2}, t\right)$, t any real number

9. $(-t, -4t - 1, t)$, t any real number

11. No solution   **13.** $(1, 2, -2)$

15. $(1, -3, -2)$ **17.** $(\frac{1}{13}, -\frac{9}{13}, \frac{18}{13})$

19. $(t, -t, 0)$, t any real number

21. $(5t, t, 4t)$, t any real number

23. $(0, 0, 0)$

25. $(3t, -2t, t)$, t any real number

27. $(\frac{11}{2}, \frac{5}{2}, -3, 1)$ **29.** No solution

31. $(2w + 4, 1 - w, -2w, w)$, w any number

33. $(\frac{2}{3}w, -w, 1, w)$

35. $(-y, y, -2y)$, y any number

37. infinitely many solutions   **39.** $(4, 3, 2)$

41. $(-1, 2, -2, 1)$ **43.** $(\frac{1}{2}, -\frac{1}{2}, -\frac{1}{2}, 2, 1)$

Answers 12-2

1. (a) $\begin{pmatrix} 1 & 3 \\ 0 & -17 \end{pmatrix}$ (b) $\begin{pmatrix} 1 & 1 & 2 \\ 0 & 1 & 1 \\ 0 & 0 & 0 \end{pmatrix}$

(c) $\begin{pmatrix} 21 & 35 & 42 \\ 0 & -32 & -30 \end{pmatrix}$ (d) $\begin{pmatrix} 1 & 0 & 1 & 0 \\ 0 & 1 & 0 & 1 \\ 0 & 0 & 0 & 0 \\ 0 & 0 & 0 & 0 \end{pmatrix}$

Answers 12-3

1. 2

3. 3

5. Compare with answers in Section 12-1

7. Compare with answers in Section 12-1

Answers 12-4

1. -1 **3.** 9 **5.** 12

7. -12 **9.** $(7, -3)$ **11.** $\left(\frac{7}{13}, \frac{2}{13}\right)$

Answers 12-5

1. 2

3. 0

5. Hint: multiply first row by $-d/a$

7. Compare with answers in Section 12-1.

Answers 12-6

1. (a) $\begin{pmatrix} \frac{1}{2} & -\frac{1}{2} & \frac{3}{2} \\ 1 & 2 & 0 \\ \frac{3}{2} & -\frac{3}{2} & \frac{1}{2} \\ 1 & 0 & 0 \end{pmatrix}$ (b) $\begin{pmatrix} 2 & -4 & 0 \\ 4 & 16 & 1 \\ 6 & -12 & 5 \\ 4 & 0 & 0 \end{pmatrix}$

(c) $\begin{pmatrix} 0 & 2 & 6 \\ 0 & -8 & -1 \\ 0 & 6 & -3 \\ 0 & 0 & 0 \end{pmatrix}$ (d) $\begin{pmatrix} 2 & 0 & 12 \\ 4 & 0 & -1 \\ 6 & 0 & -1 \\ 4 & 0 & 0 \end{pmatrix}$

3. $\begin{pmatrix} 2 & 3 \\ -2 & 4 \end{pmatrix}$ **5.** $\begin{pmatrix} 5 \\ -2 \end{pmatrix}$

7. (b) $A = \begin{pmatrix} 1 & 1 \\ 0 & 0 \end{pmatrix}$, $B = \begin{pmatrix} 0 & 0 \\ 1 & 1 \end{pmatrix}$

9. $-\frac{1}{7}, \begin{pmatrix} 1 & -2 \\ 0 & -7 \end{pmatrix}$ **13.** $\left(\frac{1}{2}, 0\right)$

11. $-\frac{60}{7} \begin{pmatrix} \frac{1}{4} & -\frac{1}{2} \\ -\frac{2}{5} & \frac{1}{3} \end{pmatrix}$

15. $\begin{pmatrix} 1 & 0 \\ 0 & 0 \end{pmatrix}$ and $\begin{pmatrix} 0 & 0 \\ 1 & 0 \end{pmatrix}$

19. $A^n = I$, when n is even
$A^n = A$, when n is odd

Answers 12-7

5. $\begin{pmatrix} 0 & 1 & -1 \\ 1 & -1 & 0 \\ 0 & 0 & 1 \end{pmatrix}$

7. $\begin{pmatrix} 1/a & 0 & 0 & 0 \\ 0 & 1/b & 0 & 0 \\ 0 & 0 & 1/c & 0 \\ 0 & 0 & 0 & 1/d \end{pmatrix}$

9. (a) $-\frac{1}{9}, \begin{pmatrix} -5 & -12 & 1 \\ 4 & 6 & 1 \\ 1 & -3 & -2 \end{pmatrix}$

(b) $x = 7, y = -2, z = -5$

11. x must be a real number for which $x^3 - 9x^2 + 15x + 25 = 0$. Hence, $x = -1, x = 5$

13. 1

Answers 12-8

1. $f(0, 1) = -1$ is the minimum
$f(1, 0) = 4$ is the maximum

3. $f(0, 2) = 16$ is the maximum
$f(-3, 0) = f(7, -5) = -12$ is the minimum

5. Minimum of f: $-\frac{1}{2}$
Maximum of f: 1

7. $1250

9. 7 oz. of meat, $1\frac{1}{2}$ pts. of cottage cheese daily

Answers 13-1

1. $\{a, b, c, d\}$　　**3.** $\{d\}, \{a, b\}, \{b, c\}$　　**5.** $\{a, c\}$

7. $\{a, b, d\}$　　**9.** $\varnothing$

11.

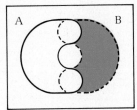

13.

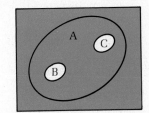

15.   **17.**

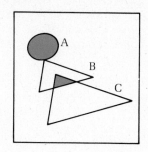   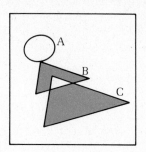

19. ∅   **21.** $(-1, 0)$ **23.** $(-5, -\frac{1}{2}]$

25. $\{1\}$, $\{2\}$, $\{3\}$, $\{4\}$, $\{5\}$, $\{1, 2\}$, $\{1, 3\}$, $\{1, 4\}$, $\{1, 5\}$, $\{2, 3\}$, $\{2, 4\}$, $\{2, 5\}$, $\{3, 4\}$, $\{3, 5\}$, $\{4, 5\}$, $\{1, 2, 3\}$, $\{1, 2, 4\}$, $\{1, 2, 5\}$, $\{1, 3, 4\}$, $\{1, 3, 5\}$, $\{1, 4, 5\}$, $\{2, 3, 4\}$, $\{2, 3, 5\}$, $\{3, 4, 5\}$, $\{2, 4, 5\}$, $\{1, 2, 3, 4\}$, $\{1, 2, 3, 5\}$, $\{1, 2, 4, 5\}$, $\{1, 3, 4, 5\}$, $\{2, 3, 4, 5\}$, $\{1, 2, 3, 4, 5\}$, ∅

27. $\{1, 2, 3, 4, 5, 6, 7\}$ **29.** $\left(\frac{7}{5}, -\frac{2}{5}\right)$

31. $\left\{0, \pi, 2\pi, \frac{7\pi}{6}, \frac{11\pi}{6}\right\}$ **33.** $\subset$

35. ∅

37. $n(A \cup B') = n(A) + n(B') - n(A \cap B')$
$= n(A) + n(B')$,
since $n(A \cap B') = \varnothing$ (this is because $A \subset B$)

Answers 13-2

1. $5! = 120$ **3.** One chance in 2^{10}

5. $2 \cdot 3 \cdot 5 = 30$ **7.** 16

9. 120

Answers 13-3

1. $\binom{10}{2} = 45$ **3.** $4\binom{10}{2} = 180$

5. $\binom{26}{3}\binom{26}{2}$

7. (a) $\binom{13}{4}\binom{13}{3}\binom{13}{3}\binom{13}{3}$

 (b) $4\binom{13}{4}\binom{13}{3}\binom{13}{3}\binom{13}{3}$

9. 2 in 1 room, 1 in another, 1 in another.

Answers 13-4

1. $(2a)^{93} \quad 3^7 \dbinom{100}{7}$ **3.** $(2a)^{49} \quad 3^{51} \dbinom{100}{51}$

5. $x^8 - \dfrac{8x^7}{y^2} + \dfrac{28x^6}{y^4} - \dfrac{56x^5}{y^6} + \dfrac{70x^4}{y^8} - \dfrac{56x^3}{y^{10}} + \dfrac{28x^2}{y^{12}} - \dfrac{8x}{y^{10}} + \dfrac{1}{y^{12}}$

7. $\sin^5 x + 5 \sin^4 x \cos x + 10 \sin^3 x \cos^2 x + 10 \sin^2 x \cos^3 x + 5 \sin x \cos^4 x + \cos^5 x$

9. $a^3 + b^3 + c^3 + 3a^2b + 3a^2c + 3ab^2 + 3b^2c + 3ac^2 + 3bc^2 + 6abc$

Answers 14-2

1. $\frac{5}{6}$ **3.** $\frac{1}{2}$ **5.** $\dfrac{\dbinom{48}{9}}{\dbinom{52}{13}}$

7. $1 - \left[\dbinom{39}{13} \; \dbinom{52}{13} \right]$ **9.** $\dbinom{3}{2} \dfrac{1}{2^3} = \dfrac{3}{8}$

11. $1 - \dfrac{1}{2^3} = \dfrac{7}{8}$ **13.** $\frac{5}{8}$

15. $1 - \dfrac{2}{36} = \dfrac{17}{18}$ **17.** $\frac{1}{36}$

19. $\dbinom{10}{1} \left(\dfrac{1}{3}\right) \left(\dfrac{2}{3}\right)^9$ **21.** $\dbinom{10}{3} \left(\dfrac{1}{3}\right)^3 \left(\dfrac{2}{3}\right)^7$

23. $\left(\frac{1}{3}\right)^{10}$ **25.** $1 - \left(\frac{2}{3}\right)^{10}$

27. 1 **29.** $1 - \dfrac{12_n}{12^n}$

Answers 14-3

1. P (Fraternal twins are both girls) $= P$ (they have the same sex and that sex is female) $= \frac{2}{3} \cdot \frac{1}{2} = \frac{1}{3}$

3. $\dbinom{5}{3} \left(\dfrac{1}{6}\right)^3 \left(\dfrac{5}{6}\right)^2 = \dfrac{250}{6^5}$

5. $\left(\dfrac{2}{3}\right)^3 + \dbinom{3}{1} \dfrac{1}{3} \left(\dfrac{2}{3}\right)^3 + \dbinom{4}{2} \left(\dfrac{1}{3}\right)^2 \left(\dfrac{2}{3}\right)^3 = \dfrac{64}{81}$

7. P [The gambler goes broke] $= P$ [LL or LWLL or LWLWLL or ...] $= \dfrac{1}{2^2} + \dfrac{1}{2^4} + \dfrac{1}{2^6} + \cdots = \dfrac{\frac{1}{4}}{1 - \frac{1}{4}} = \dfrac{1}{3}$

9. $\dfrac{1}{2^3} + \dfrac{1}{2^6} + \dfrac{1}{2^9} + \cdots = \dfrac{\frac{1}{8}}{1 - \frac{1}{8}} = \dfrac{1}{7}$

Tables

Table 1

Common Logarithms of Numbers

n	0	1	2	3	4	5	6	7	8	9
1.0	.0000	.0043	.0086	.0128	.0170	.0212	.0253	.0294	.0334	.0374
1.1	.0414	.0453	.0492	.0531	.0569	.0607	.0645	.0682	.0719	.0755
1.2	.0792	.0828	.0864	.0899	.0934	.0969	.1004	.1038	.1072	.1106
1.3	.1139	.1173	.1206	.1239	.1271	.1303	.1335	.1367	.1399	.1430
1.4	.1461	.1492	.1523	.1553	.1584	.1614	.1644	.1673	.1703	.1732
1.5	.1761	.1790	.1818	.1847	.1875	.1903	.1931	.1959	.1987	.2014
1.6	.2041	.2068	.2095	.2122	.2148	.2175	.2201	.2227	.2253	.2279
1.7	.2304	.2330	.2355	.2380	.2405	.2430	.2455	.2480	.2504	.2529
1.8	.2553	.2577	.2601	.2625	.2648	.2672	.2695	.2718	.2742	.2765
1.9	.2788	.2810	.2833	.2856	.2878	.2900	.2923	.2945	.2967	.2989
2.0	.3010	.3032	.3054	.3075	.3096	.3118	.3139	.3160	.3181	.3201
2.1	.3222	.3243	.3263	.3284	.3304	.3324	.3345	.3365	.3385	.3404
2.2	.3424	.3444	.3464	.3483	.3502	.3522	.3541	.3560	.3579	.3598
2.3	.3617	.3636	.3655	.3674	.3692	.3711	.3729	.3747	.3766	.3784
2.4	.3802	.3820	.3838	.3856	.3874	.3892	.3909	.3927	.3945	.3962
2.5	.3979	.3997	.4014	.4031	.4048	.4065	.4082	.4099	.4116	.4133
2.6	.4150	.4166	.4183	.4200	.4216	.4232	.4249	.4265	.4281	.4298
2.7	.4314	.4330	.4346	.4362	.4378	.4393	.4409	.4425	.4440	.4456
2.8	.4472	.4487	.4502	.4518	.4533	.4548	.4564	.4579	.4594	.4609
2.9	.4624	.4639	.4654	.4669	.4683	.4698	.4713	.4728	.4742	.4757
3.0	.4771	.4786	.4800	.4814	.4829	.4843	.4857	.4871	.4886	.4900
3.1	.4914	.4928	.4942	.4955	.4969	.4983	.4997	.5011	.5024	.5038
3.2	.5051	.5065	.5079	.5092	.5105	.5119	.5132	.5145	.5159	.5172
3.3	.5185	.5198	.5211	.5224	.5237	.5250	.5263	.5276	.5289	.5302
3.4	.5315	.5328	.5340	.5353	.5366	.5378	.5391	.5403	.5416	.5428
3.5	.5441	.5453	.5465	.5478	.5490	.5502	.5514	.5527	.5539	.5551
3.6	.5563	.5575	.5587	.5599	.5611	.5623	.5635	.5647	.5658	.5670
3.7	.5682	.5694	.5705	.5717	.5729	.5740	.5752	.5763	.5775	.5786
3.8	.5798	.5809	.5821	.5832	.5843	.5855	.5866	.5877	.5888	.5899
3.9	.5911	.5922	.5933	.5944	.5955	.5966	.5977	.5988	.5999	.6010
4.0	.6021	.6031	.6042	.6053	.6064	.6075	.6085	.6096	.6107	.6117
4.1	.6128	.6138	.6149	.6160	.6170	.6180	.6191	.6201	.6212	.6222
4.2	.6232	.6243	.6253	.6263	.6274	.6284	.6294	.6304	.6314	.6325
4.3	.6335	.6345	.6355	.6365	.6375	.6385	.6395	.6405	.6415	.6425
4.4	.6435	.6444	.6454	.6464	.6474	.6484	.6493	.6503	.6513	.6522
4.5	.6532	.6542	.6551	.6561	.6571	.6580	.6590	.6599	.6609	.6618
4.6	.6628	.6637	.6646	.6656	.6665	.6675	.6684	.6693	.6702	.6712
4.7	.6721	.6730	.6739	.6749	.6758	.6767	.6776	.6785	.6794	.6803
4.8	.6812	.6821	.6830	.6839	.6848	.6857	.6866	.6875	.6884	.6893
4.9	.6902	.6911	.6920	.6928	.6937	.6946	.6955	.6964	.6972	.6981
5.0	.6990	.6998	.7007	.7016	.7024	.7033	.7042	.7050	.7059	.7067
5.1	.7076	.7084	.7093	.7101	.7110	.7118	.7126	.7135	.7143	.7152
5.2	.7160	.7168	.7177	.7185	.7193	.7202	.7210	.7218	.7226	.7235
5.3	.7243	.7251	.7259	.7267	.7275	.7284	.7292	.7300	.7308	.7316
5.4	.7324	.7332	.7340	.7348	.7356	.7364	.7372	.7380	.7388	.7396

Table 1

Common Logarithms of Numbers (Cont.)

n	0	1	2	3	4	5	6	7	8	9
5.5	.7404	.7412	.7419	.7427	.7435	.7443	.7451	.7459	.7466	.7474
5.6	.7482	.7490	.7497	.7505	.7513	.7520	.7528	.7536	.7543	.7551
5.7	.7559	.7566	.7574	.7582	.7589	.7597	.7604	.7612	.7619	.7627
5.8	.7634	.7642	.7649	.7657	.7664	.7672	.7679	.7686	.7694	.7701
5.9	.7709	.7716	.7723	.7731	.7738	.7745	.7752	.7760	.7767	.7774
6.0	.7782	.7789	.7796	.7803	.7810	.7818	.7825	.7832	.7839	.7846
6.1	.7853	.7860	.7868	.7875	.7882	.7889	.7896	.7903	.7910	.7917
6.2	.7924	.7931	.7938	.7945	.7952	.7959	.7966	.7973	.7980	.7987
6.3	.7993	.8000	.8007	.8014	.8021	.8028	.8035	.8041	.8048	.8055
6.4	.8062	.8069	.8075	.8082	.8089	.8096	.8102	.8109	.8116	.8122
6.5	.8129	.8136	.8142	.8149	.8156	.8162	.8169	.8176	.8182	.8189
6.6	.8195	.8202	.8209	.8215	.8222	.8228	.8235	.8241	.8248	.8254
6.7	.8261	.8267	.8274	.8280	.8287	.8293	.8299	.8306	.8312	.8319
6.8	.8325	.8331	.8338	.8344	.8351	.8357	.8363	.8370	.8376	.8382
6.9	.8388	.8395	.8401	.8407	.8414	.8420	.8426	.8432	.8439	.8445
7.0	.8451	.8457	.8463	.8470	.8476	.8482	.8488	.8494	.8500	.8506
7.1	.8513	.8519	.8525	.8531	.8537	.8543	.8549	.8555	.8561	.8567
7.2	.8573	.8579	.8585	.8591	.8597	.8603	.8609	.8615	.8621	.8627
7.3	.8633	.8639	.8645	.8651	.8657	.8663	.8669	.8675	.8681	.8686
7.4	.8692	.8698	.8704	.8710	.8716	.8722	.8727	.8733	.8739	.8745
7.5	.8751	.8756	.8762	.8768	.8774	.8779	.8785	.8791	.8797	.8802
7.6	.8808	.8814	.8820	.8825	.8831	.8837	.8842	.8848	.8854	.8859
7.7	.8865	.8871	.8876	.8882	.8887	.8893	.8899	.8904	.8910	.8915
7.8	.8921	.8927	.8932	.8938	.8943	.8949	.8954	.8960	.8965	.8971
7.9	.8976	.8982	.8987	.8993	.8998	.9004	.9009	.9015	.9020	.9025
8.0	.9031	.9036	.9042	.9047	.9053	.9058	.9063	.9069	.9074	.9079
8.1	.9085	.9090	.9096	.9101	.9106	.9112	.9117	.9122	.9128	.9133
8.2	.9138	.9143	.9149	.9154	.9159	.9165	.9170	.9175	.9180	.9186
8.3	.9191	.9196	.9201	.9206	.9212	.9217	.9222	.9227	.9232	.9238
8.4	.9243	.9248	.9253	.9258	.9263	.9269	.9274	.9279	.9284	.9289
8.5	.9294	.9299	.9304	.9309	.9315	.9320	.9325	.9330	.9335	.9340
8.6	.9345	.9350	.9355	.9360	.9365	.9370	.9375	.9380	.9385	.9390
8.7	.9395	.9400	.9405	.9410	.9415	.9420	.9425	.9430	.9435	.9440
8.8	.9445	.9450	.9455	.9460	.9465	.9469	.9474	.9479	.9484	.9489
8.9	.9494	.9499	.9504	.9509	.9513	.9518	.9523	.9528	.9533	.9538
9.0	.9542	.9547	.9552	.9557	.9562	.9566	.9571	.9576	.9581	.9586
9.1	.9590	.9595	.9600	.9605	.9609	.9614	.9619	.9624	.9628	.9633
9.2	.9638	.9643	.9647	.9652	.9657	.9661	.9666	.9671	.9675	.9680
9.3	.9685	.9689	.9694	.9699	.9703	.9708	.9713	.9717	.9722	.9727
9.4	.9731	.9736	.9741	.9745	.9750	.9754	.9759	.9763	.9768	.9773
9.5	.9777	.9782	.9786	.9791	.9795	.9800	.9805	.9809	.9814	.9818
9.6	.9823	.9827	.9832	.9836	.9841	.9845	.9850	.9854	.9859	.9863
9.7	.9868	.9872	.9877	.9881	.9886	.9890	.9894	.9899	.9903	.9908
9.8	.9912	.9917	.9921	.9926	.9930	.9934	.9939	.9943	.9948	.9952
9.9	.9956	.9961	.9965	.9969	.9974	.9978	.9983	.9987	.9991	.9996

Table 2   *The Trigonometric Functions (at Angles, in Radians)*

t	sin t	cos t	tan t	cot t	sec t	csc t
.00	.0000	1.0000	.0000	. . . . .	1.000	. . . . .
.01	.0100	1.0000	.0100	99.997	1.000	100.00
.02	.0200	.9998	.0200	49.993	1.000	50.00
.03	.0300	.9996	.0300	33.323	1.000	33.34
.04	.0400	.9992	.0400	24.987	1.001	25.01
.05	.0500	.9988	.0500	19.983	1.001	20.01
.06	.0600	.9982	.0601	16.647	1.002	16.68
.07	.0699	.9976	.0701	14.262	1.002	14.30
.08	.0799	.9968	.0802	12.473	1.003	12.51
.09	.0899	.9960	.0902	11.081	1.004	11.13
.10	.0998	.9950	.1003	9.967	1.005	10.02
.11	.1098	.9940	.1104	9.054	1.006	9.109
.12	.1197	.9928	.1206	8.293	1.007	8.353
.13	.1296	.9916	.1307	7.649	1.009	7.714
.14	.1395	.9902	.1409	7.096	1.010	7.166
.15	.1494	.9888	.1511	6.617	1.011	6.692
.16	.1593	.9872	.1614	6.197	1.013	6.277
.17	.1692	.9856	.1717	5.826	1.015	5.911
.18	.1790	.9838	.1820	5.495	1.016	5.586
.19	.1889	.9820	.1923	5.200	1.018	5.295
.20	.1987	.9801	.2027	4.933	1.020	5.033
.21	.2085	.9780	.2131	4.692	1.022	4.797
.22	.2182	.9759	.2236	4.472	1.025	4.582
.23	.2280	.9737	.2341	4.271	1.027	4.386
.24	.2377	.9713	.2447	4.086	1.030	4.207
.25	.2474	.9689	.2553	3.916	1.032	4.042
.26	.2571	.9664	.2660	3.759	1.035	3.890
.27	.2667	.9638	.2768	3.613	1.038	3.749
.28	.2764	.9611	.2876	3.478	1.041	3.619
.29	.2860	.9582	.2984	3.351	1.044	3.497
.30	.2955	.9553	.3093	3.233	1.047	3.384
.31	.3051	.9523	.3203	3.122	1.050	3.278
.32	.3146	.9492	.3314	3.018	1.053	3.179
.33	.3240	.9460	.3425	2.920	1.057	3.086
.34	.3335	.9428	.3537	2.827	1.061	2.999
.35	.3429	.9394	.3650	2.740	1.065	2.916
.36	.3523	.9359	.3764	2.657	1.068	2.839
.37	.3616	.9323	.3879	2.578	1.073	2.765
.38	.3709	.9287	.3994	2.504	1.077	2.696
.39	.3802	.9249	.4111	2.433	1.081	2.630
.40	.3894	.9211	.4228	2.365	1.086	2.568
.41	.3986	.9171	.4346	2.301	1.090	2.509
.42	.4078	.9131	.4466	2.239	1.095	2.452
.43	.4169	.9090	.4586	2.180	1.100	2.399
.44	.4259	.9048	.4708	2.124	1.105	2.348

Table 2   The Trigonometric Functions (Cont.)

t	sin t	cos t	tan t	cot t	sec t	csc t
.45	.4350	.9004	.4831	2.070	1.111	2.299
.46	.4439	.8961	.4954	2.018	1.116	2.253
.47	.4529	.8916	.5080	1.969	1.122	2.208
.48	.4618	.8870	.5206	1.921	1.127	2.166
.49	.4706	.8823	.5334	1.875	1.133	2.125
.50	.4794	.8776	.5463	1.830	1.139	2.086
.51	.4882	.8727	.5594	1.788	1.146	2.048
.52	.4969	.8678	.5726	1.747	1.152	2.013
.53	.5055	.8628	.5859	1.707	1.159	1.978
.54	.5141	.8577	.5994	1.668	1.166	1.945
.55	.5227	.8525	.6131	1.631	1.173	1.913
.56	.5312	.8473	.6269	1.595	1.180	1.883
.57	.5396	.8419	.6310	1.560	1.188	1.853
.58	.5480	.8365	.6552	1.526	1.196	1.825
.59	.5564	.8309	.6696	1.494	1.203	1.797
.60	.5646	.8253	.6841	1.462	1.212	1.771
.61	.5729	.8196	.6989	1.431	1.220	1.746
.62	.5810	.8239	.7139	1.401	1.229	1.721
.63	.5891	.8080	.7291	1.372	1.238	1.697
.64	.5972	.8021	.7445	1.343	1.247	1.674
.65	.6052	.7961	.7602	1.315	1.256	1.652
.66	.6131	.7900	.7761	1.288	1.266	1.631
.67	.6210	.7838	.7923	1.262	1.276	1.610
.68	.6288	.7776	.8087	1.237	1.286	1.590
.69	.6365	.7712	.8253	1.212	1.297	1.571
.70	.6442	.7648	.8423	1.187	1.307	1.552
.71	.6518	.7584	.8595	1.163	1.319	1.534
.72	.6594	.7518	.8771	1.140	1.330	1.517
.73	.6669	.7452	.8949	1.117	1.342	1.500
.74	.6743	.7358	.9131	1.095	1.354	1.483
.75	.6816	.7317	.9316	1.073	1.367	1.467
.76	.6889	.7248	.9505	1.052	1.380	1.452
.77	.6961	.7179	.9697	1.031	1.393	1.437
.78	.7033	.7109	.9893	1.011	1.407	1.422
.79	.7104	.7038	1.009	.9908	1.421	1.408
.80	.7174	.6967	1.030	.9712	1.435	1.394
.81	.7243	.6895	1.050	.9520	1.450	1.381
.82	.7311	.6822	1.072	.9331	1.466	1.368
.83	.7379	.6749	1.093	.9146	1.482	1.355
.84	.7446	.6675	1.116	.8964	1.498	1.343
.85	.7513	.6600	1.138	.8785	1.515	1.331
.86	.7578	.6524	1.162	.8609	1.533	1.320
.87	.7643	.6448	1.185	.8437	1.551	1.308
.88	.7707	.6372	1.210	.8267	1.569	1.297
.89	.7771	.6294	1.235	.8100	1.589	1.287

Table 2   *The Trigonometric Functions (Cont.)*

t	$\sin t$	$\cos t$	$\tan t$	$\cot t$	$\sec t$	$\csc t$
.90	.7833	.6216	1.260	.7936	1.609	1.277
.91	.7895	.6137	1.286	.7774	1.629	1.267
.92	.7956	.6058	1.313	.7615	1.651	1.257
.93	.8016	.5978	1.341	.7458	1.673	1.247
.94	.8076	.5898	1.369	.7303	1.696	1.238
.95	.8134	.5817	1.398	.7151	1.719	1.229
.96	.8192	.5735	1.428	.7001	1.744	1.221
.97	.8249	.5653	1.459	.6853	1.769	1.212
.98	.8305	.5570	1.491	.6707	1.795	1.204
.99	.8360	.5487	1.524	.6563	1.823	1.196
1.00	.8415	.5403	1.557	.6421	1.851	1.188
1.01	.8468	.5319	1.592	.6281	1.880	1.181
1.02	.8521	.5234	1.628	.6142	1.911	1.174
1.03	.8573	.5148	1.665	.6005	1.942	1.166
1.04	.8624	.5062	1.704	.5870	1.975	1.160
1.05	.8674	.4976	1.743	.5736	2.010	1.153
1.06	.8724	.4889	1.784	.5604	2.046	1.146
1.07	.8772	.4801	1.827	.5473	2.083	1.140
1.08	.8820	.4713	1.871	.5344	2.122	1.134
1.09	.8866	.4625	1.917	.5216	2.162	1.128
1.10	.8912	.4536	1.965	.5090	2.205	1.122
1.11	.8957	.4447	2.014	.4964	2.249	1.116
1.12	.9001	.4357	2.066	.4840	2.295	1.111
1.13	.9044	.4267	2.120	.4718	2.344	1.106
1.14	.9086	.4176	2.176	.4596	2.395	1.101
1.15	.9128	.4085	2.234	.4475	2.448	1.096
1.16	.9168	.3993	2.296	.4356	2.504	1.091
1.17	.9208	.3902	2.360	.4237	2.563	1.086
1.18	.9246	.3809	2.427	.4120	2.625	1.082
1.19	.9284	.3717	2.498	.4003	2.691	1.077
1.20	.9320	.3624	2.572	.3888	2.760	1.073
1.21	.9356	.3530	2.650	.3773	2.833	1.069
1.22	.9391	.3436	2.733	.3659	2.910	1.065
1.23	.9425	.3342	2.820	.3546	2.992	1.061
1.24	.9458	.3248	2.912	.3434	3.079	1.057
1.25	.9490	.3153	3.010	.3323	3.171	1.054
1.26	.9521	.3058	3.113	.3212	3.270	1.050
1.27	.9551	.2963	3.224	.3102	3.375	1.047
1.28	.9580	.2867	3.341	.2993	3.488	1.044
1.29	.9608	.2771	3.467	.2884	3.609	1.041
1.30	.9636	.2675	3.602	.2776	3.738	1.038
1.31	.9662	.2579	3.747	.2669	3.878	1.035
1.32	.9687	.2482	3.903	.2562	4.029	1.032
1.33	.9711	.2385	4.072	.2456	4.193	1.030
1.34	.9735	.2288	4.256	.2350	4.372	1.027

Table 2   The Trigonometric Functions (Cont.)

t	sin t	cos t	tan t	cot t	sec t	csc t
1.35	.9757	.2190	4.455	.2245	4.566	1.025
1.36	.9779	.2092	4.673	.2140	4.779	1.023
1.37	.9799	.1994	4.913	.2035	5.014	1.021
1.38	.9819	.1896	5.177	.1931	5.273	1.018
1.39	.9837	.1798	5.471	.1828	5.561	1.017
1.40	.9854	.1700	5.798	.1725	5.883	1.015
1.41	.9871	.1601	6.165	.1622	6.246	1.013
1.42	.9887	.1502	6.581	.1519	6.657	1.011
1.43	.9901	.1403	7.055	.1417	7.126	1.010
1.44	.9915	.1304	7.602	.1315	7.667	1.009
1.45	.9927	.1205	8.238	.1214	8.299	1.007
1.46	.9939	.1106	8.989	.1113	9.044	1.006
1.47	.9949	.1006	9.887	.1011	9.938	1.005
1.48	.9959	.0907	0.938	.0910	11.029	1.004
1.49	.9967	.0807	12.350	.0810	12.390	1.003
1.50	.9975	.0707	14.101	.0709	14.137	1.003
1.51	.9982	.0608	16.428	.0609	16.458	1.002
1.52	.9987	.0508	19.670	.0508	19.965	1.001
1.53	.9992	.0408	24.498	.0408	24.519	1.001
1.54	.9995	.0308	32.461	.0308	32.476	1.000
1.55	.9998	.0208	48.078	.0208	48.089	1.000
1.56	.9999	.0108	92.620	.0108	92.626	1.000
1.57	1.0000	.0008	1255.8	.0008	1255.8	1.000
1.58	1.0000	−.0092	−108.65	−.0092	−108.65	1.000
1.59	.9998	−.0192	−52.067	−.0192	−52.08	1.000
1.60	.9996	−.0292	−34.233	−.0292	−34.25	1.000

Table 3   *Values of Trigonometric Functions at Angles (in Degrees)*

θ	sin θ	tan θ	cot θ	cos θ	
0°	.0000	.0000	. . . .	1.0000	90°
1°	.0175	.0175	57.290	.9998	89°
2°	.0349	.0349	28.636	.9994	88°
3°	.0523	.0524	19.081	.9986	87°
4°	.0698	.0699	14.301	.9976	86°
5°	.0872	.0875	11.430	.9962	85°
6°	.1045	.1051	9.5144	.9945	84°
7°	.1219	.1228	8.1443	.9925	83°
8°	.1392	.1405	7.1154	.9903	82°
9°	.1564	.1584	6.3138	.9877	81°
10°	.1736	.1763	5.6713	.9848	80°
11°	.1908	.1944	5.1446	.9816	79°
12°	.2079	.2126	4.7046	.9781	78°
13°	.2250	.2309	4.3315	.9744	77°
14°	.2419	.2493	4.0108	.9703	76°
15°	.2588	.2679	3.7321	.9659	75°
16°	.2756	.2867	3.4874	.9613	74°
17°	.2924	.3057	3.2709	.9563	73°
18°	.3090	.3249	3.0777	.9511	72°
19°	.3256	.3443	2.9042	.9455	71°
20°	.3420	.3640	2.7475	.9397	70°
21°	.3584	.3839	2.6051	.9336	69°
22°	.3746	.4040	2.4751	.9272	68°
23°	.3907	.4245	2.3559	.9205	67°
24°	.4067	.4452	2.2460	.9135	66°
25°	.4226	.4663	2.1445	.9063	65°
26°	.4384	.4877	2.0503	.8988	64°
27°	.4540	.5095	1.9626	.8910	63°
28°	.4695	.5317	1.8807	.8829	62°
29°	.4848	.5543	1.8040	.8746	61°
30°	.5000	.5774	1.7321	.8660	60°
31°	.5150	.6009	1.6643	.8572	59°
32°	.5299	.6249	1.6003	.8480	58°
33°	.5446	.6494	1.5399	.8387	57°
34°	.5592	.6745	1.4826	.8290	56°
35°	.5736	.7002	1.4281	.8192	55°
36°	.5878	.7265	1.3764	.8090	54°
37°	.6018	.7536	1.3270	.7986	53°
38°	.6157	.7813	1.2799	.7880	52°
39°	.6293	.8098	1.2349	.7771	51°
40°	.6428	.8391	1.1918	.7660	50°
41°	.6561	.8693	1.1504	.7547	49°
42°	.6691	.9004	1.1106	.7431	48°
43°	.6820	.9325	1.0724	.7314	47°
44°	.6947	.9657	1.0355	.7193	46°
45°	.7071	1.0000	1.0000	.7071	45°
	cos θ	cot θ	tan θ	sin θ	θ

Table 4

Powers and Roots

No.	Sq.	Sq. Root	Cube	Cube Root	No.	Sq.	Sq. Root	Cube	Cube Root
1	1	1.000	1	1.000	51	2,601	7.141	132,651	3.708
2	4	1.414	8	1.260	52	2,704	7.211	140,608	3.733
3	9	1.732	27	1.442	53	2,809	7.280	148,877	3.756
4	16	2.000	64	1.587	54	2,916	7.348	157,464	3.780
5	25	2.236	125	1.710	55	3,025	7.416	166,375	3.803
6	36	2.449	216	1.817	56	3,136	7.483	175,616	3.826
7	49	2.646	343	1.913	57	3,249	7.550	185,193	3.849
8	64	2.828	512	2.000	58	3,364	7.616	195,112	3.871
9	81	3.000	729	2.080	59	3,481	7.681	205,379	3.893
10	100	3.162	1,000	2.154	60	3,600	7.746	216,000	3.915
11	121	3.317	1,331	2.224	61	3,721	7.810	226,981	3.936
12	144	3.464	1,728	2.289	62	3,844	7.874	238,328	3.958
13	169	3.606	2,197	2.351	63	3,969	7.937	250,047	3.979
14	196	3.742	2,744	2.410	64	4,096	8.000	262,144	4.000
15	225	3.873	3,375	2.466	65	4,225	8.062	274,625	4.021
16	256	4.000	4,096	2.520	66	4,356	8.124	287,496	4.041
17	289	4.123	4,913	2.571	67	4,489	8.185	300,763	4.062
18	324	4.243	5,832	2.621	68	4,624	8.246	314,432	4.082
19	361	4.359	6,859	2.668	69	4,761	8.307	328,509	4.102
20	400	4.472	8,000	2.714	70	4,900	8.367	343,000	4.121
21	441	4.583	9,261	2.759	71	5,041	8.426	357,911	4.141
22	484	4.690	10,648	2.802	72	5,184	8.485	373,248	4.160
23	529	4.796	12,167	2.844	73	5,329	8.544	389,017	4.179
24	576	4.899	13,824	2.884	74	5,476	8.602	405,224	4.198
25	625	5.000	15,625	2.924	75	5,625	8.660	421,875	4.217
26	676	5.099	17,576	2.962	76	5,776	8.718	438,976	4.236
27	729	5.196	19,683	3.000	77	5,929	8.775	456,533	4.254
28	784	5.291	21,952	3.037	78	6,084	8.832	474,552	4.273
29	841	5.385	24,389	3.072	79	6,241	8.888	493,039	4.291
30	900	5.477	27,000	3.107	80	6,400	8.944	512,000	4.309
31	961	5.568	29,791	3.141	81	6,561	9.000	531,441	4.327
32	1,024	5.657	32,768	3.175	82	6,724	9.055	551,368	4.344
33	1,089	5.745	35,937	3.208	83	6,889	9.110	571,787	4.362
34	1,156	5.831	39,304	3.240	84	7,056	9.165	592,704	4.380
35	1,225	5.916	42,875	3.271	85	7,225	9.220	614,125	4.397
36	1,296	6.000	46,656	3.302	86	7,396	9.274	636,056	4.414
37	1,369	6.083	50,653	3.332	87	7,569	9.327	658,503	4.431
38	1,444	6.164	54,872	3.362	88	7,744	9.381	681,472	4.448
39	1,521	6.245	59,319	3.391	89	7,921	9.434	704,969	4.465
40	1,600	6.325	64,000	3.420	90	8,100	9.487	729,000	4.481
41	1,681	6.403	68,921	3.448	91	8,281	9.539	753,571	4.498
42	1,764	6.481	74,088	3.476	92	8,464	9.592	778,688	4.514
43	1,849	6.557	79,507	3.503	93	8,649	9.644	804,357	4.531
44	1,936	6.633	85,184	3.530	94	8,836	9.695	830,584	4.547
45	2,025	6.708	91,125	3.557	95	9,025	9.747	857,375	4.563
46	2,116	6.782	97,336	3.583	96	9,216	9.798	884,736	4.579
47	2,209	6.856	103,823	3.609	97	9,409	9.849	912,673	4.595
48	2,304	6.928	110,592	3.634	98	9,604	9.899	941,192	4.610
49	2,401	7.000	117,649	3.659	99	9,801	9.950	970,299	4.626
50	2,500	7.071	125,000	3.684	100	10,000	10.000	1,000,000	4.642

Formulas from Trigonometry

$$\sin(-x) = -\sin x \qquad\qquad \cos x = \cos(-x)$$

$$\cos(\pi - x) = -\cos x \qquad\qquad \sin(\pi - x) = \sin x$$

$$\cos(x \pm \pi) = -\cos x \qquad\qquad \sin(x \pm \pi) = -\sin x$$

$$\cos\left(\frac{\pi}{2} - x\right) = \sin x \qquad\qquad \sin\left(\frac{\pi}{2} - x\right) = \cos x$$

$$\cos(x \pm y) = \cos x \cos y \mp \sin x \sin y$$

$$\sin(x \pm y) = \sin x \cos y \pm \cos x \sin y$$

$$\cos 2x = \cos^2 x - \sin^2 x$$
$$= 1 - 2\sin^2 x$$
$$= 2\cos^2 x - 1$$

$$\sin 2x = 2 \sin x \cos x$$

$$\cos^2 x = \frac{1 + \cos 2x}{2}$$

$$\sin^2 x = \frac{1 - \cos 2x}{2}$$

$$\cos^2 \frac{x}{2} = \frac{1}{2}(1 + \cos x)$$

$$\sin^2 \frac{x}{2} = \frac{1}{2}(1 - \cos x)$$

$$\tan x = \frac{\sin x}{\cos x}, \ \cot x = \frac{1}{\tan x},$$

$$\sec x = \frac{1}{\cos x}, \ \csc x = \frac{1}{\sin x}$$

$$\tan(x + y) = \frac{\tan x + \tan y}{1 - (\tan x)(\tan y)}$$

$$\tan 2x = \frac{2 \tan x}{1 - \tan^2 x} \qquad\qquad \tan^2 \frac{x}{2} = \frac{1 - \cos x}{1 + \cos x}$$

$$1 + \tan^2 x = \sec^2 x \qquad\qquad 1 + \cot^2 x = \csc^2 x$$

Index